STUDIES IN PHYSICAL GEOGRAPHY

Geomorphological Processes

STUDIES IN PHYSICAL GEOGRAPHY
Edited by K. J. Gregory

Also published in the series

Man and Environmental Processes
Edited by
K. J. GREGORY and D.E. WALLING

Ecology and Environmental Management
C. C. PARK

Forthcoming volume

Applied Climatology
J. E. HOBBS

STUDIES IN
PHYSICAL GEOGRAPHY

Geomorphological Processes

E. DERBYSHIRE
Reader in Physical Geography, University of Keele

K. J. GREGORY
Professor of Geography, University of Southampton

J. R. HAILS
Director of Environmental Studies, University of Adelaide

Routledge
Taylor & Francis Group

LONDON AND NEW YORK

First published 1979 by Westview Press

Published 2018 by Routledge
52 Vanderbilt Avenue, New York, NY 10017
2 Park Square, Milton Park, Abingdon, Oxon OX14 4RN

Routledge is an imprint of the Taylor & Francis Group, an informa business

British Library Cataloguing in Publication Data

Derbyshire, Edward
    Geomorphological processes. - (Studies in physical geography; 2
    ISSN 0142-6389).
    1. Geomorphology
    I. Title   II. Gregory, Kenneth John
    III. Hails, John R
    551.4   GB401.5   79–5285
    ISBN 0-7129-0833-1
    ISBN 0-7129-0923-0 Pbk

    ISBN 0-89158-695-4 (Westview)
    ISBN 0-89158-864-7 Pbk (Westview)

ISBN 13: 978-0-36702194-8 (hbk)

ISBN 13: 978-0-367-17181-0 (pbk)

Contents

Preface

This book originated from a proposal by one author (J. R. H.) who was subsequently joined by a second (E. D.) and then by a third (K. J. G.). It has taken longer to produce than we expected because of the complications imposed by the distances which the authors have succeeded in putting between themselves during the past three years. The basic objective was to produce a short book which would introduce geomorphological processes to students in the first or second year of their higher education courses. We believed that there was a need for such a book reviewing a range of geomorphological processes which would offer a prelude to the symphonies which are available in books devoted to specific processes and their effects, many of which are signposted in the lists of further reading at the end of each chapter. We are aware that the range of suitable preludes is wide, but we have endeavoured to compose one which expresses at least some of the recent achievements in the study of geomorphological processes. Emphasis is placed on the nature of processes and upon their controls but the effects of processes in creating landforms are not reviewed in any detail. In addition to the selected references at the end of each chapter, we have collected a bibliography of works cited at the end of the book but this is not intended to be as exhaustive as the references collated in more advanced works.

To continue the musical analogy, the first movement must be a sound understanding of landscape processes and mechanics and this book is designed to represent this stage. The second movement can subsequently elaborate several themes including the influence of man and the theoretical principles underlying the operation of individual processes, whereas the third movement may be concerned with the chronology of landscapes in the past and with applications of knowledge gained as a basis for understanding future tendencies of processes. Grateful acknowledgement is due to the cartographic units in the Universities of Keele, Exeter, Southampton and Adelaide. It is a pleasure to acknowledge the assistance given by Mrs R. Flint, Mrs P. Cornes and Mrs S. Howell, and to record our thanks to three families, and especially to Maryon, Chris and Josie, for continuing their stalwart toleration during the production of this book.

January 1979

Ed. Derbyshire
Ken. Gregory
John Hails

Illustrations

Tables

1
Introduction

Geomorphology, as the scientific study of landforms and land form change, was founded more than a century ago. The word literally means the study of the form of the earth. It has frequently been acknowledged however, that a complete understanding of geomorphological processes is necessary for advancement of our understanding of landforms. Although the necessity for the study of processes has often been stated throughout the history of the science (Chorley, Dunn and Beckinsale 1964, 1973), the record of achievement of the subject shows that until 1960 there was little substantive investigation of processes by geomorphologists. Instead, geomorphology concentrated upon the investigation of landforms and only when the needs arose to estimate the nature of landform change in areas that had not been surveyed in detail, and to predict the future course of change, did a greater concern for the study of processes become more generally accepted.

Since 1960 there have been a number of approaches to, and trends within, geomorphology. These can be characterized as a geomorphological bandwaggon parade: J. N. Jennings asked the question, 'any millenniums today lady?' in an address in which he envisaged a series of impacts or bandwaggons, shown as approaches to the study of geomorphology. The existence of diverse approaches has also been recognized by Butzer (1973) who concluded that pluralism was inevitable and other authors have recognized alternative conceptual approaches. Chorley (1971) found an analogy for physical geography in the tight-rope walker confronted by two tightropes. One tightrope was largely

Table 1.1
SOME APPROACHES IN GEOMORPHOLOGY WITHIN PHYSICAL GEOGRAPHY

Fashions in geomorphology have included (Jennings, 1973):

Denudation chronology	elucidation of stages of evolution; particularly the study of stage in the Davisian trilogy of 'structure, processs and stage'.
Climatic geomorphology	Recognition of morphogenetic systems which acquire a particular character and evolve in their own way under combinations of exogenic processes varying particularly with climate.
Morphometry	Use of quantitative techniques to define and to describe the nature of landforms and their spatial pattern.
General systems approach	Use of concepts adopted from thermodynamics to focus upon process-form relationships.
Process study	Quantification of present processes and understanding process in physical or chemical terms.
Structural geomorphology	Acknowledgement of the influence of rock type and rock disposition.

Current Trends in geomorphology were suggested (Dury, 1972) to include:

Quantification	Techniques introduced from various directions.
Random variation	A general interest in models and particularly in analog models.
Climatic geomorphology	Interest in the extent to which terrain is in harmony with the existing climate.

Slope studies
Measurement of the speed of geomorphic processes
Remote sensing

Systems basic to an approach to physical geography may be viewed (Chorley and Kennedy, 1971), as:

Morphological systems	Morphological physical properties integrated to form a recognizable operational part of physical reality, with the strength and direction of their connectivity often revealed by correlation analysis.
Cascading systems	A chain of subsystems, often characterized by thresholds, having both spatial magnitude and geographical location, and dynamically linked by cascade of man or energy.
Process-response systems	Formed by intersection of morphological with cascading systems.
Control systems	Intelligence can operate valves which influence distribution of energy and mass within cascading systems, and bring about changes in equilibrium relationships involving the morphological variables linked with them in process-response systems.

The content and relationships of physical geography (Brown, 1975) suggested:

Physical geography is too internally unbalanced and centrifugal in character—geomorphology plays too dominant a role.

Physical geography as an integrated subject has been rediscovered by non-geographers under the guise of environmental science.

Does human geography need to pay attention to the physical world?

Could physical geography cease to be an integral part of geography as environmental sciences are developing?

If physical geography were incorporated within environmental science and human geography featured within the spatial aspects of social science then someone would reinvent geography!

Theory in geomorphology can be envisaged (Chorley, 1978) through a number of phases:

Teleological	Studies of terrain related to views of overall design of nature.
Immanent	Explanation of landforms in terms of inherent nature, e.g. rocks and relief.
Historical	Interpretation of landforms by reference to a series of events.
Taxonomic	Explanation of landforms by grouping into spatially associated classes.
Functional	Form and function thought to be related in repeated and predictable regulation.
Realist	Explanation involves more than prediction based upon observed regularities.
Conventionalist	Include belief that no useful distinctions drawn between observation and theory and that observation precedes theory.

expressive of the investigation of process-response systems by the study of contemporary processes, their character as a response to external constraints, and their effects. The second tightrope was visualized as representing the study of the evolution of landscape and landforms and was therefore concerned primarily with historical development and chronology. Although these two tightropes were portrayed by Chorley in 1971 as ever-diverging, it is perhaps inevitable that the two tightropes will eventually come much closer together. These and some other alternative approaches to geomorphology are proposed in Table 1.1.

1.1 A GEOMORPHOLOGICAL EQUATION

Progress in geomorphology has been achieved by a diversity of approaches. A simple equation may be envisaged as a vehicle for the explanation of recent trends in geomorphology (Gregory, 1978a). This equation involves three main terms which are results or landforms (F), processes (P) and materials (M). The landform or result (F) may be regarded as the dependent variable whereas processes and materials may be thought of as independent variables. The equation may therefore be expressed as

$$F = f(PM) \, \mathrm{d}t$$

where f denotes a function of, and $\mathrm{d}t$ is the mathematical way of denoting change over time.

This equation gives a simple basis for summarizing the various approaches available to geomorphology and there are four levels of study of the equation which may be stated as follows:

LEVEL 1 Study of *elements of the equation*, i.e. investigation of form, processes or materials independently.

LEVEL 2 *Balancing the equation*, i.e. obtaining relationship between form, processes and materials for a particular area at a specific time.

LEVEL 3 *Differentiating the equation*, involving an examination of the way in which the relationship between the three ingredients, form, process and materials, varies over time.

LEVEL 4 *Applying the equation* is where we endeavour to utilize the results from the preceding three levels to apply to contemporary environmental problems.

Studies of individual elements can be undertaken at level 1, and this embraces study of form, process or materials independently for their own sake. Of these, studies of form have been undertaken most frequently by geomorphologists, and the branch of geomorphology called landform geography (e.g. Zakrzewska, 1967) was devoted to the description of specific landforms and landform assemblages. In this way the shape of cirques, of drumlins, or of slopes has been described often employing quantitative techniques. Analysis of networks has been undertaken using topological methods obtained from combinatorial analysis in mathematics. Although the study of form is a necessary pre-requisite to later geomorphological analysis it has been argued that it should not be an end in itself because it is very difficult to understand the past development of form, the present significance or future character, from morphology alone. However, the form of a drumlin, particularly its shape, orientation and angles of surface slope are features which are much more pertinent to the use of the drumlin for cultivation and rural land use than is the age or stratigraphy of the drumlin.

Hence the investigation of form as the appropriate starting point for much geomorphological analysis may be vindicated. The description of form has been enhanced with greater availability of information from satellite platforms and with the potential of data retrieval systems so that, in addition to maps of conventional style, automated cartography can be employed to provide maps of many dimensions of the form of the land, and data storage systems offer scope for the assembly of large amounts of terrain information and for its retrieval.

Materials studied by the geomorphologist are those upon which processes may act to produce landforms. Traditionally, rock, weathered material, superficial deposits and soils have been regarded as those materials which are of basic interest to the geomorphologist. Initially, study of the influence of materials upon landforms and upon landscape was achieved largely qualitatively by structural geomorphology but, particularly since 1960, it has been appreciated that the geomorphologist needs to be more aware of the physical characteristics of materials, often expressed in a quantitative way, as a basis for understanding the behaviour of particular materials when they are subjected to the influence of specific processes. The improved understanding of earth materials was advocated by Yatsu (1971) when he proposed a landform-materials science which would require a knowledge of their physical properties including shear strength, porosity and mineral composition. The essential features of materials are outlined in Section 1.2.

Processes provide the third element of the geomorphological equation and the theme for this book. It has often been argued that until 1950, the study of geomorphology was characterized by a tendency to ignore the detailed investigation of geomorphological processes. This has now been remedied by studies which have elaborated the way in which specific processes operate, the controls which affect the operation of processes, and the rate at which processes change the environment. This greater emphasis on the study of processes has become feasible with the development of monitoring systems which can record the character and rate of process operation and so provide the basis for analysis of the causes of process response. Process response models have therefore featured prominently in demonstrating how a particular stimulus provides a specific response which may have predictable effects.

The trend towards a greater concern with processes has necessitated both a shift in the content of geomorphological studies and a greater familiarity with measurement methods, with techniques of quantitative analysis of process response, and with the principles governing the movement of water, ice, air, and particles fundamental to exogenetic, or surface, processes and the movement of earth relative to subsurface, or endogenetic, processes. Geomorphological processes can be thought of as those occurring on slopes and in river channels which are embraced in drainage basins (Chapter 2), those involving glacier and ground ice (Chapter 5), those occurring in coastal situations (Chapter 3), and those depending upon air movement (Chapter 4). Each of these processes can be resolved into production of material, often by erosion; transport of material; and deposition. The weathering processes which are a fundamental prelude to the impact of other geomorphological processes are briefly outlined in Section 1.3. Endogenetic processes have not been studied by physical geographers with the same degree of interest accorded to exogenetic processes. This is because the endogenetic processes originating within the earth have been the subject of research by geologists and geophysicists although the geomorphologist should be aware of the effects which these processes have.

Geomorphological analysis is achieved at level 2 by establishing relationships between the three elements of the equation and by showing how a specific process operating upon a particular material can produce a certain result or landform. The systems approach is ideally suited to the identification of the relationships between the elements of the equation and has been instrumental in clarifying the diverse ways in which indices of materials, of process, and of form are related. In this way it is essential for the geomorphologist to understand the interaction of materials beneath a glacier and the movement of ice and, in the same way, it is necessary to know how the process operating upon a sandy beach during a tidal cycle will interact to generate a particular beach form. Such interactions, which may be regarded as *balancing the geomorphological equation*, offer the most stimulating problems for geomorphological analysis.

Many ideas have been propounded to apply to this level of the geomorphological equation, and equilibrium concepts have drawn attention to the way in which a balance is often realized by the interaction between processes and materials to produce equilibrium forms. Such a balance may depend upon the occurrence of a process of a particular magnitude. Thus a storm beach profile may be an equilibrium form which is generated by a previous high tide. There is thus a dependence upon some events which may be more significant than others and there may be a delay between the occurrence of the formative process and the appearance of the resulting form. In a classic study of the magnitude and frequency of geomorphic processes (Wolman and Miller, 1960) it was agreed that the meso-scale events which occur at least once or perhaps several times each year, are often more significant than those processes which occur more frequently, or those which occur very rarely. In the investigation of geomorphological processes much can be learnt from the effects of a simple event such as a large infrequent storm (see Chapters 2 and 3).

Whereas studies which involve balancing the geomorphological equation may often be applied at the local scale, it is also necessary to develop an understanding of how processes vary on a world scale and this is the domain of climatic geomorphology which is reviewed in Section 1.4.

Two further levels at which the geomorphological equation may be applied both relate to the way in which geomorphological systems change or adjust over time. The third level, that of *differentiating the equation*, may be visualized as the one at which consideration is given to the nature of geomorphological systems in the past and this necessarily requires reference to the way in which climate, sea levels and earth movements have operated in the past and to the way in which man is a potent influence upon contemporary geomorphological systems. It has been shown that time scales used by the geomorphologist may be thought of at three levels. Short periods of study time refer to periods of tens of years, graded time pertains to hundreds or perhaps a few thousands of years, and cyclic time applies to the millions of years which would be necessary for an erosion surface to be produced (Schumm and Lichty, 1965). Geomorphological analysis may be undertaken at any of these three scales and the variables considered will be chosen according to the time scale which is being employed. Short-term investigations of geomorphological processes are undertaken at the scale of steady time; the effects of these processes in producing a graded river profile can be considered at the level of graded time; and much longer time spans apply to the evolution of the entire land surface at the scale of cyclic time.

A fourth level of consideration is offered by *applying the geomorphological equation* to contemporary environmental problems. This involves the use of a knowledge of the relations between form, process and materials obtained by balancing the equation (level 2), and of the differentiation of the equation to allow for past development (level 3), as a basis for estimation of the behaviour of geomorphological systems either in locations where processes have not been measured (spatial prediction) or in the future (temporal prediction). This introduces the field of applied geomorphology (e.g. Hails, 1977) the great potential of which is becoming realized partly through the development of geomorphic engineering (Coates, 1976).

We can think of the environment as a machine which we need to control. However, such control can be achieved only if we fully understand how the geomorphological machine works. This book is dedicated to the workings, or processes, of the geomorphological machine in the firm belief that this is an essential prerequisite for understanding either changes in the past or the potential for developments in the future.

1.2 EARTH MATERIALS

Rock, the material of the solid earth, is an aggregate of mineral particles, the composition of which varies according to primary (formation) processes and secondary (diagenetic) processes. Rocks may be classified into three broad groups on the basis of their formation, namely igneous, sedimentary and metamorphic (Table 1.2).

Igneous rocks originate within the earth's crust and solidify from magmatic bodies. Those which solidify to rock before reaching the surface are *intrusive*. Slow crystallization at great depth produces plutonic igneous rocks with relatively large, interlocking crystals and a massive, coarser-grained structure. At shallower depth, hypabyssal igneous rocks are finer-grained and occur commonly in subhorizontal sheets called sills and subvertical walls called dykes. Mineral composition may vary to produce a wide range of intrusive rock types but granite is a common acidic type and gabbro a basic type of plutonic rock, equivalent hypabyssal types being rhyolite and dolerite. When magma erupts or flows at the surface, it is extrusive and produces volcanic rocks. It may be ejected as showers of ash or rock lumps (volcanic 'bombs') or flow out as lava. The former give rise to rock types such as ash and tuff, while the latter vary from basalt (fine-grained) to obsidian (volcanic glass) depending upon the rate of cooling. When lava cools in water it produces pumice.

Most igneous rocks consist of several types of minerals, inorganic substances of determinable chemical composition occurring most commonly in crystal form. For example, granite contains the constituent minerals quartz, feldspar (orthoclase) and mica (biotite and muscovite). Gabbro, on the other hand, lacks quartz and is composed predominantly of plagioclase feldspar and pyroxene with some olivine. Decrease in the content of aluminium and of silicon, usually in the form of quartz and increase in the ferromagnesian minerals are the criteria used in the gradational series of igneous rocks from acid to basic (Table 1.3).

Rocks comprising particles derived from the breakdown of other rocks, followed by transport and deposition by rivers, wind, the sea or glacier ice are called *clastic sedimentary rocks*. Examples, listed in order of decreasing particle size, are conglomerate, sandstone and shale. Slow interstitial infilling with quartz cement in a quartz-sandstone produces a sedimentary quartzite. Deposition of sediments may also result from chemical reactions followed by precipitation. Some limestones are of this origin, together with rock-salt and some nitrates, for example. Rocks resulting from accumulation of salts at the surface in conditions of strong evaporation are called evaporites; gypsum is an example. Deposits of

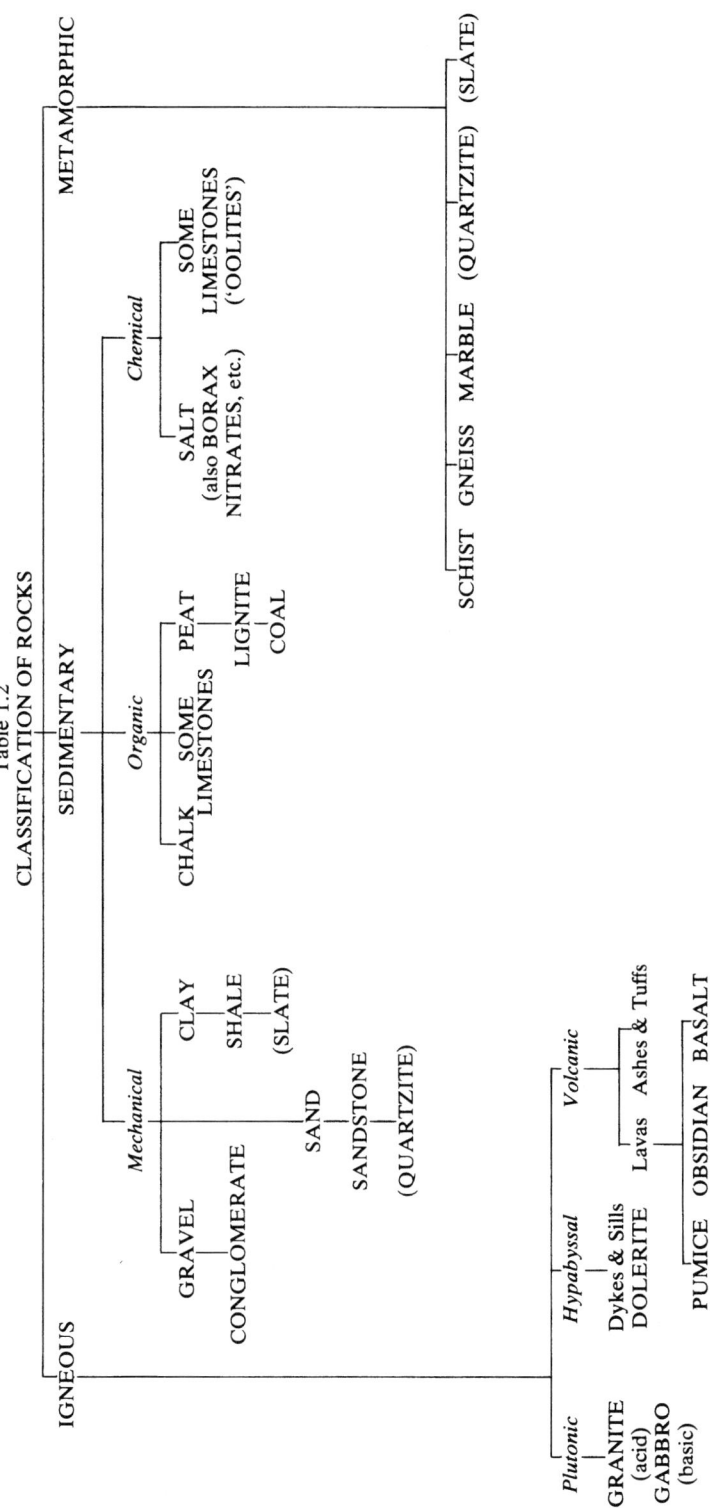

Table 1.2
CLASSIFICATION OF ROCKS

The rock family. After S. H. Beaver. *The Geology of Sand and Gravel*, (Sand and Gravel Assoc. Gt. Britain), 1968, 14.

Table 1.3
GRADATIONAL SERIES OF IGNEOUS ROCKS FROM ACID TO BASIC

	ACID	INTERMEDIATE		BASIC	ULTRABASIC
PLUTONIC Coarse grain size	GRANITE GRANODIORITE	SYENITE	DIORITE	GABBRO NORITE	PERIDOTITE
HYPABYSSAL Medium grain size	MICRO-GRANITE	MICRO-SYENITE	MICRO-DIORITE	DOLERITE	
	⟵ PEGMATITES ⟶				
	⟵ VARIOUS PORPHYRIES & APLITES ⟶				
VOLCANIC (Extrusive) Fine grain size	RHYOLITE	TRACHYTE PHONOLITE	ANDESITE	BASALT SPILITE	OLIVINE-RICH BASALTS
VOLCANIC Glassy Rocks	PITCHSTONE FELSITE			TACHYLITE	
	⟵ OBSIDIAN ⟶				
	Orthoclase feldspar				
PRINCIPAL	Quartz	Sodic-Plagioclase feldspar	Sodic-Plagioclase feldspar	Calcic-Plagioclase	
MINERAL	Orthoclase feldspar				Ferromagnesians predominant
CONSTITUENTS	Micas	Some Ferromagnesian minerals	Ferromagnesian minerals	Ferromagnesian minerals	
	Little Quartz				

An elementary classification of igneous rocks, following Sparks, B. W., *Geomorphology*, London, Longmans, p. 30.

organic origin include the lime-rich parts of the skeletons of corals, snails and shells and carbonaceous material (peat and coal) derived from terrestrial vegetation. A distinctive structural feature of sedimentary rocks is their bedding or stratification. Initially, most bedding planes are horizontal but earth movements tilt and fold them.

The internal form or fabric of rock materials may be altered during earth movements by high confining pressures and high temperatures to produce metamorphic rocks. Sedimentary rocks thus modified are called metasediments. Metamorphosed coarse-grained clastic sediments such as conglomerate and sandstone become quartzite following the filling of interstices with quartz to produce one of the most resistant of all rocks. Metamorphism of limestone produces marble, and the metamorphism of intrusive igneous rocks may give rise to gneiss, a distinctively banded, coarse-grained rock. Extreme metamorphism of finer-grained rocks produces the very fine foliation characteristic of schist.

In the main, rocks vary in their resistance to chemical processes according to their mineral composition, and in their mechanical strength according to their mineral composition and their *rock fabric*, which includes the disposition and especially the interlocking of grains and the extent and frequency of joints and bedding planes. For example, quartzites are resistant, both chemically and mechanically, in most climates, while limestones may be mechanically strong but subject to relatively rapid chemical attack. On the other hand, shale is chemically resistant but mechanically weak. Such differences are reflected in the landscape, with both weathering and erosional activity tending to be more effective on the mechanically weaker units and less effective on the stronger units of a stratified succession.

This process of differential erosion is reflected in the major forms (or geomorphic units: see Currey in Hails 1977) of the landscape (Figure 1.1). The term structural landforms is applied to these features and to those arising directly from tectonic events such as faulting (e.g. fault scarps) and volcanic action (e.g. cones, craters, calderas).

Sometimes, the properties of a particular rock type may be the predominant factor in the landforms produced. The best example of this is probably provided by limestones, many of which are both soluble in the mild acid solutions present in the soil, and characterized by bedding planes and frequent joints. The result is a highly distinctive landscape in which streams sink and emerge as resurgences along master joints, solution produces a variety of closed rock depressions on the surface and caves at depth within the rock, and groundwater-controlled solution maintains broad areas of gentle slope which contrast with the steep slopes of the unconsumed residual hills (Figure 1.2). Such *karst* landscapes display considerable variety in detail but all have in common the residual hills and enclosed rock depressions.

Unconsolidated earth materials derived from rock weathering, erosion and re-deposition are called soils. *In situ* rock weathering in which the weathered products suffer little or no surface transportation produces a distinctive series of layers or horizons with depth below the surface known as the solum or soil. Unconsolidated sediments derived from erosion are also called soils by engineers. The term is used in a generic way here to include both *in situ* weathering profiles and engineering soils.

Soils which have been dried in the laboratory consist of air (in voids) and solid particles, in contrast to completely saturated soils which consist of water and solid particles. In the very common condition of partial saturation, air, water and solids are present. The ratio

24

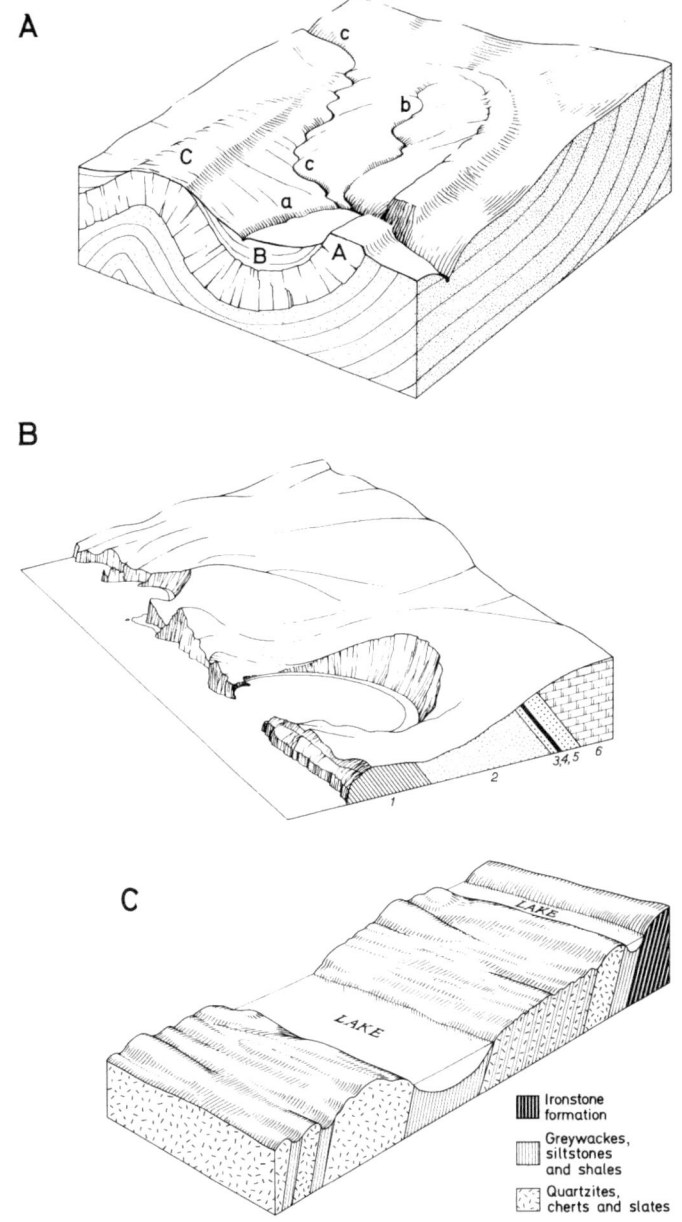

FIG. 1.1. DIFFERENTIAL EROSION

A – fluvial etching of gently folded weak and strong rocks showing a cuesta (A), synclinal valley in a softer stratum (B), strike streams in this valley (a,b), discordant river (c) cutting through the cuesta, and anticline crest in process of erosion (C). (Modified after Tricart, 1974.)

B – Breaching of resistant bed of Portlandian and Purbeckian limestones (1) by the sea, to produce Lulworth Cove (Dorset, England), in soft Weald clays and sands (2). Main ridge made up of Lower Greensand (3), Gault Clay (4), Upper Greensand (5) and Chalk (6).

C – Ice-smoothed ridges of resistant rock and ice-gouged lake basins in weaker rocks. Steeply-dipping metasediments of Proterozoic age, central Labrador-Ungava.

FIG. 1.2. LIMESTONE TOWERS OF SUB-TROPICAL KARST LANDSCAPE NEAR KWEILIN, PEOPLE'S REPUBLIC OF CHINA (Based on photograph by E. Derbyshire.)

between the volume of the voids in a soil (V_v) and the total volume of the soil (V_o) is called the porosity (n), thus

$$n = \frac{V_v}{V_o} \qquad (1.1)$$

the ratio of the volume of the voids to the volume of the soil solids (V_s) being termed the voids ratio (e),

$$e = \frac{V_v}{V_s} \qquad (1.2)$$

Porosity in soils can be expected to vary with factors such as the amount of cement between grains, particle size, particle shape and packing, and the presence of discontinuities such as fractures and bedding planes. Variations in packing are best illustrated by considering a deposit of uniformly-sized spherical grains (Figure 1.3). With a cubic packing arrangement, porosity (n) is at its theoretical maximum ($n = 0.476$). The closest packing type is rhombohedral in which $n = 0.26$. Many sands made up of rounded or angular grains have porosities which are lower than the porosity of clays. Clays are made up of very small platy particles characterized by strong molecular forces between the particles and between the particles and the soil water, thus giving them the property of cohesion. Therefore, clays may have higher porosity values than sand but their ability to *transmit* water is less (see illustrations in Chilingarian and Wolf, 1975).

The ability of a soil to transmit water through its voids is termed the hydraulic conductivity or permeability (K). This is a function of the specific weight of fluid (γ), its viscosity (μ), and the average pore size (d). Also to be considered is a shape factor (C) introduced as a

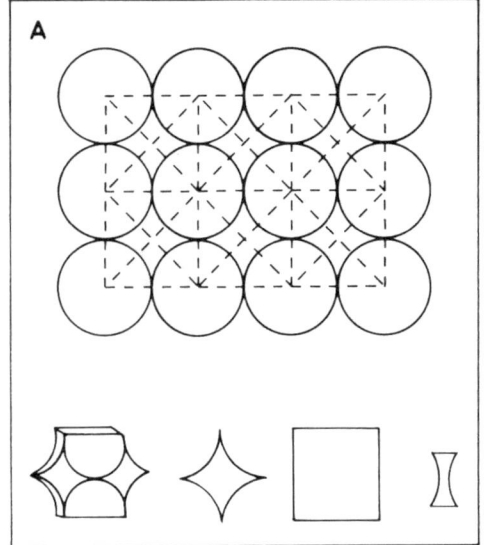
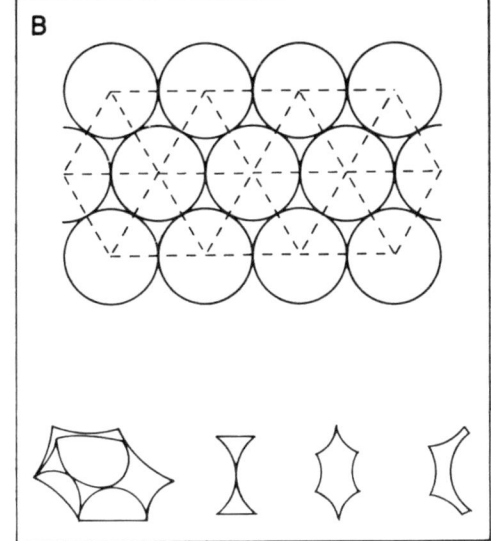

FIG. 1.3. TWO-DIMENSIONAL SKETCH SHOWING REGULAR PACKING OF UNIFORM GRAINS
A – cubic packing, showing (below) three-dimensional sketch of the pore space shapes, left, and different cross-sectional shapes of this three-dimensional pore.
B – Rhombohedral packing, showing (below) three-dimensional sketch of the pore space shapes, left, and different cross-sectional shapes of this three-dimensional pore. Pore space diagrams based on Chilingarian and Wolf, 1975, by permission of the publishers.

constant to take account of features such as packing, stratification, grain-size and porosity.

Thus
$$K = Cd^2 \frac{\gamma}{\mu} \tag{1.3}$$

in which K is expressed as a rate of flow per unit area of soil in millimetres per second. This relationship, Darcy's law, assumes chemical and mechanical stability of water and earth material. A variety of processes render this unlikely, notably ionic exchange on clay surfaces (affecting mineral volume and therefore pore size and shape). Variations in the rate of ground water movement, solution and deposition will also play a part, as well as changes in temperature and pressure affecting the gas/non-gas state of soil solutions.

Water in voids of a porous medium experiences pressure referred to as the *pore water pressure*. In the phreatic condition (that is, in the saturated condition below the water table), this is positive but becomes negative in the unsaturated zone. Pore water pressure (P_w) at a point is directly proportional to the distance of that point below the water table, thus

$$P_w = \gamma H_p \tag{1.4}$$

where γ is the unit weight of water and H_p is the pressure head (Figure 1.4a). In the unsaturated zone of variable water content (the *vadose zone*) pore water pressures are negative and are termed pore water tension (i.e. $P_w = -\gamma H_p$: Figure 1.4b).

The percentage water content at which soil flows under its own weight (i.e. it becomes

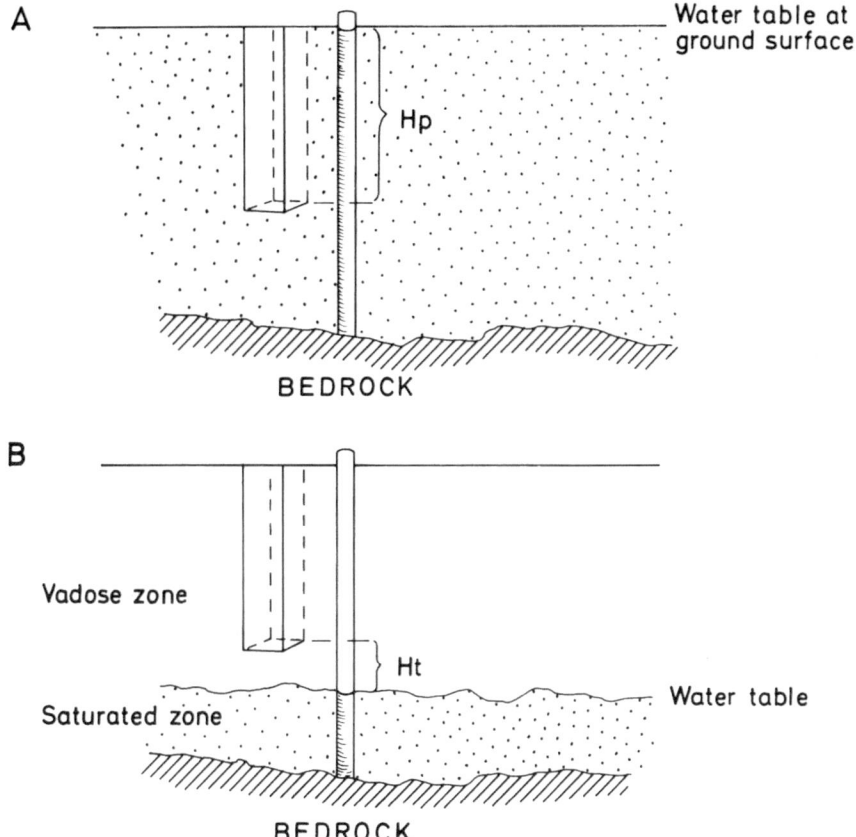

FIG. 1.4. PIEZOMETRIC POTENTIAL AT BASE OF COLUMN OF SOIL IN TWO DIFFERENT SITUATIONS
A – with watertable at ground surface, pore-water pressure at base of the soil column measured in a tube is positive and derives from pressure head *Hp*.
B – at same point in the soil in unsaturated conditions above watertable, pore-water pressures at same point are negative (pore-water tension)—*Ht* is equivalent to negative *Hp*.

liquid) is known as the *liquid limit*. The minimum moisture content at which soil can be rolled into a thread 3 mm in diameter without breaking is known as the *plastic limit*. The highest liquid limits are to be expected in clays and are found to average about 50 per cent compared with an average of 25 per cent in silts. When a soil has a field moisture content which is the same as the liquid limit determined in the laboratory it is said to have a *liquidity index* of 1.0. Moisture contents below this level give lower liquidity indices. The difference between the liquid limit and the plastic limit moisture percentages is termed the *plasticity index*, also expressed as a percentage. An example illustrating some of these differences is provided in Table 1.4. It will be noted that the silty soil from central Breida has a plasticity index of 0. In behavioural terms, this means that the soil has some bearing strength until it reaches 13 per cent moisture when it passes into the liquid state, there being no transition to plastic behaviour. It should also be noted that the plasticity index rises in the samples as the clay percentage increases.

Table 1.4
COMPARISON OF SOIL PROPERTIES

Soil Properties	Central Breida Iceland	West Breida Iceland	Taylor Valley Antarctica
Handling description	Silty	More Clayey	Most Clayey
Liquid limit (%)	13	20	27
Plastic limit (%)	13	15	11
Plasticity index (%)	0	5	16
Natural water content (%)	9	7.4	<1
Dry density (g/ml)	2.37	1.9	2.14
Angle of internal friction (degrees)	46–48	46–48	45

Some geotechnical properties of till from two locations in Iceland (Breidamerkursandur) and Taylor Valley, Antarctica. By courtesy of A. McGown.

In assessing the likelihood of rock or debris moving under gravity on sloping surfaces, we are concerned with the strength of the material. The strength of earth materials consists of two major components: namely, those due to frictional and to cohesive forces. As already mentioned, fine-grained sediments (fine silts and clays) possess the property of cohesion. Sediments coarser than this are little affected by electrostatic forces and may be termed non-cohesive. These materials owe their strength to the internal frictional resistance between one grain and another. The degree to which grains are forced together reflects several factors including the amount of overburden at the point under study. This affects the steepness of the surface slope which can be developed in any debris and which is known as the *angle of repose* of the material. Such surface slopes range between about 25° and 40°. The angle is steeper for angular-shaped grains, the interlocking of the particles yielding a higher internal frictional angle. Sometimes, non-cohesive materials may maintain surface slopes which are greater than those to be expected from measured internal friction angles. In this case the presence of water in thin films and under considerable surface tension acts as a cohesive force. The tensional or stretching force along the water surfaces serves to hold the soil particles together so that this apparent cohesion is an important factor in maintaining slopes in partly saturated non-cohesive soils.

The strength of a soil may be exceeded if sufficient *force* is applied to it, force being defined as the mass of a body multiplied by its acceleration. Force per unit area acting at a point is termed *stress*. Stress is relieved by *strain* or the amount by which a body is deformed. Strain may be elastic, viscous, or plastic in type. Finite strain under a given stress, followed by complete recovery from deformation is known as elastic strain. If the elastic limit is exceeded, brittle failure occurs. When a body undergoes strain indefinitely under a given stress, the *rate* of strain being proportional to the stress, it is said to behave in viscous fashion. When the rate of strain is related to the stress in a linear fashion, the material is called a fluid. Unlike these two conditions, rigid plastic material shows little or no strain until the stress reaches a threshold value or *yield stress*. Stress acting at right angles to a surface is called a normal stress. When such a normal stress is applied to a body with lateral confining stresses the soil may ultimately fail along a planar surface at some intermediate angle between the initial plane and planes at right angles to it (Figure 1.5).

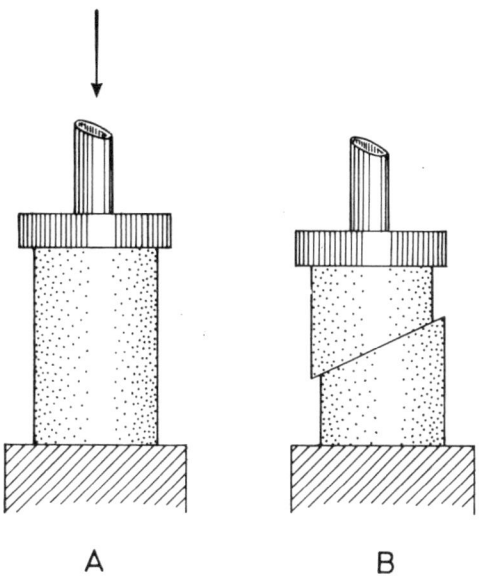

A   B

FIG. 1.5. COMPRESSIVE NORMAL STRESS APPLIED TO COLUMN OF SOIL BY PISTON (A) MAY BE RESOLVED INTO A SHEAR FAILURE IN THE SPECIMEN (B)

Such a failure involves the relative movement of one part of the soil against the other at an intermediate angle. In other words, the normal stress is resolved into a shear stress which is relieved by failure along a shear plane. Samples of soil tested in this way in the laboratory provide quantitative indications of the shear strength of the materials.

Failure in earth materials is described by the Coulomb equation,

$$\tau_f = c' + \sigma_n \tan \varphi \qquad (1.5)$$

where τ_f is the shear strength, c' is the soil cohesion and φ is the angle of internal friction of the material. Three types of soil behaviour may be distinguished by plotting the shear strength (τ_f) against normal stress on the shear plane (σ_n). In the first type, only the frictional component is involved in sands, for example, and is explained entirely by the grain-to-grain forces (slope in Figure 1.6a). In the second case, only cohesion is involved, as in a pure clay, with no frictional component (slope in Figure 1.6b). In the third case, both frictional and cohesion forces are involved (slope in Figure 1.6c). This is, in fact, the commonest condition owing to the widespread presence of soils of mixed grain size.

When a saturated soil is loaded, the weight is carried not only by the inter-granular contacts but also by the water present in the voids. Thus, an increase in the normal stress will result in an increase in the pore-water pressure. This pore-water pressure will serve to

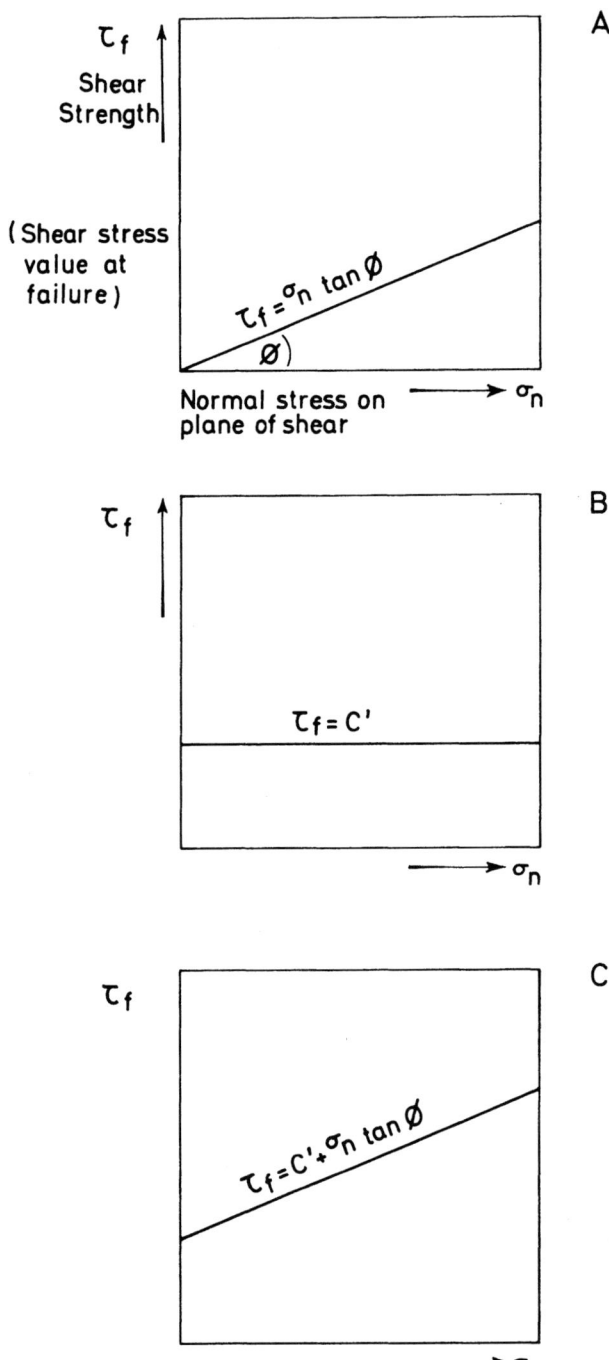

FIG. 1.6. SHEAR STRESS PLOTTED AGAINST NORMAL STRESS FOR THREE TYPES OF SOIL
A – in the absence of cohesive materials, the slope is determined by internal friction angle ϕ; B – the relationship is determined solely by cohesion in pure clay soil; C – in mixed soils, relationship is function of both cohesion and internal friction

reduce the total applied stress because it takes part of the load. The total applied stress minus the pore-water pressure is referred to as the *effective stress*. If the Coulomb equation is re-written to take account of effective stress,

$$\tau_f = c' + (\sigma_n - P_w) \tan \varphi \tag{1.6}$$

it is evident that very high values of pore-water pressure (P_w) can greatly reduce the shear strength (τ_f) of the materials. Conversely, negative pore-water pressures can increase the shear strength as in the partial saturation condition described above.

The extent to which particles in a soil interact is influenced by their size and shape and also by their disposition or *fabric*. Fabric, which refers to the directional properties of a sediment, consists of several sub-fabrics. On the largest scale, features such as bedding planes, laminations and joints make up the *macrofabric*. The most widely studied directional properties of soils are the orientation and dip of the broken fragments or *clasts* (*cf.* clastic, above). These measurements may be applied to the longest axis (or *a* axis) of elongate particles and to the largest plane within the particle (defined by its *a* axis and its intermediate or *b* axis and so known as the *a/b* plane of the clast: Figure 1.7). Such measurements are referred to as the clast fabric or *mesofabric*. By the use of microscopy, particles of the *matrix* (silts and clays) may also be measured: this is the *microfabric*. The rock particles in a soil show varying degrees of parallelism to one another. When a preferred average orientation is clearly evident, the soil fabric is said to be anisotropic. The fabric properties of a soil may be the product of depositional processes (e.g. bedding and particle overlapping or *imbrication*) and the result of post-depositional processes (such as joints produced by desiccation or contraction on freezing).

Other properties affecting soil strength which arise from formative processes and subsequent stress history include compaction, bulk density and degree of consolidation. The accommodation of one particle to another as a soil is loaded reduces the volume of pore space, so reducing the bulk volume. This is the process of *compaction*. The degree to which a soil will compact under a standard load (i.e. change its packing density, see above) depends on factors such as grain size and shape, soil with many platy particles showing greater degrees of compaction than those dominated by equidimensional (equant) grains of the same volume. The *bulk density* (or unit weight) of a soil is the weight of the soil particles and any contained water per unit of volume. The major determinants of bulk density are voids ratio (or porosity) and moisture content. Bulk density also varies with the degree of *consolidation*, defined as the gradual decrease in the water content of a soil under a constant load.

When sediments are subjected to loading, for example by overlying sediments or an overriding glacier, water is expressed from the voids. The degree of consolidation of a sediment increases with the amount of overburden pressure. Thus, if the degree of consolidation ($C1$) of a sample of clay is greater than that to be expected from the thickness of its present overburden or depth below the surface ($C2$), the sediment is said to be *overconsolidated*. The ratio $C1/C2$, the overconsolidation ratio, has been used to assess former thicknesses of overburden or glacier ice. This approach must be viewed with caution, however, because consolidation may also result from desiccation without additional loading.*

*The process of consolidation is distinct from compaction, and the two terms should not be confused.

32

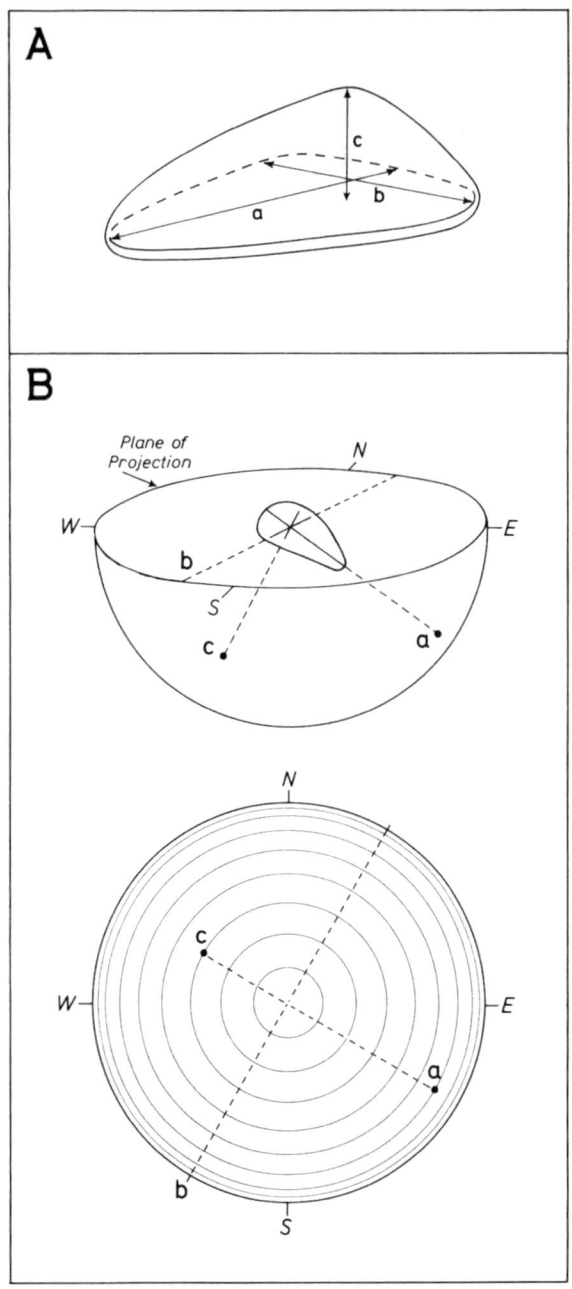

FIG. 1.7.

A – THE THREE PRINCIPAL AXES OF A PARTICLE: *a*, LONG; *b*, INTERMEDIATE; *c*, SHORT.

B – THREE-DIMENSIONAL DISPOSITION OF ROCK PARTICLES SHOWN ON A PROJECTION OF A LOWER HEMISPHERE.

In upper (perspective) diagram, the prolongation of the *a* axis of a pebble centred on the plane of the projection cuts the hemisphere at a, and a prolongation of the *c* axis cuts it at point c. Lower diagram shows conventional projection plot of axial data for these two points. When axes of 25 or more particles are plotted in this way, density of points provides indication of degree to which particles in a soil have preferred orientation. Cf. Figure 5.48.

If overburden pressure on a soil is relieved (for example, by erosion or by thinning of an overlying ice sheet) water may re-enter the voids so that the stressed particles will themselves expand in elastic fashion resulting in a bulk swelling or *dilation*. This process results in tension at right angles to the ground surface which may result, in turn, in the generation of sub-horizontal fissures, large scale fabric elements called *dilation joints*. Geomorphologically active areas may suffer cyclic consolidation and dilation.

1.3 ENDOGENETIC PROCESSES

Landforms are fashioned from earth materials by geomorphic processes. These *exogenetic* processes derive their energy from Earth's internal (*endogenetic*) mobility (tectonism) and from climate (and hence, ultimately from the sun). As a first approximation, landforms may thus be thought of as products of the extremely complex interaction of earth material resistance on the one hand and tectonically- and climatically-derived forces on the other.

Endogenetic processes are energy forces which emanate from within the earth's crust and include crustal or non-isostatic warping within the mantle (epeirogenesis), earthquakes, folding (orogenesis), faulting, metamorphism due to heat flow and vulcanism. In 1959, L. C. King introduced the concept of *cymatogeny*, whereby a landscape, often hundreds of kilometres wide, is either arched or domed to thousands of metres with minimum rock deformation. According to King, modern mountain ranges are usually cymatogenic, not orogenic, and any local rock deformation is presumed to antedate the simple cymatogeny that, followed by erosion and valley cutting, has created modern mountainous terrain. The disparity between present rates of denudation and orogeny is related to the critical factor of time. For example, the European Alps have been uplifted over a period of approximately 4×10^6 years, and are still rising at a rate of about 1 mm per year at the same time as the mean denudation rate is around 0.1 mm per year. Ultimately, the irregular jagged relief of the Alps will be reduced to a degree of flattening characteristic of the shield areas of the world.

Although endogenetic processes have often been ignored in geomorphological studies (see Chapter 2, p. 104), it is increasingly appreciated that effects of such movements, expressed in earthquake activity and in vulcanicity, are important influences in particular areas of the world including the circum-Pacific belt and areas of young-fold mountains. It is in such areas that the endogenetic influences can provide additional energy to alter the rate of exogenetic processes. Although the mechanisms of endogenetic processes are the province of the geologist and the geophysicist, the effects of endogenetic movements should be kept in mind as they influence the incidence of certain geomorphic processes such as mass movement as analysed in New Guinea (see p. 105). Endogenetic processes are not considered further in this volume but volcanic forms are reviewed by Rittman (1962), Ollier (1969) and Francis (1976) and relationships between the major morphological elements of the continents and oceans and the global tectonic framework are considered at an introductory level by Tarling and Tarling (1971).

1.4 EXOGENETIC PROCESSES

1.4a Weathering Processes

The complex interaction between chemical, physical and biological factors ultimately results in the disintegration of bedrock to form new products. This process is known simply as *weathering*, which is both selective and differential, and represents the interaction between the lithosphere, atmosphere, hydrosphere and biosphere. The various ways in which rocks and minerals are broken down mechanically are well documented in the literature together with the processes of chemical alteration and, therefore, only a summary of the salient points is given here.

The rate of chemical weathering is highly variable and is controlled in any region by the composition of the parent material, morphology, climate, biological activity and perhaps the most significant factor of all, time. Solution, hydration, hydrolysis, oxidation, reduction and carbonation are the main reactions in *chemical weathering*. Generally, the first stage of chemical weathering is *solution*, the degree of which depends upon the solubility of a mineral in water and weak acids. Common salt, for example, is extremely soluble in water compared with gypsum. In contrast, limestone is insoluble in water but reacts with carbonic acid to form calcium bicarbonate which, in turn, is soluble in water:

$$CaCO_3 + H_2O + CO_2 \rightarrow Ca(HCO_3)_2$$

Volume changes, which may enhance physical weathering, are often caused by solutions precipitating chemicals. Also, the products of other chemical weathering processes are transported as carbonates, sulphates or chlorides in solution in rivers and groundwater. Solution is probably the dominant weathering process on limestone coasts but the chemical reactions involved are exceedingly complex and present many problems that still require examination under controlled conditions.

Hydration, the chemical combination of water with another substance, is an exothermic reaction involving a significant change in volume which is also an important process in the formation of clay minerals and may be important in physical weathering with regard to exfoliation and granular disintegration.

Hydrolysis is a chemical process of decomposition involving the addition of water which may be expressed by the equation

$$Na_2CO_3 + H_2O \rightarrow HCO_3^- + OH^- + 2Na^+$$

It can be seen that the above chemical reaction is between the H or OH ions of water, and the ions of the mineral sodium. During the decomposition of orthoclase, the dominant alkali-feldspar of granite, for example, two types of hydrolysis reactions occur. In the first, shown by the equation

$$K(AlSi_3O_8) + H^+ + OH^+ \rightarrow H(AlSi_3O_8) + K^+ + OH^-$$
$$\text{(water)} \qquad \text{(insoluble)} \quad \text{(soluble)}$$

the products of hydrolysis are an undissociated insoluble fraction and a double base KOH. Thus, in this case, orthoclase reacts like a salt of a weak acid and a strong base. Following an interchange reaction the insoluble fraction $H(AlSi_3O_8)$ reacts like a salt of a weak acid (silicic acid) and a weak base (aluminium hydroxide). As both the H^+ and the OH^- ions of water are adsorbed by the mineral fraction, the second type of hydrolysis occurs, resulting in the formation of silicic acid and aluminium hydroxide:

(a) $H(AlSi_3O_8) \rightarrow Al_{1/3}(HAl_{2/3}Si_3O_8)$

(b) $Al_{1/3}(HAl_{2/3}Si_3O_8) + 3H^+ + 3OH^-(H_2O) \rightarrow Al(OH)_3 + (H_4Si_3O_8)$

Hydrolysis is believed to be the main type of chemical process associated with *spheroidal weathering*, a term which is used for a constant volume alteration type of exfoliation whereby weathering products are removed in groundwater.

Oxidation, or the process of combining with oxygen, results in an increase in positive valence or a decrease in negative valence as illustrated in the example:

$$Cu^+ \text{ is oxidized to } Cu^{++} \text{ and } S^{--} \text{ is oxidized to } S$$

In other words, this process represents the removal of one or more electrons from an ion or an atom. Also, oxidation is important in the complete and partial decomposition of the mafic minerals, such as amphiboles, biotite, olivines and pyroxenes which contain Fe_2^+ iron, because of an increase in positive electrical charge in the crystal structure. In this way, the partial decomposition of biotite, for example, results in the formation of mica clays.

Reduction is simply the opposite process of oxidation and generally occurs under anaerobic conditions in the absence of free oxygen.

Carbonation is another process of chemical weathering by which minerals containing lime, soda, potash or other basic oxides are changed to carbonates by the action of carbonic acid in water or in air.

Physical weathering, the *in situ* breakdown of material, is caused by thermal expansion and pressure release. These stresses may originate within rocks or be applied externally. Because the geological time factor cannot be simulated experimentally it is not surprising that detectable physical weathering has not been recorded in laboratory work hitherto.

The arched fractures which transgress rock structures and divide them into a series of massive sheets or exfoliation slabs are believed to result from the release of pressure consequent upon the erosion of superincumbent strata. This form of physical weathering is invariably termed *pressure release unloading*. Unloading or exfoliation sheets in different rock types give rise to particular landforms. It is pertinent to mention here that exfoliation is commonly used to describe processes such as spheroidal weathering which involves constant volume alteration as already mentioned; unloading, or expansion due to innate forces within rocks; and flaking, a type of expansion resulting from external forces.

Differential weathering reflects lithological and structural variations in rocks. Cavernous weathering pits, as well as many caves in rocks other than limestone, are examples of features formed by the differential weathering of less resistant layers underlying more resistant beds. Crystal growth with freezing of water or from salt-rich solutions is an important source of mechanical stress in rocks. The principal types of physical weathering and the related geomorphic processes are summarized in Table 1.5.

During weathering the rock-forming minerals (quartz, feldspar, mica, olivine, pyroxene, amphibole and the carbonates) are altered into secondary minerals. The rate at which this change occurs determines the rate of rock weathering. Mineral weathering, of course,

Introduction

Table 1.5
GENERALIZED PHYSICAL WEATHERING PROCESSES

Primary Process	Mechanism	Geomorphic Process
Stress relief (unloading)	Dilation jointing	Sheeting, exfoliation
Differential expansion of mineral crystals	Intergranular and intragranular stress	Granular disaggregation
Ice crystal growth	In rock: Thermoclasty	Gelifraction Granular disaggregation
	In soil: Frost heaving	Ground heaving and cracking Disruption of bedding (involutions)
Salt crystal growth	Haloclasty	Granular disaggregation
Adsorption of water (hydration)	Volume increase of hydrated minerals	Granular disaggregation Spheroidal weathering
	Swelling of mixed-layer clays (e.g. montmorillonite)	Disruption of bedding (involutions), ground heaving and cracking

depends upon several properties, the more important of which include structure, composition, and crystal size, shape and perfection. It is generally acknowledged that the mechanical stability of accessory minerals (non-essential rock constituents) like tourmaline, rutile and zircon, for example, is controlled largely by their physical properties. It seems likely, though, that differences in the stability of a particular mineral vary from one area to another according to variations in climate, topography, and vegetation which are known collectively as the environment of the mineral. In addition, it is believed that the resistance of a mineral to weathering depends upon its particular variety.

As a result of many detailed weathering and soil profile studies (see references at the end of this Chapter), a *weathering series* has been constructed for the commoner minerals as shown in Table 1.6.

Table 1.6
ORDER OF PERSISTENCE OF MINERALS IN THE GEOLOGIC COLUMN

Least persistent minerals are listed first. Read down columns

Olivine	Topaz	Apatite
Actinolite	Andalusite	Biotite
Diopside	Hornblende	Garnet
Hypersthene	Epidote	Monazite
Sillimanite	Kyanite	Tourmaline
Augite	Staurolite	Zircon
Zoisite (epidote group)	Magnetite	Rutile
Sphene	Ilmenite	Anatase

However, because of the innate structure and composition of a mineral, as well as the environmental factors mentioned above, the reader should be aware that there is no absolute scale of degree of weathering.

1.4b Climate and Surface Processes

The term 'weathering' clearly recognizes the general importance of weather and climate in the alteration of rock surfaces. Various attempts have been made to specify this relationship using mean annual values of temperature and precipitation, an example of which is shown in Figure 1.8A and B. The result is a broad generalization and rests on certain

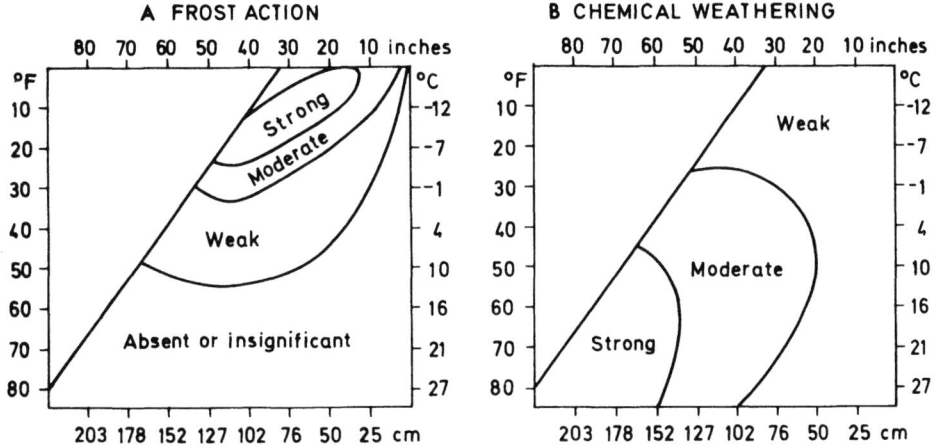

FIG. 1.8. INCIDENCE OF STRONG, MODERATE AND WEAK FROST ACTION (A) AND CHEMICAL WEATHERING (B) IN TERMS OF MEAN ANNUAL PRECIPITATION AND TEMPERATURE
redrawn by permission from *Annals Assoc. Amer. Geog.*, 40, L. C. Peltier.

assumptions which cannot always be supported. For example, mechanical breakage of rock is effected by a range of processes and is by no means limited to climates in which frost is frequent. Furthermore, while chemical weathering is enhanced by readily available water and high temperatures, it occurs to some extent in all climates. Nevertheless, the differentiation of weathering types on the basis of gross climatic parameters is a valuable first step in understanding process variations in space and time.

A similar approach has been applied to the classification of landforms on a continental or regional scale. The origins of this approach are closely related to the scientific acceptance of the glacial theory and the recognition that many of the larger landforms found in areas of the world now experiencing a temperate climate owe their origin to glacial erosion and deposition, e.g. cirques and moraines. Similar inferences have been applied to large landforms in other climatic zones, such as the extensive plateau-like surfaces of intertropical Africa *(pediplains)* with isolated residual hills (known by the German word *inselberg*) rising above them, together making up the essential elements of the *inselberg-landschaft* particularly associated with the seasonally wet savannas.

The generalized climatic boundaries (in terms of mean annual temperature and precipitation) of some geomorphic process realms (mass movement, surface runoff and wind action) were set out by L. C. Peltier in 1950. These were combined with the climatic boundaries of the weathering types to yield nine *morphogenetic* regimes. Classifications of this kind are superficially attractive but they are dangerous because they divert attention from the functional complexity of the relationships so portrayed. One well-known example which demonstrates this fact is the work of Langbein and Schumm (1958). These authors have shown that the intensity of fluvial erosion as expressed in the sediment yield from a drainage basin, is not related to climatic parameters in a simple way. Specifically, erosion tends to increase as precipitation increases up to a point at which the vegetation cover becomes multi-layered and so protects the surface. The sediment yield and fluvial erosion are greatest in climates with seasonal aridity including climates classified on an annual basis as semi-arid, sub-humid and monsoonal. Lee Wilson (1973) used this seasonal factor in classifying geomorphic process suites on the basis of dynamic climate types. The six resultant climatic regimes (Figure 1.9) were regarded as the basis of *climate-*

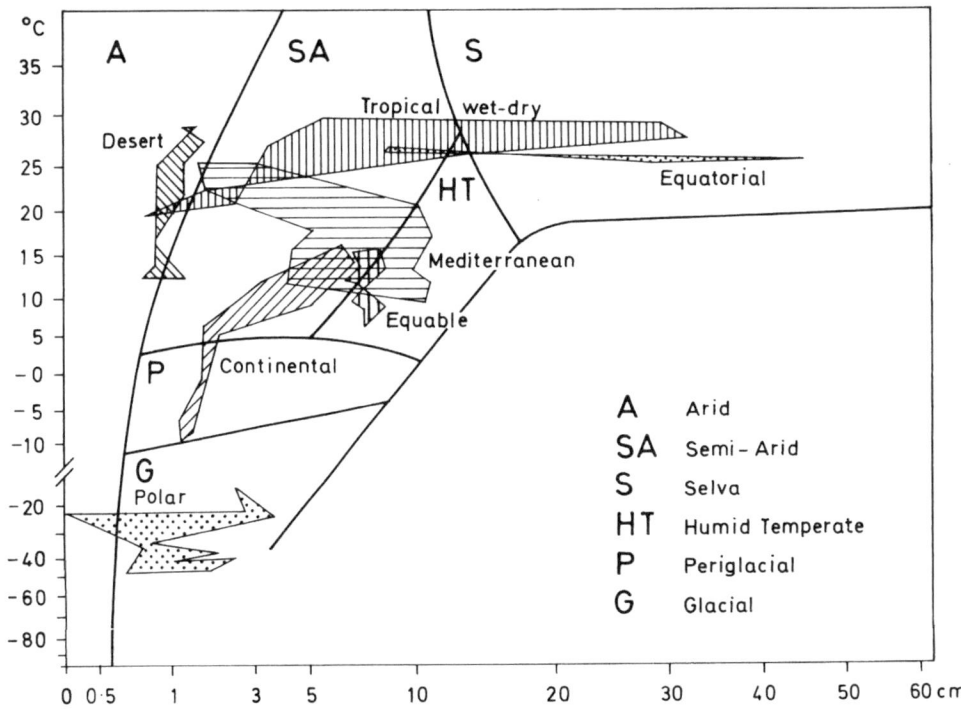

FIG. 1.9. CLIMATIC REGIMES AND CLIMATE-PROCESS SYSTEMS
The shaded areas are climographs defined by plotting the mean temperature and precipitation for each month of the year for a location representative of the six climatic regimes. The environmental ranges of the six climate-process systems is shown by solid boundary areas. (After Wilson in Derbyshire 1973, 279.)

process systems, although the poor level of understanding of how these climatic inputs are expressed in landforms was recognized.

Variations in the rate and incidence of physical, chemical and biotic reactions at the Earth's surface form the basis of the *morphoclimatic* regions of the French geomorphologists Jean Tricart and André Cailleux. The emphasis here is on the distribution of present-day bioclimatic and geomorphic *processes* and not on regional assemblages of *landforms* as in the work of Julius Büdel (Figure 1.10). The problem of scale was specifically recognized by Büdel who pointed out that the climatically diagnostic landforms tend to be the smaller features, the more extensive forms being the product of non-

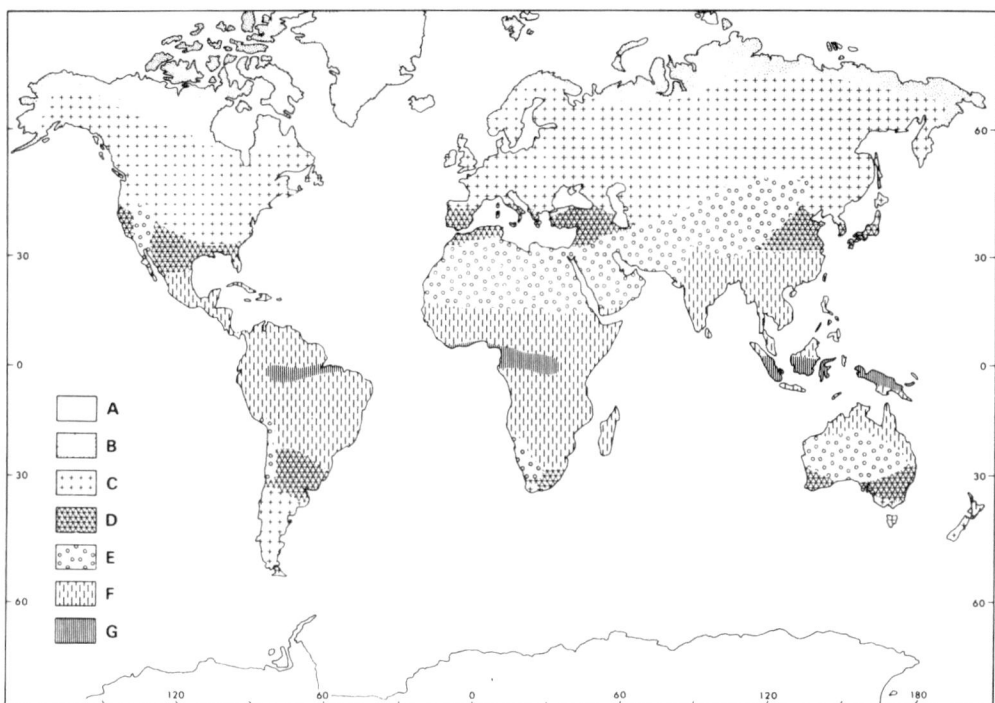

FIG. 1.10. CLIMATOGENETIC ZONES OF THE WORLD ACCORDING TO BÜDEL (1969).
A – glaciated areas; B – zone of pronounced valley formation; C – extra-tropical zone of valley formation; D – subtropical zone; E – arid zone; F – circum-tropical zone of excessive plain formation; G – intertropical zone of partial plain formation.

climatic, especially tectonic (mountain building) and epeirogenic (continent forming) factors.

The basic assumption of climatic geomorphology is that climate, varying over space and time, controls the nature and incidence of the geomorphic processes which, in turn, combine to produce a range of distinctive landform units. This paradigm is extremely difficult to test in the case of many individual landforms, let alone whole landform assemblages or regional landscape types. A major problem is to discriminate between the effects of present climatic inputs and those of the past. Occasionally, of course, the relationship may be clear as, for example, in the case of recently deglacierized slopes upon which current processes are imposing a sequence of gully, fan and avalanche landforms (Rapp, 1960) but, in other than waning glacial conditions, such clarity is less common. Study of the literature of the residual rock masses called *tors* may be used to exemplify the fundamental problem posed by climatic geomorphology. Tors are composed of joint-bound rock units surrounded by uniform slopes covered with boulders but containing residual masses of deeply weathered soil, and have been explained as originating in two different ways. In the first model, a period of deep weathering acts along joints of variable spacing so that rock units are decomposed and an uneven *weathering front* forms. Subsequently, the weathered layer is progressively removed by mass movement under a frost climate, by etching as rivers incise their channels following a fall in sea level, or both, and the tors are revealed at the surface. In the second model, denudation and weathering proceed together: slope retreat aided by rock fall and frost action consumes all but the residual rock masses on crests and summits.

These models, often referred to as the two-cycle and one-cycle hypothesis respectively, specify a series of processes to explain the resultant forms, and associate the processes with broad climatic types. However, tors appear to be *azonal*, in that they occur over a great range of present-day climatic types from polar desert to equatorial. It follows that they cannot be regarded as climatically-diagnostic landforms. If it is true that, with certain exceptions, landform assemblages are not diagnostic of particular, narrowly-defined climatic regions, the assumed relationship between climate and geomorphic processes and between geomorphic processes and landforms must be assessed. A good deal of current research is concerned with the second of these relationships and much of this book is devoted to summarizing it. Even so, it will be evident from Chapters 2–5 that an understanding of the relationship between processes, landforms and sediment types is still somewhat tenuous. Such understanding tends to advance in spasmodic fashion, as is well-illustrated by the development of fluvial geomorphology since 1890. In contrast, the relationship between climatic inputs and geomorphic processes remains for the most part poorly documented. As this is the fundamental relationship upon which dynamic work in climatic geomorphology must rest, the climatic-geomorphic approach remains an acceptable descriptive scheme on the world scale but is insecurely founded in dynamic terms.

Despite the relatively weak dynamic basis of climatic geomorphology, there is abundant evidence of superimposition of new landforms upon older surfaces formed under a different process regime. The study of the whole complex of landscape made up of elements of past and present environmental conditions was called *climatogenetic geomorphology* by Büdel. Many geomorphologists have drawn attention to the rapid alternations of climate characteristic of the Quaternary (the last 2.5 million years) as an important source of variation in the landform components in a region. Conditions in parts of the middle latitudes, for example, shifted from warm temperate to polar desert several times during this period. As a result, the landforms of a notable percentage of the cool temperate lands of the earth are of glacial and periglacial origin.

In warmer climates such as sub-humid to semi-arid parts of eastern Australia, present day river channels are contained within much larger channels derived from past climates which were conducive to much higher runoff yields. Great aeolian dune systems in intertropical Africa have been 'overprinted' by soil development and sheet and gully erosion with a shift toward more humid climatic conditions. The time scale of such shifts varies and serves to underline the problem posed by *relaxation time*, that is the time-lag between the arrival of a new stimulus derived from a change in the climatic input and the development of a suite of landforms in equilibrium with the new environment. The magnitude of a climatic event may itself influence the relaxation time. A catastrophic flood may result in erosional and depositional modifications of a floodplain which will persist with only minor modifications until the next catastrophic event of similar magnitude.

On the other hand, a climatic change such as a fall in average temperatures will result in persistence and accumulation of snow at lower altitudes in mountain areas. If this persists, glacier ice will form and modification of the high level valley heads will begin, leading to the development of cirques. The relaxation time associated with such a long-term change in the climatic balance may be of the order of centuries. Successful recognition of a series of landforms in an area as the result of one or more shifts in the climatic environment is dependent on relaxation time in each phase being sufficient for landforms to reach an approximate equilibrium with the climatically-induced processes. It may be objected that climate, changing continuously in amplitude and duration, produces threshold conditions for landform change of varying lengths so that any one landscape may preserve only those landforms which were produced by long climatic episodes or by the presence of coincidental non-climatic factors such as resistant rock or tectonic uplift. It is for such reasons that landforms are regarded as the weakest basis for stratigraphical studies.

Nevertheless, given that the level of resolution of the climate-landform sequence is rarely very high, the imprint of past climates on the landforms of a region is always present and may be very strong. In some cases, the gross landform units (at the scale of valleys and interfluves) are in all essentials the product of geomorphic processes which are no longer dominant or even present in an area. If this were not so, the knowledge of landforms and of geomorphic processes would be far less advanced than it is today. Present knowledge of landform genesis remains founded on the study of relict rather than currently active landforms, in many cases because modern correlatives of some ancient landforms are either unknown or inaccessible. Moreover, modern analogues of many Pleistocene landforms, where they exist, are unlikely to be precise because, for example, glaciation of lowland, mid-latitude areas occurred under a climatic regime not replicated anywhere on earth to-day. Despite its deficiencies, notably its poor level of resolution, the climato-genetic approach remains, therefore, an important framework for the study of landforms.

1.5 PROCESSES AND SYSTEMS

An approach to geomorphological processes within the context of a geomorphological equation (Section 1.1) necessarily requires attention to be devoted to the way in which processes, materials and resulting forms are interrelated. Relationships have been attempted at the scale of climatic control of geomorphological systems as reflected in the progress achieved towards a climatic geomorphology reviewed in Section 1.4b. In addition, it is desirable to adopt a fundamental approach to the analysis of relationships between processes, materials and forms which is applicable at all scales. Such an approach was sought during the decade of the 1960s when the investigation of geomorphological processes was increasing in significance, at a time when appropriate statistical and mathematical methods were becoming of greater import due to the advent of improved computational facilities, and when a greater range of methods for the measurement of geomorphological processes was becoming available.

General systems theory provided a basic conceptual approach which was advocated generally by Chorley in 1962. By focusing upon interaction between variables and upon equilibrium situations this stimulated a fundamental shift in the direction of geomorphology. In 1971 Chorley and Kennedy provided a systems approach to physical geography and published an important text which was structured on the basis of the types of systems which they distinguished. *Morphological systems* were proposed as those which depend upon the interrelationships between properties of form. *Process-response systems* were designated as occurring in situations in which a particular parameter of process could be a dependent variable related to independent variables of form, process or materials. Thus the amount of mass movement on a specific slope might be envisaged as the response to the processes occasioned by input of precipitation to a slope with particular intrinsic features of angle, height, and underlying material. Such an approach depends upon a knowledge of the input of energy, transfer of energy and output of energy from a system.

It is expedient to distinguish systems and subsystems so that a *cascading system* is linked by a cascade of energy in the way that the output of energy or material reaching the base of a slope constitutes one of the inputs to the stream channel system at the slope base. Similarly, over a coastal slope the output from the slope base is one input into the coastal system. All three types of system may be subject to the influence of prominent variables which act as regulators upon the operation of process-response, of cascading, or of morphological systems. Such regulators can promote changes in the distribution of mass and energy and human influence is one variable which has been particularly important in determining the character of such *control systems* (Gregory and Walling, 1979).

Use of a systems approach has been widely advocated in many branches of geomorphology since 1960. In the field of glacial geomorphology, Andrews (1975) and Sugden and John (1976) developed approaches founded upon systems ideas. This involved (Sugden and John, 1976) the identification of geological environment, climate, and regional relief and slope forms as independent variables and glacier morphology and glacier size as dependent variables. The complex interrelations between the variables determines how a single glacier flows and operates. In a comparable way King (1974) reviewed processes in the coastal system by considering waves, tides, and winds as processes which operate on beach materials to determine the response variables which include the form and plan of the beach and the coast.

Some criticisms have been directed towards the systems approach. It has been said, for example, that it merely formalizes what is already known and that it affords a method which is difficult to apply completely because it depends upon very complex methods of analysis. However, the systems approach is important because it directs our attention to the complete range of dependent and independent variables which are involved in process-response systems and as such ensures that we identify the components involved in contemporary landscape dynamics. Subsequent chapters of this book are intended to introduce some of these components and if we succeed in providing at least a partial basis for an appreciation of the nature of contemporary geomorphological systems, that may be a foundation for further progress.

FURTHER READING

A useful outline of rock composition and type can be found in ERNST, W. G., 1969, *Earth Materials* (Prentice Hall) and an introduction to the behavioural properties of rocks and soils in WHALLEY,

W. B., 1976, *Properties of Materials and Geomorphological Explanation* (Oxford). The processes and products of weathering are considered in OLLIER, C. D., 1969, *Weathering* (Oliver and Boyd). A collection of some of the classic papers in climatic geomorphology is available in DERBYSHIRE, E., 1973, *Climatic Geomorphology* (Macmillan). The systems approach is employed in stimulating manner in CHORLEY, R. J. and KENNEDY, B. A., 1971, *Physical Geography: A Systems Approach* (Prentice Hall) and a valuable general reference is FAIRBRIDGE, R. W. (ed.), 1968, *The Encyclopedia of Geomorphology*, Vol. 3 of Encyclopedia of the Earth Sciences (Van Nostrand Reinhold).

2
Drainage Basin Processes

All landsurfaces are composed of drainage basins which are the units of the land surface which collect, concentrate and promote the movement of water and sediment (Figure 2.1). The drainage basin is the unit which is basic to the transmission of energy from climate through geomorphological processes to give morphological results or landforms. Although the drainage basin can be visualized as a unit of study in ice-covered areas or in

FIG. 2.1. HEADWATER AREA OF TRIBUTARY OF RIVER EXE, DEVON
Snow covers the interfluve areas and the slopes of many of the valley sides are snow-free. Fluvial characteristics of such drainage basins may be investigated employing one of several approaches (Figure 2.2), and the process operating on several types of slope element can also provide focus for study. Maximum slope angles of this type of area have been investigated in relation to character of the subsurface material (p. 77) and, in addition to an expanding drainage network, subsurface stormflow (p. 56) contributes to the generation of stream hydrographs.

deserts it is most useful where weathering, mass movement and water flow provide the basis of landscape morphogenesis. It is difficult to establish when the significance of the basin was first appreciated. This depended upon the realization that water from precipitation generates streamflow and Pierre Perrault in 1674 had measured rainfall over the Seine basin in France and had shown that as the river flow of the Seine was only one sixth of the rainfall amount then precipitation was ample to sustain river flow. This deduction therefore produced the concept of the basin as an area of the land within which precipitation could generate streamflow.

Drainage basins are composed of slopes but not all slope studies have used the drainage basin as a frame of reference. Although studies of drainage basins and rivers have often been conducted independently from investigations of slopes, there are good reasons for an integrated study of the two themes. This integration is attempted here by outlining the perspectives for contemporary understanding according to the aims of previous studies (2.1); by describing the processes which occur in the basin (2.2); by enumerating the physical controls upon drainage basin processes (2.3); and by proceeding to outline spatial patterns of processes (2.4).

2.1 PERSPECTIVES ON THE DRAINAGE BASIN

At the end of the nineteenth century geomorphology was provided with two alternative models and strategies for the development of research. The one widely adopted had been proposed by W. M. Davis as the normal cycle of erosion in which landscape development was seen as a function of structure, process and time. The cycle approach was adapted for nearly half a century and produced a geomorphology which was essentially timebound and concerned with temporal evolution of landscape over periods of time as much as millions of years. The significance of rivers and slopes within the cycle of erosion was based upon an assumed knowledge of processes. Although the demerits of the cycle have been noted to include lack of a quantitative approach and of recent knowledge of recent earth history, and of its general evolutionary overtones following Darwinian ideas of evolution, it did provide a useful vehicle for the expansion and crystallization of geomorphology.

The alternative model available in the early twentieth century was that advocated by G. K. Gilbert. After working in many parts of western USA and undertaking laboratory experiments he proposed that landscape processes should be studied at a more fundamental level than was possible under the Davisian cycle. This approach, which required a greater understanding of the principles of engineering and the methods of mathematics, did not readily find favour. Instead the Davisian approach dominated the study of rivers until 1950 and it retarded the study of slopes because insufficient attention was given to slope processes. Occasional diversions included the development of slopes as proposed by Penck (1924) in his particular style of landscape development.

Until 1950 therefore, slopes found little place in geomorphological writing and rivers were usually approached in terms of long-term landscape development. Since 1950, and beginning before that time, the geomorphological study of hillslopes and rivers in drainage basins took on a new complexion as a consequence of developments internal and external to geomorphology.

2.1a Drainage Basin Studies

The rise of interest in the drainage basin may be perceived as the consequence of growth in seven areas of study (Figure 2.2). These all offered alternatives to the Davisian approach and yet each tended to be pursued rather independently (Gregory, 1976). Studies of the *morphometry* of stream networks (Figure 2.2) were advocated by R. E. Horton in 1945

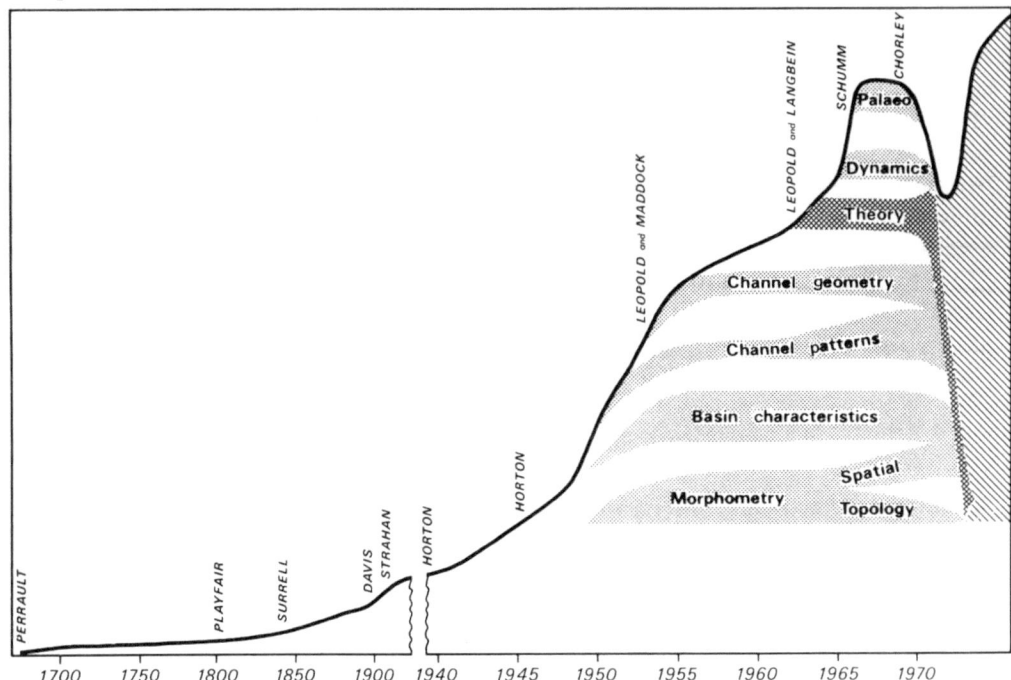

FIG. 2.2. DEVELOPMENT OF APPROACHES TO STUDY OF THE DRAINAGE BASIN
Increase in studies after 1940 depicted in terms of seven approaches described in text (p. 46). The *y* axis of the graph is suggested to represent degree of interest in studies of drainage basins, and use of the drainage basin unit may have declined slightly in the 1970s as specific approaches were investigated in parts of the basin, prior to further increase in basin studies catalyzed by greater understanding of the inter-relationships of the seven themes (After Gregory, 1976).

and 'laws' of drainage composition were tested and eventually rejected as it was appreciated that they were the consequences of the statistics underlying the methods of ordering. More recently such studies of morphometry have become concerned with the mathematical topology of stream networks and with the spatial patterns displayed by values of morphometric parameters.

A second type of approach was also founded by Horton. The basis for this second theme was that *drainage basin processes* are the consequence of the way in which drainage basin characteristics of relief, soil, land use and rock type modify inputs of water and radiation. The objective of such studies was the development of equations which could express the relations between inputs, basin characteristics and output, so that outputs could be predicted for areas with no records of river flow or sediment production. Considerable progress was made in this field although it was not necessary to know how the landscape worked to model its performance.

A third type of approach was represented by studies of the *hydraulic geometry* of stream channels. For example, work by engineers on irrigation canals in India, had already demonstrated that it was possible to design and construct a stream channel which was 'in regime'. This channel would have its size, slope and shape just capable of conveying the water and sediment supplied to it. Such a design situation was desirable because it meant that the irrigation canals did not erode by scouring or become smaller through siltation. The design of these canals required a knowledge of the relationships between the form and processes in a river channel cross section and such relationships were identified by Leopold and Maddock (1953) as the hydraulic geometry (Figure 2.2) of stream channels. Whereas regime theory was developed for controlled situations, like irrigation canals with fairly constant flows, it is much more difficult to apply in the case of natural channels where the water flow rates vary considerably. Also developed for practical reasons were studies of *river channel patterns* particularly in areas where rapid changes had consequences for irrigation and human activity. Early studies of river channel pattern were concerned with the dimensions of meandering channels and with their relations to stream discharge.

These four themes figured in research on stream channels before 1960 but they were then complemented by three others which were concerned with *drainage basin dynamics*, with theory, and with palaeohydrological objectives. Whereas the drainage basin had long been thought of as a static unit which transformed climatic inputs into water and sediment yield, it was appreciated in the 1960s that this transformation could be achieved only by a basin which was itself dynamic. Thus not only does the water table fluctuate over the year but the amount of moisture in the soil, the extent of the drainage network and of saturated areas near the streams all vary over time. *Theoretical approaches* also appeared in the 1960s as a number of workers attempted to propose theories that accounted for relationships and patterns of landscape development. Whereas the earlier studies of basin characteristics, of morphometry, of hydraulic geometry, and of channel patterns had necessitated the use of statistical methods, such techniques produced relationships which were not necessarily explanatory. It was therefore necessary to develop theory (Figure 2.2) which was often based upon physical principles.

Six rather disparate strands of fluvial research may be discerned up to 1965 and an integrating conceptual basis for further developments had not clearly emerged. This was remedied by the proposal of S. A. Schumm that geomorphology should be concerned with palaeohydrology and with the metamorphosis of rivers and river channels. This *palaeo approach* (Figure 2.2) seeks to use an understanding of present processes as a basis for interpretations of past changes of rivers, river channels and drainage basins. The physical geographer studying rivers is therefore primarily concerned with the variations in water and sediment production over time and with the effects of processes upon landforms at present, in the past and also in the future. More recently greater attention has been devoted to applied studies (e.g. Schumm, 1977).

2.1b Slope Studies

Studies of slopes cannot be resolved into such definite categories but at least four types of study may be distinguished. The earliest approaches were concerned with field measurement and morphology and de la Noë and Margerie (1888) advocated the measurement of slopes. This approach was intensified after 1950 when a number of workers developed the technique of morphological mapping (Savigear, 1965) and others employed detailed analysis of slope profiles (Young, 1972). A difficulty with this essentially *morphological approach* is that the form of slopes is not necessarily a guide to the processes which fashioned the slope and which still operate today. Because slopes are so complex and occur in a great diversity over the entire land surface of the earth, therefore it is not easy to establish statistical correlations between data on slope form and values of climatic, rock type and soil parameters as a basis for useful *models of slope equilibrium and development.*

Whereas such empirical methods using large amounts of field data are essentially inductive in character, an alternative approach was based upon deductive models. Such deductive approaches assume a knowledge of slope processes; they are illustrated by models of slope development proposed by W. Penck in 1924 and by other workers in the 1940s and 1950s. A third type of study developed around the *analysis of mass movements* and particularly of landslide behaviour. This was initially of particular interest to engineers and it required a knowledge of the engineering characteristics of the materials on a slope, but ultimately it has been a type of study followed profitably by a number of geomorphologists.

In addition to the investigation and analysis of mass movements on slopes, the study of the *processes of water, sediment and solute movement* over a slope has engaged a number of geomorphologists as a fourth focus for research. The third and fourth types of study, both primarily concerned with slope processes, have given the basis for more elaborate and physically meaningful models of slope processes and slope development, including derivation from physical theories in the same way that a theoretical approach may be discerned in studies of rivers (Figure 2.2).

The types of study employed thus highlight two particular problems which may be referred to as the purpose of geomorphology and the methods of the subject. Two alternative purposes for geomorphology are provided by either the study of present processes or of landform evolution as a reflection of past and changing processes. The first alternative has been followed by researchers who have concentrated upon the measurement, analysis and investigation of the processes at present fashioning the landscape. The second, longer-established, alternative has been adopted by those workers wishing to establish the sequence of stages of landscape evolution and so leads to the chronology of landscape development. These two alternatives were depicted by R. J. Chorley in 1971 as two ever-diverging tightropes (see Chapter 1).

2.1c The Geomorphological Equation applied to the Drainage Basin

The purpose of drainage basin geomorphology may be presented very simply in terms of a geomorphological equation (Chapter 1, Section 1.1). Studies of hillslopes and drainage basins are concerned with the three elements of the equation: the study of form or morphology; of processes; and of earth materials.

Within a specific drainage basin a particular slope between interfluve and stream can be measured morphologically in terms of its slope angles and the form of the slope can be plotted. For the same slope it is possible to identify each of the processes operating, to measure their frequency and intensity. It is also possible to take samples of the soil, superficial deposits and rock beneath each element of the slope and then to analyse these samples in the laboratory to reveal their particle size characteristics, shear strength and infiltration capacity. Each of these items could be investigated individually but as a second stage it is necessary to show how the processes operating at present on the materials of specific character are modifying or creating the form of the slope. It may transpire that the morphological characteristics of the slope cannot be explained entirely in terms of present processes and materials because for example, climate has changed or man has affected the slope by cultivation so that past climates or man's influence have to be taken into consideration. Only by attempting to solve the relation between present and past processes and landform can we hope to indicate how the slope will alter or adjust in the future.

If a small drainage basin is homogeneous in character it is possible to derive indices of the form of the basin which include its relief and the density of the drainage network. Similarly the character of the soils, superficial deposits and underlying rocks can be described with particular reference to their ability to transmit water. The easiest way to analyse basin processes is to measure the water and sediment discharged from the outlet of the basin and this can provide an expression of the way in which the streamflow rate varies over time and this can identify the *discharge hydrograph*. Within the single basin it is possible to consider the materials, the morphological aspects and the processes separately, but it is desirable to proceed to see how the climatic input to the basin is transformed by the materials and morphology available to produce the characteristic processes. If we can achieve this balancing of the equation (p. 19) it is then feasible to see how much water is removed over time and with what frequency, and also to analyse the rate and frequency of sediment and solute movement.

Such an analysis will indicate how the small basin has operated in the recent past but it could also be used to show how changes may take place in the near future. Thus if the basin remained unchanged we could calculate how much water or sediment could be removed in the next ten or twenty years. However if a reservoir were constructed near the outlet of the basin we could attempt a calculation of the rate at which sediment may accumulate in the reservoir. This illustrates the application of the geomorphological equation (see p. 20) because the rate of sediment accumulation could afford some idea of the rate at which the capacity of the reservoir might be reduced.

2.1d Types of Model

Achievement of relationships between form, process and materials is dependent upon the use of several types of model. To relate the three elements it is often necessary to collect data from a number of slopes or from a number of basins and to relate them by reference to a particular type of model. It is possible to distinguish three types of model namely empirical, theoretical and experimental.

Empirical models are essentially inductive in nature, whereby a large amount of field data is collected and is analysed by reference to an assumed model which is often statistical in character. This approach may be illustrated generally by Figure 2.3 which shows the streamflow hydrographs from contrasted catchments. If the four small basins have a similar climatic input then their differing responses may be ascribed to their different catchment characteristics. We may therefore anticipate differences in streamflow from a single storm and the hydrographs would differ from permeable and impermeable rocks, from basins with different shapes, from basins with contrasting soils and from catchments with distinct vegetation cover (Figure 2.3). Similarly we might envisage several different slopes each on the same type of material and in a comparable position but subjected to different amounts of rainfall. This difference could then be related to the intensity and frequency of mass movements on the slope.

Such empirical methods, although frequently used, are often confronted by the complexity of reality. It is often difficult to find a number of slopes which are identical in every

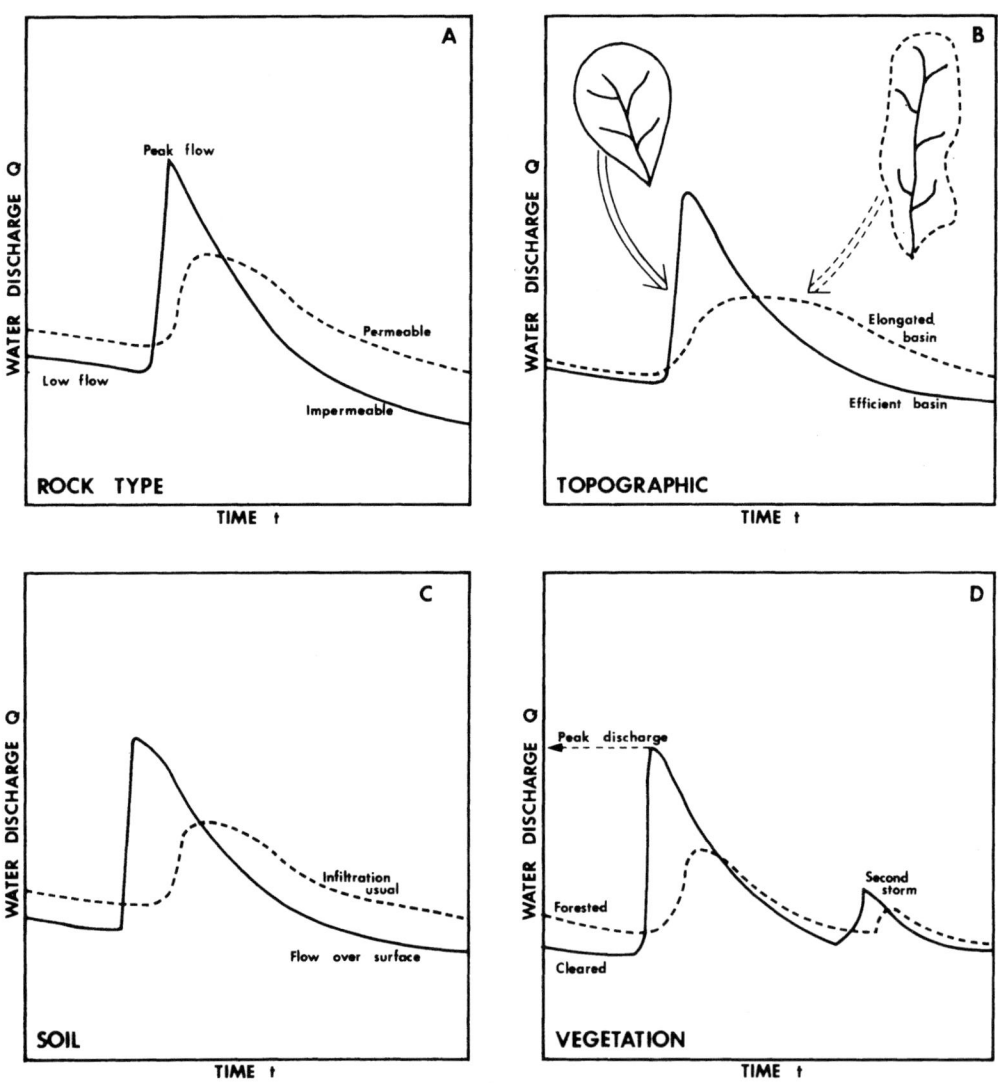

FIG. 2.3. HYDROGRAPHS RELATED TO DRAINAGE BASIN CHARACTERISTICS
Input of uniform precipitation to two basins identical in all but one respect is used to show effect of (A) rock type, (B) topographic factors, (C) soil character and (D) vegetation.

respect except climate and it is equally difficult to find a set of drainage basins which have a uniform climatic input. A further problem arises because it is not easy to express all characteristics of the slope or of the basin in quantitative terms. An alternative is therefore to employ a *theoretical model*. In this case the relation between form and process is assumed usually in simple mathematical terms, and the model is then employed to show how changes could take place over time. Thus we could assume that slope material was perfectly homogeneous and that slope denudation was directly proportional to height above the slope base. Although a gross oversimplification, this could then be the basis for the suggestion of how slope form could change over time. The model of slope evolution could then be compared with slopes that exist in nature.

A third alternative is to make a physical or *experimental model* of the landscape and to use this simplification of reality to proceed towards the understanding of form, process and materials and to how they may change over time. A slope could be modelled in the laboratory and subjected to different artificial rainfall amounts to show when and how it would fail. A slightly larger version could be provided by investigating a simple slope provided by a slope in a sand pit or on a slag heap. Similarly a laboratory river channel, or flume, could be constructed to show how a particular input of water could change the artificially created river channel. Considerable progress has been achieved using such scale models but it is not easy to scale down the size of materials and it is difficult to relate such simplified models of nature to the real world which is necessarily much more complex.

Predictably we can learn a lot by using all three types of model (see Section 2.3d) but together they offer the best hope of a complete understanding of the way in which slopes and drainage basins function.

2.2 PROCESSES IN THE DRAINAGE BASIN

The qualitative understanding of drainage basin processes is a necessary prerequisite for the quantitative appreciation of landscape mechanics. Although it is undesirable to perpetuate the artificial separation of slope and fluvial processes it is expedient to outline the ways in which water and sediment move through a drainage basin, and then to elaborate the types of slope failure that can occur.

2.2a Water Movement

Movement of water in a drainage basin necessarily involves appreciation of the stages of the hydrological cycle. A drainage basin anywhere in the world will receive precipitation of which some may be intercepted by vegetation. Different kinds of vegetation will obviously intercept different amounts of precipitation because trees with a large leaf surface area can intercept more precipitation than grass cover. However if precipitation continues the initial amount of precipitation that can be intercepted and retained by leaves, branches or blades of grass will be exceeded and water will proceed down to the ground surface.

During a rainstorm a film of water may be retained on vegetation surfaces but water will reach the ground surface by running down tree trunks as stemflow and by falling between the tree crowns as throughfall. A proportion of precipitation received will reach the ground surface and this will initially accumulate upon the ground surface as surface detention. Water reaching the ground surface by throughfall and stemflow has to pass through the litter layer before it reaches the mineral soil. A number of workers have summarized the interception process (Dunne and Leopold, 1978) as follows:

$$R = P-I$$
$$= P-(C+L)$$
$$= (T+S-L)$$

where R = net rainfall entering the mineral soil below the litter layer, P = gross rainfall, I = total interception from canopy (C) and (L), T = throughfall, and S = stemflow.

The percentage of the rainfall that reaches the ground surface will depend upon the extent of the vegetation cover, and upon the amount and intensity, or rate, of the rainfall. Measurements of the amount of interception indicate that forests may intercept about 10 to 15 per cent of gross rainfall although interception may be larger than this and there is some evidence to indicate that coniferous forests intercept more precipitation than deciduous forests. Whereas grass and some crops can intercept substantial amounts of gross precipitation during the months of maximum growth they are responsible for smaller annual interception amounts than are trees. Although we think of interception applying largely to vegetation surfaces it is important to remember that in urban areas there will be some retention of water on the surfaces of buildings and man-made structures. When precipitation is in the form of snow there may be a delay between the fall of snow and receipt of water at the ground surface because until the snow pack melts the water will not be released to be available to the ground surface.

Such temporary storages in the form of snow pack or interception are not the only losses to occur. Evapotranspiration includes evaporation from water on the surface of vegetation, soil or land, and transpiration which is the loss of water from the stomatal openings or from the cuticles of plants. Loss of water by evapotranspiration is very substantial and accounts for approximately two thirds of the precipitation falling on the United States, for as much as 90 per cent of the gross rainfall over parts of Africa, and in many areas potential evaporation can exceed the annual precipitation total.

It is difficult to assess evapotranspiration for a particular area (Gregory and Walling, 1973) but some of the variables involved are indicated by the equation proposed by Penman (1963). He suggested a formula useful for estimating evapotranspiration from surfaces completely covered by short green crops under temperate conditions where the supply of soil moisture is abundant. The equation expressed E as the total energy available for evaporation and evapotranspiration as:

$$E = \frac{\left(\dfrac{\Delta}{\gamma}H + E_a\right)}{\left(\dfrac{\Delta}{\gamma} + x\right)}$$

where Δ = a constant depending on temperature,
γ = the constant of the wet- and dry-bulb psychrometer equation,

so that $\underline{\Delta} =$ a dimensionless factor weighting the relative effects of energy supply and
γ ventilation or evaporation,
$H =$ the total heat budget
$E_a =$ an aerodynamic expression for the drying power of the air, involving wind
    speed and saturation deficit,
$x =$ a factor depending on the stomatal geometry of the plant cover and on the
    length of day.

The value calculated from this equation may be compared with the potential evapotranspiration (E_t) and the ratio (E_t/E) varied with the length of daylight and in southeast England averaged 0.75.

The proportion of precipitation that reaches the ground surface will initially accumulate on the surface as surface detention and evaporation may take place from the pools or ponds. Where soils or the material below the surface is permeable however, water will not persist for very long upon the ground surface but it will infiltrate into the ground. Water is drawn into the soil by the forces of gravity and by capillary attraction. The rate at which water infiltrates depends upon the rate at which rain is falling, that is the rainfall intensity; the characteristics of the material immediately below the surface and particularly the size of the pore spaces; and the amount of water already present. Material that is coarse-grained will allow water to infiltrate rapidly whereas material that is fine-grained will be typified by a lower infiltration rate.

The maximum rate at which a specific material can allow water to infiltrate has been described as its infiltration capacity. It has been shown that where water is supplied at a rate greater than the infiltration capacity then the upper layers are saturated so that a wetting front develops where the instantaneous infiltration rate (f) can be expressed (Kirkby, 1969) by an equation including two flow components:

$$f = A + Bt^{-1/2}$$

where $A =$ conductivity flow under gravity expressed as transmission constant involving flow through a continuous network of large pores,
$B =$ diffusion term representing slow filling of soil air pores from surface downwards by capillary flow, declining with time as water enters the soil,
$t =$ time.

The infiltration rate at a specific point will decrease during rainfall as the wetting front advances and the soil moisture storage is filled. Final infiltration rates therefore vary very substantially according to the character of the soil but a range of values collected for sites in the USA (Dunne, 1978) shows that they can be as much as 8 cm.h^{-1}.

An early model of runoff production proposed by Horton (1933) envisaged that once the infiltration capacity was exceeded then water would be retained on the surface in pools, and when these were sufficiently large then water would flow from one pool to another in sheets as sheet-flow or overland flow which is defined as the flow of water over the land surface towards a stream channel. On natural slopes Horton (1945) proposed that mixed flow takes place because areas of turbulent flow are interspersed with areas of laminar flow. As the slope angle increases the water will become increasingly concentrated and eventually a stream channel will be formed. Thus water flowing over slopes will be motivated by the force of gravity acting downslope and retarded by the frictional effects of vegetation and of the roughness of the ground surface over which it moves. When the first of these forces exceeds the shear strength of the surface material and its associated vegetation cover, then a stream channel may be formed. The depth of overland flow may be estimated, for turbulent flow, by combining the continuity equation and the Manning flow equation (Emmett, 1970) (see p. 57) and by assuming depth is equivalent to hydraulic radius, because the cross section of overland flow is very wide and shallow.

Continuity equation $q = DV$

Manning flow equation $V = \dfrac{1}{n} D^{0.67} S^{0.5}$

where q = unit discharge, D = mean depth, V = mean velocity, n = Manning roughness coefficient, and S = slope.

Peak rates of overland flow vary very substantially according to surface topography, precipitation characteristics, and character of material immediately below the surface but velocities ranging between 0.3 and 15 cm.sec^{-1} have been quoted (Dunne, 1978).

Whereas Horton envisaged such overland flow to be a feature of many areas of the world it has now been realized that it is a process which is confined to some specific areas or restricted to small portions of a drainage basin. This conclusion followed the observation that maximum infiltration rates often exceeded precipitation intensities particularly in humid temperate areas; that overland flow is seldom if ever observed in many humid areas although it has been described from semi-arid areas where vegetation density is low; and that lateral flow occurs in the unsaturated zone between the water table and the ground surface.

The water that infiltrates below the surface will first move vertically downwards directly under the influence of gravity (Figure 2.4). The infiltration rate will depend not only upon the rate and intensity of water supplied to the surface but also on the characteristics of the material. The spacing of particles and the extent to which voids are connected will both influence water movement. The porosity of a material is a measure of the volume of its pore spaces whereas the permeability of a material reflects the rate at which water can be transmitted through the material. Water infiltrating vertically may continue uninterrupted to the water table to recharge the ground water supply which is the zone below the water table in which all pore spaces and voids are filled with water. Water from the water table will gradually be discharged in springs or seepages.

A distinction is frequently made between those rocks that contain large quantities of water, called aquifers, and those which do not allow water to circulate or to be stored, which are termed aquicludes. Sandstones and limestones are usually good aquifers whereas clays are usually aquicludes. The rate of groundwater movement relates to differences in head and is retarded as it moves through the voids of an aquifer. Darcy's law expresses the rate of flow Q of water in an aquifer as:

$$Q = w\, d\, K \left(\frac{\Delta h}{\Delta l} \right)$$

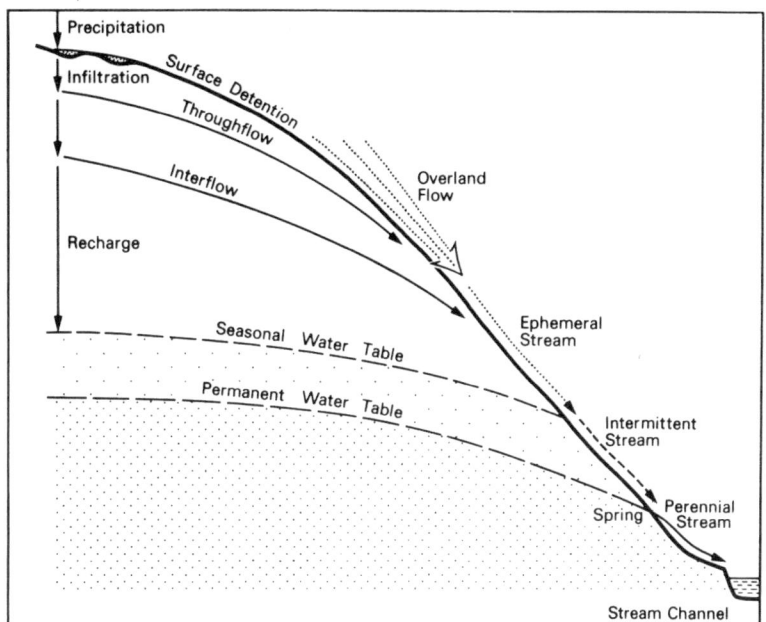

FIG. 2.4. GENERALIZED SECTION OF WATER FLOW ROUTES THROUGH DRAINAGE BASIN

where W = width of flow, d = depth of flow, K = coefficient of permeability or hydraulic conductivity, Δh = difference in head, and Δl = distance between measurement points in the direction of flow.

A subdivision into a phreatic zone below the water table and a vadose zone above the water table is necessarily a gross over-simplification. Where there is an alternation of aquifer and aquiclude it is likely that there may be several local or perched water tables above the main one. There are also variations in the vadose zone above the water table where throughflow, interflow, and pipeflow have been identified (Figure 2.4). The Horton model of runoff production visualized streamflow as derived either from groundwater flow from springs and seepages or from surface overland flow which occurred when the infiltration capacity was exceeded by the precipitation intensity. Research undertaken particularly after 1960 demonstrated that overland flow was less common than would be expected if the Horton model was applicable to all areas and in some humid regions such surface flow was never observed. This conclusion posed a problem because the rate of movement of water through the groundwater zone to springs was insufficient to account for the stream hydrographs that could be measured along streams and rivers immediately after a rainstorm.

This anomaly was largely solved when field measurements in small drainage basins in the north eastern part of the USA revealed the existence of flow taking place laterally beneath the ground surface but above the water table. This led to the distinction of throughflow taking place laterally within the soil, and of interflow occurring laterally through rock or regolith above the water table. Pipeflow is the mode of flow which occurs through interconnected voids or pipes in soil and which can range in diameter from a few millimetres to as much as several metres. Water flow in pipes is turbulent in character whereas the movement of water as throughflow, interflow or groundwater flow is laminar in character and is diffuse or matrix flow (Atkinson, 1978). Matrix flow can be directed vertically, as anticipated in the Horton model for water percolating down to the groundwater table (Figure 2.4), or it may move in a downslope direction as is the case with throughflow and interflow (Figure 2.4).

In each case water movement may take place under unsaturated or saturated conditions and if it is laminar in character it will move according to Darcy's Law (p. 54). Movement of water will occur according to a gradient of hydraulic potential and the value of this at a specific point will reflect the combined influence of the gravitational potential which arises from the height of the point above a fixed location such as the slope base, the pressure of capillary potential which is the difference between the pore-water pressure and the atmospheric pressure, and the osmotic potential which occurs due to the presence of solutes in the water and which therefore lower the vapour pressure.

Research undertaken since 1960 has demonstrated the variety of flow routes which may be followed by water passing through a particular basin. This has involved consideration of the dynamics of the drainage basin (Figure 2.2) in relation to changes which occur seasonally and which also occur during, and after, each rainstorm. Just as the water table will fluctuate seasonally (Figure 2.4) so on the surface there will be an expansion and contraction of streams, of saturated areas and of water flow types. In the zone above the water table there is really a continuum of flow types and although it is convenient to think of pipe flow as turbulent, and of matrix flow as laminar, in character there are many intermediate types because of the arbitrary distinction made between soil voids which permit matrix flow and those voids which interconnect and are large enough to allow pipe flow.

The outcome of research since 1960 has been to visualize a variety of flow types through the drainage basin and these are summarized in Table 2.1. This emphasizes the fact that whereas the Horton model envisaged surface runoff and groundwater flow as the only two flow types these are in fact complemented by a third group which is collectively referred to as subsurface runoff whereby water moves at a more rapid rate than by groundwater flow. Of the flow types included within the category of subsurface runoff some water may emerge from the soil as a result of new water added from precipitation higher up the slope and this is termed translatory flow; some may occur in saturated zones as saturated throughflow or saturated interflow and this is collectively termed subsurface stormflow; some may take place more rapidly as pipeflow; and some may move laterally as throughflow or interflow through the unsaturated zone. Because it is desirable to distinguish the flow sources which maintain the baseflow of a stream from those which contribute to the storm discharge following a storm, some workers have proposed that a more meaningful distinction is between quickflow or quick return flow, which is directly attributable to a precipitation event, and delayed flow which is delayed by transmission through groundwater or through the unsaturated zone.

The flow types summarized in Table 2.1 are not discrete but grade into one another and at a particular point on a slope the type of water movement may vary during the two days which succeed a specific precipitation event. Work by the Institute of Hydrology has shown how a slope in the Nant Gerig catchment of the upper Wye basin in central Wales is

Table 2.1
TYPES OF FLOW IN HEADWATER AREAS OF DRAINAGE BASINS

			Type of Flow	Character	Location
QUICKFLOW	SURFACE RUNOFF/FLOW		Overland flow	Surface flow of water because rainfall intensity exceeds infiltration rate. Referred to as Horton overland flow or infiltration overland flow by some workers.	Semi-arid areas where rainfall intensities high and vegetation cover sparse. In humid areas may occur adjacent to stream channels or in topographic hollows where water converges
QUICKFLOW	SURFACE RUNOFF/FLOW		Saturated overland flow	Surface flow of water which occurs because soil is saturated and infiltration capacity has not been exceeded	Locations usually close to stream channels or hollows where water table rises rapidly to surface during storm event
QUICKFLOW	SUBSURFACE RUNOFF	RETURN FLOW	Throughflow	Movement of water downslope in soil profile usually under unsaturated conditions. Referred to as unsaturated throughflow by some workers	Slopes with well-drained soils and often encouraged by discontinuities in soil profile. Lateral flow will occur in soil if this meets less resistance than vertical percolation of water
QUICKFLOW	SUBSURFACE RUNOFF	RETURN FLOW	Saturated throughflow	Lateral flow in soil under saturated conditions	During storm a saturated wedge will extend upslope in soil profile and saturated throughflow occurs immediately above this
QUICKFLOW	SUBSURFACE RUNOFF		Translatory flow	Lateral flow in soil occurring by displacement of stored water due to addition of 'new' water	Slope with soil with saturated zone
QUICKFLOW	SUBSURFACE RUNOFF		Interflow	May be used synonymously with throughflow. Some workers describe lateral flow above water table but below soil as interflow which could thus be through unsaturated rock or regolith	Slopes having permanent water table at depth and any lithological discontinuities may encourage lateral flow of water as interflow
QUICKFLOW	SUBSURFACE RUNOFF		Saturated interflow	Interflow occurring under saturated conditions	Affected by extension of saturated wedge beneath surface in upslope direction
QUICKFLOW	SUBSURFACE RUNOFF		Pipe flow	Flow through subsurface network of interconnected, anastomosing pipes or tubes, larger than other soil voids and may be up to 1 m in diameter	Variety of areas including steep slopes, where erodible layer lies above less permeable layer, or on flood plains marginal to channel banks
DELAYED FLOW	GROUNDWATER FLOW		Groundwater flow	Water that has infiltrated into ground, has reached groundwater and is discharged to surface from spring or seepage at rate determined by hydraulic head	Areas where groundwater storage is possible due to character of subsurface materials

underlain by a network of pipes and that these pipes occur at several levels below the surface. After a rainstorm not all pipes will convey water and all slopes demonstrate a similar variation in response according to the character of the storm, the nature of the slope particularly in relation to possible flow types, and the antecedent conditions which influence the moisture status of the slope when the storm begins.

It is necessary to know the rates at which water can move through the drainage basin by following one of the several flow routes (Table 2.1) because this is significant in relation to our interpretation of streamflow hydrographs (p. 60). Recent measurements in small basins have provided information about different flow velocities (e.g. Kirkby, 1978) and some typical ranges of velocities are collected in Table 2.2. Although velocities will vary,

Table 2.2
EXAMPLES OF VELOCITIES OF SOME TYPES OF WATER FLOW IN DRAINAGE BASINS

Type of flow	Velocity of water flow	Reference
Overland flow	3–15 cm.sec^{-1} on slope of 0.40. Less than 0.1 cm.sec^{-1} on low slopes with thick vegetation cover	Dunne, 1978
Vertical percolation	Less than 7.5 cm.day^{-1} in Whitehall Watershed, Georgia	Tischendorf, 1969
Saturated throughflow	20 cm.hr^{-1}, 0.2–37.2 cm.hr^{-1} saturated hydraulic conductivity values collected from various field measurements	Whipkey, 1965 Dunne, 1978
Throughflow	80 cm.day^{-1} in B horizon in East Twin Brook catchment, Somerset. 50 cm.day^{-1} in B/C horizon	Atkinson, 1978
Pipe flow	10–20 cm.sec^{-1} in Nant Gerig catchment, central Wales	Atkinson, 1978
Stream channel flow	Average 45 cm.sec^{-1}	Pilgrim, 1966

according to local conditions, moisture and rainfall received, these show that it is generally thought that the most rapid water movement will take place in open channels, and that this will be succeeded by pipeflow, overland flow, saturated subsurface flow, throughflow, and vertical percolation.

During a period of dry conditions water flow through the headwater area of a drainage basin will follow some of the routes listed in Table 2.1. Water flow along the stream channel will be at low flow or base flow and in the absence of any further input of water from storm rainfall or snow melt this will continue to decline exponentially. The base flow component of stream flow will be supplied from springs and seepages and will derive from groundwater and from subsurface flow from unsaturated areas. If a single rainstorm then occurs over the basin, then if this storm is sufficient in amount to exceed interception and to provide sufficient water at the ground surface to infiltrate or to flow over the surface, then quickflow will reach the stream and produce a hydrograph. This is the term given to variations in discharge over time. The hydrograph obtained by plotting volume of flow per unit time or discharge against time is a storm hydrograph if it is plotted for a period following a specific storm but we could also plot the annual hydrograph to represent the variations in discharge during a year (see p. 104) at a single location or gauging station. Soon after the storm rainfall the discharge of the stream will begin to increase, it will continue to increase until the peak discharge is achieved, and then it will decrease in discharge at a rate less than the rate of rise. The rate of decrease is termed the recession curve of the hydrograph. If the quickflow can be separated from the baseflow, or delayed flow, component of the hydrograph then it is possible to calculate the volume of quickflow and so to express this as a percentage of the total storm rainfall.

The percentage of storm rainfall which reaches the stream channel will depend upon the character of the storm, upon the nature of the basin or basin characteristics, and upon the prevailing and antecedent conditions. The character of the storm is important especially the amount of precipitation, its duration and its intensity. The most intense storms are usually the ones responsible for the most significant hydrograph rise and peak discharge. However, the conditions prevailing in the basin at the time are also pertinent because if temperatures are high then losses by evapotranspiration will also be high and if there has not previously been much rainfall then much of the storm rainfall may be utilized in filling the deficiency in soil moisture. Conversely, if the storm succeeds a number of significant precipitation events then there may be a high rate of runoff. Antecedent conditions thus play an important role in determining the extent to which stream discharge will increase, and at what rate, following a specific rainstorm. The percentage of runoff from a particular catchment is shown in Table 2.3 varying with antecedent conditions as well as with storm rainfall amount and intensity.

Table 2.3
EXAMPLES OF ANALYSIS OF DISCHARGE HYDROGRAPHS FROM A SMALL BASIN
(AREA 0.25 km²)

Event	Precipitation			Prior conditions	Peak discharge	Runoff volume	Percentage of precipitation output as runoff
	Storm amount (mm)	Average intensity $(mm.h^{-1})$	Maximum intensity $(mm.h^{-1})$	Preceding discharge $(l.sec^{-1})$	$(l.sec^{-1})$	(mm)	
12 Feb.	34.0	0.25	2.03	68	991	10.6	31
3 March	11.8	3.05	10.16	57	1034	4.7	39
13 June	10.7	2.54	3.56	70	694	3.4	32
19 June	30.5	3.05	4.06	9	1019	4.5	15
19 Dec.	16.0	1.52	3.56	79	665	5.9	37

In general, the greater the preceding discharge and the larger the precipitation amount and the more intense the storm, the higher the peak discharge of the hydrograph. Runoff volume is calculated by separating quickflow from delayed flow and measuring the area of quickflow on the hydrograph, which can then be converted to a volume of water which is divided by drainage area to give a runoff amount which is directly comparable to the depth of precipitation measured.

In addition to the influence of storm rainfall characteristics and antecedent conditions, there is also the influence of drainage basin characteristics. These are the features of the relief, the rock type, soil type, and vegetation and land use which determine the rate and manner by which water is transmitted through the drainage basin. The incidence of the possible types of flow (Table 2.1) will be determined at least in part by the permeability of the rock type, by the size of voids within the soil, by the density of the vegetation cover and by the slope of the basin. The hydrograph produced by a particular basin following a specific storm will have a number of distinctive features and some of these attributes may reflect particular basin characteristics more than others. Hydrograph characteristics include peak discharge, the lag time which is the time elapsing between maximum precipitation and hydrograph peak, and the total runoff volume which is the volume of water passing a particular location during the passage of a single hydrograph. The way in which some of these parameters vary in a single basin is illustrated in Table 2.3.

Because drainage basin characteristics influence hydrograph character we can suggest examples of ways in which hydrographs may differ between catchments. In Figure 2.3 the effect of identical storms on four pairs of basins is depicted. A permeable rock underlying a basin tends to produce lower peak flows but to have higher base flows than a basin underlain by an impermeable rock (Figure 2.3A). The topographic characteristics of the basin including shape, relief and density of stream channels will also exercise an influence and a pear-shaped basin will be more efficient and will produce a higher peak discharge than a more elongated basin (Figure 2.3B). Where overland flow occurs the peak discharge may be greater and occur earlier than the peak flow from a basin with more permeable soils where infiltration occurs more frequently (Figure 2.3C). This contrast in soil character may dictate the mechanisms of water flow through the basin (Table 2.1) and in addition to a lower and later peak discharge, the low flow may be higher from the basin with soils where water is released more slowly from subsurface flow. A contrast may also be expected between a forested basin which will delay the peak discharge and give a lower peak flow than from a comparable basin which has been deforested (Figure 2.3D).

These examples are necessarily grossly simplified because hydrographs are often more complex and may have more than one peak; because two basins will seldom be identical in all characteristics except the one being scrutinised; and because rainstorms over two basins will rarely be identical in amount, intensity and duration. It is therefore necessary to endeavour to unravel the ways in which these multivariate controls determine the character of hydrographs and this has to be based upon an appreciation of the ways in which drainage basin characteristics may influence specific hydrograph parameters (e.g. Gregory and Walling, 1973).

A major reason for interest in the influence of drainage basin characteristics upon streamflow is the need to estimate streamflow at sites along particular rivers where flow has not been measured and this can be done by equations relating hydrograph indices to measures of storm precipitation and drainage basin characteristics. Furthermore, the influence of human activity can induce changes in drainage basin characteristics and we need to know of the likely consequences of deforestation for example. Removal of the forest from a particular basin could lead to higher peak discharges (Figure 2.3D) which give an increase in the flood hazard. The most dramatic increase in flood hazard can arise after building activity and urbanization. The introduction of impervious surfaces of houses, drives and roads to replace the previous land use leads to a reduction of infiltration and hence to an effective potential increase in overland flow. Because rates of overland flow are more rapid than rates of subsurface flow (Table 2.2) then hydrographs may have higher peak discharges and shorter lag times from urban catchments than from rural areas.

Furthermore, urbanization often involves the installation of a network of storm water drains so that this provides additional channel flow. Near Iowa City, Iowa, the density of all channels in the South Branch basin was 50 per cent greater after suburbanization than it was in the natural condition (Graf, 1977) and the density of channels in a small basin on the margin of Exeter, Devon increased from 5.8 to 17.6 km.km^{-2} during urbanization of 30 per cent of the basin area (Gregory, 1976).

Movement of water through a drainage basin can be visualized as movement between several stores and only if one store is full will water proceed to the next. Such a model is illustrated in Figure 2.5 and this can be the basis for computer models which can be developed to evaluate the streamflow discharge for a particular river location, draining a specific basin at a point in time. For a specific time period it is also possible to evaluate the water balance as an equation for the particular drainage basin. This can be achieved by relating water discharge as runoff by volume *(Q)*, to precipitation *(P)*, to evapotranspiration *(E$_t$)* and to changes in storage (Δ*S*) in the water balance equation expressed as:

$$P = Q + E_t \pm \Delta S$$

2.2b Mass Movement

Transfers of solid material take place within the drainage basin in a number of ways. Weathering processes operate (Chapter 1) to release material, water moving along one of the routes in the basin (Figure 2.5) may convey material as sediment or as solutes (Section 2.2c), and in addition there is the movement of material on slopes by a group of processes collectively termed mass movement. The movement of dry rock or weathered material is one extreme example and the flow of water containing a few sedimentary particles is another extreme case and between these two extremes is a range of mass movement types which could be classified according to the relative proportions of sediment and water. An important distinction between slow and rapid mass movements was made by C. F. S. Sharpe (1938) and whereas the slow category includes several forms of creep, the rapid group includes movements that occur by sliding. This emphasizes the distinction between those movements that occur for much of the year such as creep, and the rapid movements which occur as a failure of a slope in a relatively short period of time.

An alternative approach to mass movement can be based upon the three modes of movement namely flow, slide or subsidence/heave. These three modes are used in Figure 2.6 to illustrate the ways in which displacement may occur. In each case the displacement of a rectangular block is represented and in flow the amount of displacement or movement varies with depth. When movement occurs by slide there is no such variation in amount of movement with depth, and the third type involves vertical displacement either downwards by subsidence or upwards by heave (Figure 2.6). By representing these three types by a triangular diagram with 100 per cent of one of the three types at each apex, we can position several types of mass movement according to the relative proportions of slide, flow and heave/subsidence involved (Figure 2.6). An ideal classification is difficult to realize because of the way in which different proportions of flow, slide and vertical displacement may be involved in many mass movement processes. On the basis of mechanism, morphology, materials, and rate of movement Hutchinson (1968) distinguished four types of mass movement phenomena namely creep, frozen ground phenomena, landslides, and subsidence.

Creep takes place as a result of the slow downslope movement of superficial materials, of soil, or of rock debris and it is often indicated by the accumulation of material at the upslope side of field boundaries or trees, by the bending of posts, the leaning of trees and

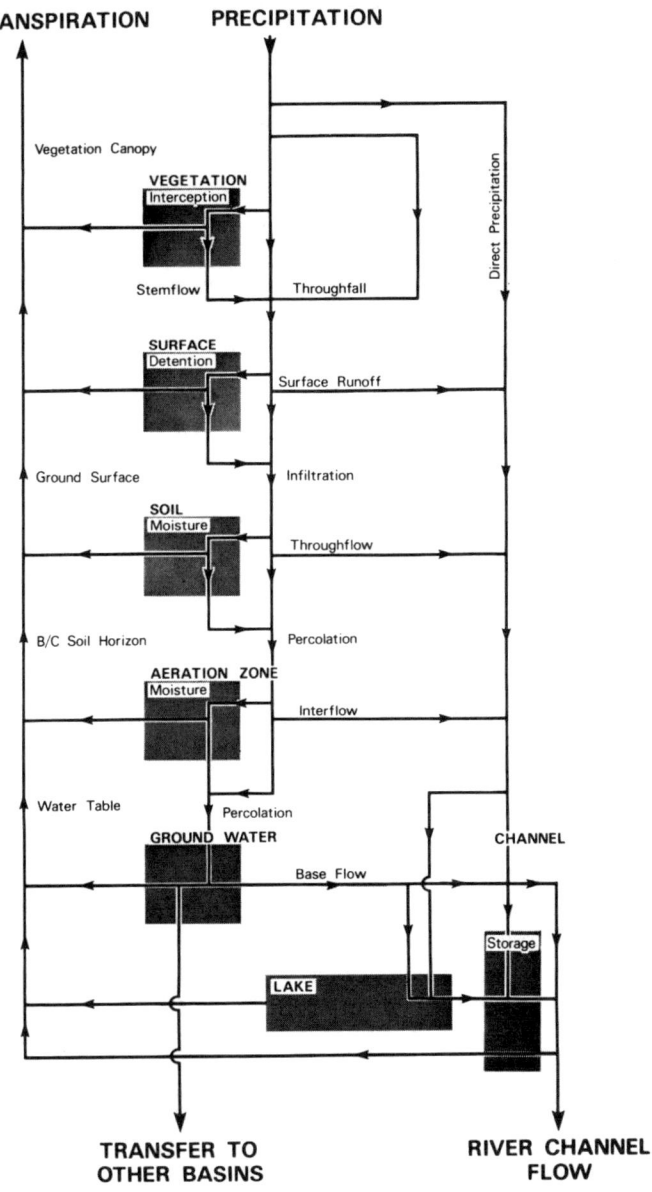

FIG. 2.5. MODEL OF COMPONENTS OF HYDROLOGICAL CYCLE
Compare with Figure 2.3 and Table 2.1. Shading indicates that each component can store moisture or transmit moisture to a subsequent component.

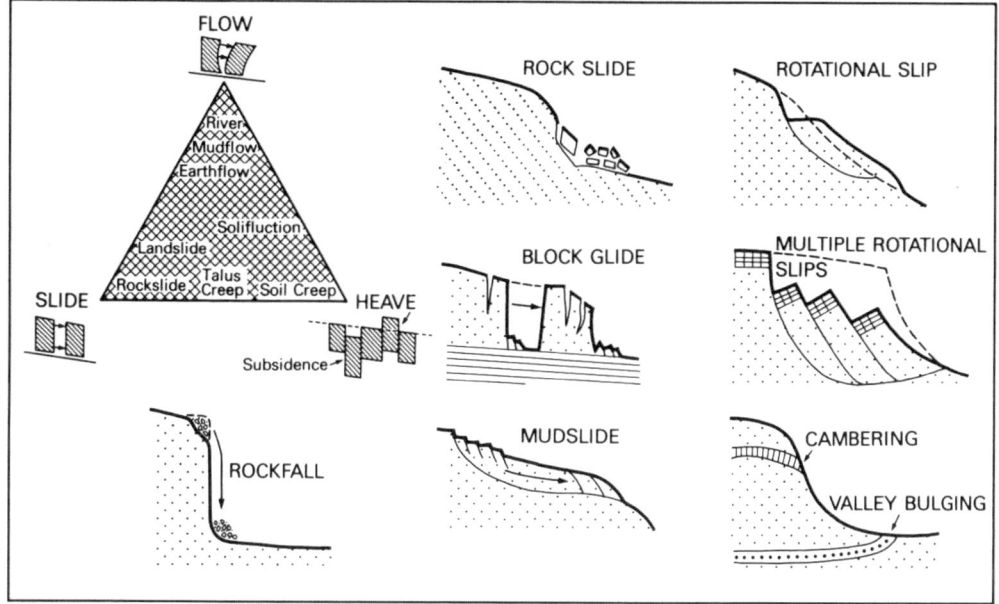

FIG. 2.6. SOME TYPES OF MASS MOVEMENT
Triangular diagram indicates types of mass movement according to relative importance of flow, slide and heave/subsidence. Sketch sections illustrate several types of rapid mass movement.

by turf rolls. It is therefore described as soil creep, or if affecting rock fragments as talus creep. The movement is caused by the upward displacement of particles often at right angles to the slope so that the particle may subsequently move downwards under the influence of gravity and thus experience a net down slope movement. Rates of movement are usually of the order of several millimetres per year and may continue throughout the year as continuous creep, or be restricted to part of the year in seasonal creep which is confined to those depths over which seasonal fluctuations are expressed.

Although this process is slow it does operate widely over slopes so that it has been suggested that it is a very important process of mass movement. Soil creep in a basin 15 km² in area in Oregon was estimated to provide 99 per cent of the sediment lost from the basin (quoted in Dunne and Leopold, 1978). The movement of particles contributing to creep may be inspired by the action of needle ice which freezes under particles and raises them normal to the slope; by the effect of alternate wetting and drying of colloidal clay particles which expand and contract; and by the production of desiccation cracks which, once formed, allow material to accrue from the upslope side of the crack. Creep is particularly common under forested slopes and is a frequent form of mass movement in humid temperate areas.

Frozen ground phenomena include cambering and valley bulging (Figure 2.6). Where rocks of different character are interbedded then the saturation of fine-grained sediments such as clays may lead to the squeezing of more massive beds involving cambering of free faces exposed in slope profiles and the bulging of such beds beneath valley floors. The result of these movements was first identified from structures in Northamptonshire, UK and appreciated to be due to mass movement under Quaternary periglacial conditions.

Solifluction is the slow downslope movement largely by flow, of masses of soil and weathered material usually containing a large amount of water and flowing over an impervious layer. First described from periglacial areas, solifluction is a characteristic mass movement process extensively affecting slopes of the Arctic landscape where the permafrost table (p. 219) provides the impervious layer. In permafrost regions the ground is frozen from the surface downwards throughout the winter months but after the spring thaw, a thawed layer, the active layer, is developed to a depth of one or two metres above the permafrost table. Because water movement is impeded by the permafrost table the active layer will have a high water content and the movement of this layer, which can take place on slopes with angles as low as 1½ degrees has been referred to as congelifluction by Dylik (1966). Although movement is largely by flowage, a surface crust of material can be rafted along above the more mobile material beneath and sliding may take place in the lower layers above the permafrost table. A similar type of movement process has been reported from tropical slopes beneath rainforest. In these situations a deep layer of chemically-weathered material may mantle the slopes beneath the dense rainforest and this may become unstable after intense tropical storms (e.g. Figure 2.20) so that movement of material by tropical solifluction occurs and this may raft the vegetation cover downslope.

Landslides include several types of phenomena which depend upon shear failure along a plane below the ground surface. Translational slides (Hutchinson, 1969) are usually shallow (e.g. mudslide in Figure 2.6) and the depth to the shear plane is usually of the order of one tenth of the distance from the toe to the head of the slide. Such shallow slides include the perceptible downward sliding or falling of masses of soil or rock or a mixture of the two. A range of types exist from those such as rock slide and block glide (Figure 2.6) where the water content is very low, to debris slides, mudslides and bog bursts with a much greater water content. In many shallow mass movements both slide and flow will occur. A debris slide may involve a sliding movement at the head of the failure but the material accumulating at the toe may be moved largely by flowage. Such flow movements resemble a viscous fluid and there is a more uniformly distributed internal deformation of the material rather than shearing along a slide plane. Earthflows usually occur in fine-grained materials with a high water content and they are particularly common in rocks, including volcanic ones, that weather to give a soil with a high clay content. Such earthflows may occur below slumps or slides where the failure provides viscous material that spreads out as a lobe of debris. A mudflow (Figure 2.20) is a faster movement than an earthflow. Mudflows may occur in existing valleys amalgamated with water flow, and these fluid movements may take place on the sides of volcanoes.

Where a curved shear plane exists material may move by rotation as a rotational slip or a slump (Figure 2.6). This type of movement is responsible at the small scale for the production of terracettes but at a larger scale is particularly significant in areas where there are particular sequences of contrasting rock types superimposed and in some cases where deep cohesive and uniform material occurs. In north east Ireland, basalt, chalk and Upper Greensand overlie Lias clays which are squeezed so that large rotational slips can be produced (Figure 2.6) and in some areas multiple slumps may be produced over a slope. Outward rotation of rock masses is known as 'toppling'.

Subsidence of the ground surface can arise from a variety of causes including mining, marine erosion, subsurface solution, melting of ground ice or subsurface erosion. The settlement of surface layers may also lead to sinking of the ground surface as a result of consolidation due to the surface loading produced by a man-made reservoir, of lowering of the ground water level, or by underdrainage. A number of mass movements also involve predominantly vertical movement and these are rockfalls (Figure 2.6), where material often derived from steep cliffs moves downwards with little contact between the particles by falling, saltating and rolling.

Mass movements are not easily categorized into discrete classes because many movements are made up of several types. The inclusion of several types of movement is illustrated by avalanches which can include rock fall material, debris, snow and enclosed air and which can move downslope at speeds greater than 150 km.hr^{-1}. Ground avalanches involve wet snow together with earth, stone and debris sliding over the ground, whereas powder avalanches are dry powdery snow and great clouds fill the air as they descend.

One of the most disastrous avalanches this century occurred in western Peru on 10 January 1962 when a vast mass of ice broke away from the northern peak of the highest mountain, the Nevado de Huascarán, and led to the rapid movement of 4.6×10^6 m^3 of ice, granite and slate. This great mass moved down valley with a great roar, crushed and buried four villages and at one part of the valley covered the valley floor with mud-rock and ice to a depth of up to 40 m and to a width of 500 m. The avalanche moved more than 17 km in 15 minutes and its final toll was 7 villages, 1 town, 10 000 livestock and the death of 3500 people (Lane, 1966). Most avalanches are smaller than this but may occur more frequently and in the inhabited parts of the Swiss Alps nearly 10 000 avalanche paths have been recognized.

2.2c Sediment and Solutes in Water

Material may be transferred within the drainage basin by the mass movement phenomena reviewed in Section 2.2b but in addition, material may be transported in water as mechanical sediment or as solutes which are dissolved in water as it follows the several possible routes through the basin (Table 2.1 and Figure 2.5).

Solutes may be present in precipitation derived from atmospheric sources. Rainfall may contain dissolved ions so that precipitation near the coast of the USA has been shown to have concentrations of 8 mg.l^{-1} of chloride ions although this decreases rapidly inland away from the marine source. Contact with vegetation can give further increases in dissolved ions, as K, Ca and Na for example increase as they are derived from vegetative surfaces and particularly from leaves. Further supplies of solutes may be collected from the litter layer and during movement of water through the soil eluviation can mobilize solutes as water moves as throughflow and interflow. The removal of solutes below the surface along percolines can lead to gradual lowering of the surface by a process described as suffosion. At lower levels below the surface the supply of solutes from weathered material and from the country rock will reflect the chemical composition of the rock. Although one would expect that many of the solutes would be derived from the groundwater reservoir, where the residence time of the water in the rock is high, it has sometimes been shown that the solute content of water in areas of limestone may be attained by the time that water leaves the soil profile.

The movement of solutes in a drainage basin should not be seen as a purely physical process because solute transfers are intimately related to plant growth. Not only does vegetation release solutes to water passing through the drainage basin but there is also an uptake of nutrients by plants, and experiments in nutrient cycling have demonstrated how much material is moved through the various components of woodland ecosystems. Modification of the solute transfers can be achieved by man's influence, and the removal of vegetation cover, and particularly of forest, can lead to an increase in the nutrients available to river water because of the reduced uptake by vegetation. Conversely, man's influence may provide new sources of solutes from fertilizers which are flushed through some of the pathways of the drainage basin, and from atmospheric or river pollution.

Variations in water discharge over time can be plotted as the stream discharge hydrograph (Figure 2.3) and if we plot the way in which solutes vary over time this gives a chemograph showing the total dissolved solids or the variations in concentration of individual ions. Concentrations of four ions during the passage of two stream hydrographs are illustrated in Figure 2.7 for a small basin in Devon studied by Walling (1974). This

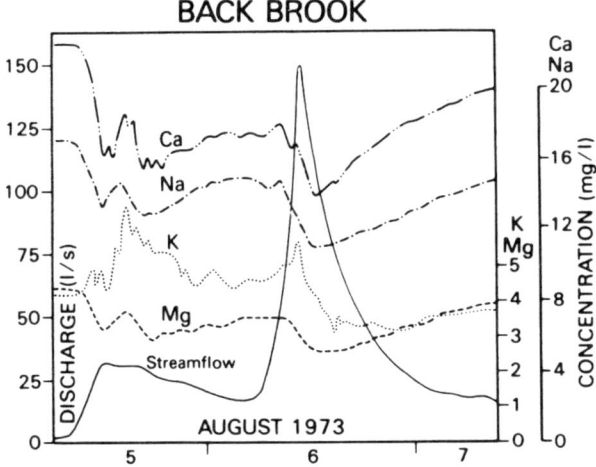

FIG. 2.7. SOLUTE PRODUCTION IN RELATION TO DISCHARGE
Event illustrated is for Back Brook, Devon (after Walling, 1974; and Walling and Foster, 1975).

illustrates the way in which concentrations of solutes decrease during the passage of a flood peak. During low baseflow discharges water is derived from ground water sources or from sources which have had a comparatively long residence time. The water hydrograph is produced by the movement of quickflow through the basin and this tends to be more dilute than the baseflow. A complicating influence may be provided where quickflow reaches new sources of nutrients and these may provide short duration pulses of solutes. Such a flushing effect may be illustrated by the behaviour of potassium in Figure 2.7 because this cation concentration increases in sympathy with the water discharge, whereas the other cations are diluted and indicate a dilution effect.

Sediment transported in the drainage basin may be released from fewer sources than those providing solutes because the major sources depend upon water flow over the surface or flowing in channels. Some dust may be present in precipitation but on the ground surface rainsplash and surface wash provide two significant sources of sediment. Raindrops possess considerable kinetic energy, proportional to the product of raindrop mass and the square of their velocity, and if this energy is expended on bare soil then particles may be detached and moved downslope. Although the vegetation cover will resist the influence of the raindrops the effect may be evident beneath forest where the openings in the canopy allow raindrops to reach the surface and the effect will be pronounced where there is little protective litter.

Surface wash over the surface will be significant where overland flow occurs and whereas the upper parts of slopes may exhibit what Horton termed a belt of no erosion, at locations lower down the slope material may be entrained as the downslope force exerted by the moving water is greater than the resisting force provided by the shear strength of the soil and its associated plant cover. In a semi-arid area of New Mexico hillslope erosion measured for nearly a decade (Leopold, Emmett and Myrick, 1966) on unrilled slopes accounted for 98 per cent of the sediment production from all sources. On badlands of Hong Kong, measurements of slope wash were made at 449 sites over a 15 month period from 16 July 1971 (Kin-Che Lam, 1977) and during this time there was a mean ground retreat of 2.17 cm when a total of 3076 mm of rainfall was received with a kinetic energy of 69 489 Joules. m^{-2}.

The movement of sediment by surface wash is not uniform but breaks up into rills and this concentration of water can be responsible for rill erosion. If rills are developed and widened then gullies will be developed which are too deep to be eliminated by ploughing and which may provide large amounts of sediment for water transport. Sediment may be released by mass movement processes on the side of the gully, by recession of the headcut at the upper end of the gully, or by deepening by mobilizing material from the gully floor. Such gully erosion will be greatest where infiltration rates are low, where vegetation resistance to surface flow is small, and where slopes are steep. Gully erosion has been found to be characteristic of semi-arid areas where annual precipitation amounts are low but where the intensities during particular storms may be high. Man's influence may be significant (Figure 2.8) particularly in areas where removal of the original vegetation cover reduces the resistance to surface erosion or where livestock densities lead to overgrazing also reducing the effectiveness of the surface vegetation to bind the soil, to mitigate the influence of rainsplash erosion and to retard surface erosion. Although classically developed in the USA, China, South Africa and Australia, gully erosion has also been identified on a smaller scale elsewhere especially due to the influence of man. In a small example in Devon a small stream channel was coverted to a gully as a result of the direction of stormwater drainage from a road into a stream channel which was shown to incise and to produce a significant gully feature (Figure 2.9) during a period of thirty years (Gregory and Park, 1976).

Along stream channels sediment may be obtained by erosion of the bedrock or the superficial deposits outcropping in the bed and banks. The force of the water may remove particles of material by the process of evorsion, by compressing air against the channel margins inducing hydraulic action, by the force of water charged with particles by corrasion, or by solution as corrosion. Material derived from the channel banks may be released by slumping where slabs of material slide down to the water level (Figure 2.10), or by sloughing where thin layers of material are displaced from the banks. The influence of

FIG. 2.8. GULLIED VALLEY FLOOR, TASMANIA
Development of such active gullies was instigated along Tea Tree Rivulet following forest clearing during
settlement expansion of the area. (Photograph: K. J. Gregory)

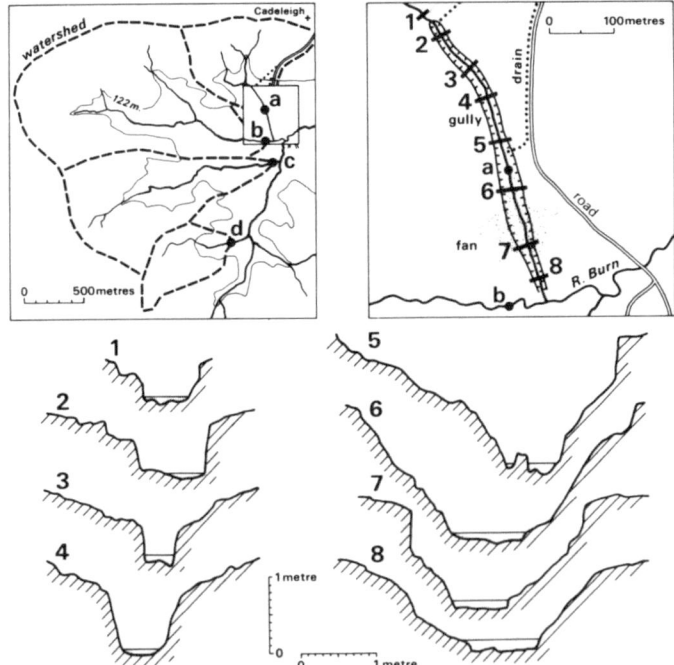

FIG. 2.9. DEVELOPMENT OF SMALL GULLY ALONG TRIBUTARY TO RIVER BURN, DEVON

FIG. 2.10. BANK COLLAPSE ALONG BANKS OF OULANKA RIVER, NORTHERN FINLAND
Bank recession on concave side of bend is indicated by fallen trees; influenced particularly by frost action and high flood discharges in Spring. (Photograph: K. J. Gregory)

water action undercutting banks and of moisture or of ice in the banks can be important influences. As material becomes available for transport in a river channel it will be moved either in suspension, as the suspended load, or as bed load which is rolled or jumps along the bed by the process of saltation. This distinction between suspended sediment and bed load is to some extent an arbitrary one because at very low velocities no material will be moved; above competent discharge for the particular river channel cross section material will move in suspension or in contact with the bed; but at higher discharges particles first moved as bed load may subsequently become suspended in the flow.

During movement along the channel the size and shape of particles will change and the composition of material will alter. An individual particle will rapidly achieve a shape characteristic of the environment (see Figure 1.7), but downstream transport may lead to particles which become progressively smaller in size and as the least durable become smaller most rapidly, so the composition of fluvial sediments will change downstream. Thus it has been suggested (Fox, 1976) that the lifetime of a sedimentary particle will vary according to the positions in which it rests. Some particles will move each time river discharge exceeds the competent discharge whereas a particle may sometimes reside in a location which becomes vegetated so that the particle is unmoved for a decade or more. In the same way that it is possible to visualize solute concentrations over time, providing a chemograph, it is also possible to relate sediment concentrations to discharge either by rating curves (see Figure 2.15) or in terms of suspended sediment hydrographs (Figure 2.16) which are explained in Section 2.3b.

Production of sediment and solutes in a drainage basin combined with mass movement processes provide material that may be transported out of the basin. The relative significance of material moved by these several methods will vary according to character of the area and the energy available particularly as expressed through the energy of particular climatic events. Table 2.4 provides a summary of examples of the supply of material from different sources.

Table 2.4
EXAMPLES OF RELATIVE SIGNIFICANCE OF EROSION COMPONENTS

2.4a   Denudation processes in the Karkevagge Valley, Sweden 1952–60 (Rapp, 1960)

Process	Tonnes.km^{-2} yr^{-1} (approx.)	
Solution	23.6	Removed in streams; 9 tonnes.km^{-2} input in precipitation.
Earth slides and mud flows	63.0	Local transfers of material, often do not reach stream.
Avalanches	14.0	Sporadic and large ones occurred in 2 years.
Rockfalls	7.9	Mainly in the form of big boulder-falls.
Solifluction and talus creep		

Other processes include creep due to needle ice, wind erosion, slope wash all very slight.

2.4b   Erosion measured for limestone basins in Mendip Hills, South West England (D. I. Smith, 1976 in Derbyshire, 1976).

Erosion derivation	Percentage of annual total erosion	Seasonal distribution percentage	
		October–March	April–September
Swallet catchments	1.3	69.3	30.7
Soil profile	10.1	87.5	12.5
Uppermost bedrock	57.1	57.7	42.3
Main mass of bedrock	31.4	77.9	22.1
Cave passages	0.1	55.7	44.3
Total	100.0	67.4	32.6

Overall erosion rate 81 m^3.km^{-2}.yr^{-1}

2.4c   Stream channel load of five small basins in south east Devon (Walling, 1971)

Suspended load	Dissolved load	Bedload
20.7–53.6%	45.4–76.8%	0.5–2.5%

2.4d   Tentative sediment budget for arid basin (0.5 km^2 area) in southern Negev, Israel (Schick, 1977).

Suspended load	Sediment in solution	Bedload
115 tonnes	1.1 tonnes	60 tonnes
65.3%	0.6%	34.1%

This mean annual sediment yield was suggested by Schick (1977) to exceed considerably the accepted norm for arid environments.

2.2d Water and Sediment in River Channels

A river channel is designed by the movement of water and sediment in relation to the material locally available in bed and banks. Therefore a subject for many empirical investigations has been the relation between the processes operating in a particular river channel, the materials locally available, and the form of the river channel which results. Research has demonstrated that a purely straight channel rarely occurs in nature because even where the channel appears to be straight, the flow of water along the channel may follow a sinuous path. This may arise because along the river channel there is an alternation of shallows or riffles, and deeper sections or pools. Successive riffle bars may occur on opposite sides of the channel so that the deepest part of the channel or the thalweg tends to follow a sinuous course. Pools tend to scour at high flow and to fill at low flows whereas riffles conversely may scour at low flow and aggrade at high flows (Keller, 1971). Such a pattern of scour and fill can maintain the pool riffle sequence in which the distance between successive pools or riffles tends to approximate four to eight times the channel width. A succession of pools may also occur along the beds of channels cut in rock where there is insufficient material available to form riffles. Material may accumulate near the convex bank of a channel, and a point bar is the deposit which is formed around the convex bank of a channel bend and this is counterbalanced by erosion taking place on the opposite concave bank of the bend (Figure 2.10).

When viewed from above, reaches of river channels may be either single thread or multithread. Where a single thread channel occurs the channel is usually sinuous and the sinuosity can be described as the length of river channel divided by the length of the reach and on this basis a clearly meandering pattern has a sinuosity value of at least 1.5. The meandering pattern develops by accumulation of point bar deposits and by erosion on the opposite bank and these processes combine to account for the downstream migration of meanders. Where more than one channel exists as in a braided river (Figure 2.11), across an alluvial fan, or in a delta, the streamflow paths separate around bars or islands. Such multithread planforms develop in areas where there is an ample supply of coarse sediment available, where discharge changes rapidly and the rate of rise of the hydrograph is rapid, and where slope is high or changes abruptly from high to low, as for example when a river emerges from a mountain range.

The flat belt of land adjacent to the river channel is described as the flood plain and this plain should be inundated by flood flows which may occur as often as once each year on average. Not all flood plains will be inundated as frequently as this because the river flow may have been regulated, the channel may have been excavated or protected to prevent flooding, or the channel may have become incised so that floods rarely exceed the capacity of the channel. The flood plain is built up by the accumulation of sediment which may be deposited by channel or lateral accretion (Figure 2.12) whereby sediment is accumulated as point bars, as channel bars between braided channels (Figure 2.11), or as alluvial islands. An additional mode of floodplain accretion is provided by overbank sedimentation. Material can be deposited as ridges adjacent to the channel margins as levées; it may be accumulated as fine material as the suspended sediment settles from flood waters over much of the floodplain; or it can accumulate as fine clays in swamps and lakes on the flood plain.

FIG. 2.11. BRAIDED STREAM CHANNEL
Langdon Brook, Forest of Bowland, illustrating multithread channel and existence of former channels now vegetated. (Photograph: K. J. Gregory)

FIG. 2.12. SEDIMENTATION ALONG SMALL STREAM CHANNEL IMMEDIATELY AFTER PEAK DISCHARGE
Coarser material is deposited on channel margin (left) and finer sediments have been deposited on adjacent part of the small flood plain, Back Brook near Hawkerland, Devon. (Photograph: K. J. Gregory)

In addition to these processes responsible for the accumulation of material in the flood plain, at high discharges a river may break through the levées by avulsion and produce crevasse splays which lead to further aggradation of the flood plain, or cutoffs may develop when the river cuts through the neck of a meander. During migration of the meanders of a single thread channel the flood plain will be constructed including the traces of former channel bends indicated by point bar ridges separated by depressions or swales (Figure 2.13). Shifting of braided multithread channels will also associate with floodplain construction and movements of large braided rivers can be as much as 100 m per day. The Yellow River (Hwang Ho) of China with a very high silt load changes its channel where it is not

FIG. 2.13. FLOOD PLAIN DEPOSITS OF MISSISSIPPI (US GEOL. SURVEY)
Former channel positions indicated by pattern of point bar ridges and swales (with trees) with lakes surviving in some cutoffs.

controlled, and Chien-Ning (1961) suggested that lateral channel movement could be
expressed by an index of wandering (θ) which was a function of hydrograph rise during
passage of a flood (ΔQ), bankfull discharge (Q), depth of flow at bankfull discharge (h),
channel slope (J), grain size of bed material (D_{35}), maximum (Q_{max}) and minimum (Q_{min})
discharges of the flood season, surface width of rare flood (B), and surface width at
bankfull discharge (b).

The relation was of the form:

$$\theta = \left(\frac{\Delta Q}{0.5\ T.Q_n}\right) \left(\frac{hJ}{D_{35}}\right)^{0.6} \left(\frac{Q_{max}-Q_{min}}{Q_{max}+Q_{min}}\right)^{0.6} \left(\frac{B}{b}\right)^{0.45} \left(\frac{b}{h}\right)^{0.3}$$

Such equations can be developed for the channel patterns in an area because the size of
channel planform will be related to discharge and therefore to size of drainage area, to the
sediment transported, to the characteristics of the bed and bank material and their
associated vegetation, and to the longitudinal slope.

Inter-relationships between the several types of channel pattern have been elucidated
by laboratory investigations. Schumm (1977) investigated the way in which slope affects
planform by showing that in a flume at low slopes a straight channel existed; this was
succeeded by a progressively more meandering channel as slope increased, until at the
highest slope a braided channel was produced.

If the river channel is considered in cross section it is possible to establish relationships
between the processes operating in the section and measures of channel form to provide a
hydraulic geometry of stream channels. In a particular channel cross section, relationships
can be established by making measurements at different discharges of width *(w)*, depth
(d), velocity *(v)* and discharge. These relationships have been expressed as power functions in the form:

$$w = a\ Q^b, d = c\ Q^f, v = k\ Q^m$$

where *a, b, c, f, k* and *m* are numerical values. Because *w, d, v = Q* then $wdv = aQ^b.cQ^f.kQ^m$
$= ack.(Q)^{b+f+m}$. Therefore $ack = 1$ and $b + f + m = 1$.

When these equations are developed for a single cross section this provides the 'at-a-
station' hydraulic geometry. The downstream hydraulic geometry may be plotted by
obtaining measurements from a number of sections along a river channel. In this case it is
assumed that the sections measured at different sites are all equivalent and relate to similar
positions on the hydrograph. An additional approach to the investigation of the channel
cross section is based upon a consideration of the factors that determine the size, shape and
cross sectional area of the river channel, as explained in Section 2.4c in relation to the
classification of channel types.

2.3 PRINCIPLES APPLIED TO THE DRAINAGE BASIN

To understand geomorphological processes it is necessary to appreciate the principles
governing mass movement and the flow of water and sediment in stream channels. The
need for such an appreciation of physical principles was first demonstrated by A. N.

Strahler in 1952 and subsequently has been emphasized by A. E. Scheidegger in his *Theoretical Geomorphology* and by M. A. Carson in *The Mechanics of Erosion*. In these and other recent works there have been attempts to outline the principles of soil mechanics and fluid dynamics which are fundamental in geomorphology and particularly relevant to the study of slopes and streams.

2.3a Particles on Slopes

Forces acting on materials provide the essence of the theoretical approach and Strahler (1952) approached the problem on the basis of types of material, types of stress and types of strain. Three types of material were recognized. *Rigid or elastic solids* are masses of soil or rock which possess elasticity and yield by elastic strain. Such a solid will not yield until a critical threshold stress is applied but subsequently the resulting movement or strain will follow Hooke's Law in which strain is proportional to stress. *Plastic solids* are those types of material which deform by distributed or intermolecular or intergranular shear. If shear stresses exceed a certain critical limit flowage will occur similar to fluid flow and the rate of strain is proportional to the stress applied. *Fluids* provide the third type of material; they offer little resistance to a change of form, and they are unable to adjust internally when subjected to shear stress. The behaviour of these three groups of materials is illustrated in Figure 2.14 which shows how the rate of strain varies with applied stress in each type.

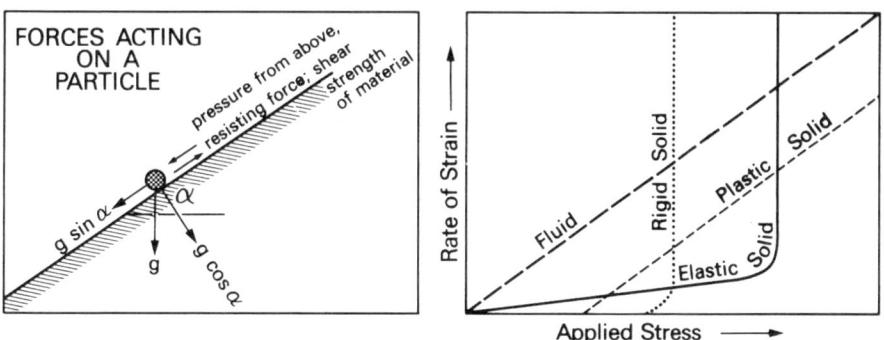

FIG. 2.14. FORCES ACTING ON A PARTICLE AND STRESS-STRAIN RELATIONSHIPS

The forces applied to a body are termed *stresses* and these may be gravitational, molecular, biological or inter-particle in character. Gravitational stresses are induced by the influence of gravity and these stresses will influence mass movement and water flow on slopes and fluid flow in channels. A simplified picture of mass movement is represented in Figure 2.14 by depicting a particle on a slope. The gravity force acts vertically downwards but as the particle cannot often subside into the slope, the gravitational component of stress will act down the slope with a value of $g \sin \alpha$ (Figure 2.14). This stress may be complemented by the pressure of material from higher up the slope and if these were the only forces operating the particle would move downslope. However there is a resisting force represented by the shear strength of the material, and for the particle to remain static the shear strength of the material and the frictional force between the particle and the slope must be greater than, or equal to, the gravitational component acting downslope.

If the shear strength of the material is reduced, for example by an increase in water content, or if the frictional force between the particle and the slope is decreased due to lubrication by water on the surface, then it is possible that the resistance will be less than the gravitational downslope component. In this case the particle would move downslope until the gravitational downslope component no longer exceeded the shear strength of the material and/or the frictional resistance. Molecular stresses act in any direction as a result for example of the swelling and shrinking of colloids, thermal expansion and contraction of ice crystals. Some stresses can be induced by plant roots or by animals including for instance the stress occasioned by roots during plant growth. Some workers (e.g. Young, 1969) have also distinguished interparticle stresses which are the forces operating between particles and so are distinguished from molecular or colloidal stresses.

The several types of stress which operate may act normally to the surface of the material, and such normal stresses which involve compression contrast with shear stresses which act parallel to the surface of material. Many of the geomorphological causes of stress have been mentioned already. Gravitational stresses include the gravitational component acting on slope materials (see Figure 2.14) and also fluid flow forces which occur as fluids flow over the slope or in a stream channel. Rainfall impact can create stresses which will include a consolidating stress acting normal to the ground surface and a shear stress acting along the slope or ground surface. Expansion or heaving forces are not responsible for moving material downslope but give molecular stresses which lift material so that gravity can then exercise an influence. Such expansion forces can occur due to changes in moisture content, thermal expansion of rock, expansion of water on freezing, and the effects of crystallization of salts in arid areas. The effect of water below the surface may be expressed in water pressure and tension forces where the head of water exerts a pressure equal to the overlying weight of water per unit area.

Two distinct types of *strain* or movement resulting from stress may be distinguished. A rock or material behaving as a rigid or elastic solid will fail, after the threshold value of stress is reached (Figure 2.14) by shear fractures or by a widening tension fracture. A shear fracture develops as a slip plane where the shear stresses are greater than the resistance of the material to shear. Such shear fractures develop in rotational slumps and may occur in translational slides (Figure 2.6). A widening tension fracture is a vertical fracture which when produced gives elongation of the mass, and crevasses in a glacier or tension cracks at the head of landslides are examples of this type of fracture. A plastic solid or a fluid will strain by one of two types of flow. In laminar flow, movement takes place as a series of layers or sheets one above another whereas in turbulent flow there is mixing of the layers and eddies will occur. During turbulent flow the particles move in irregular paths and it is possible to distinguish laminar and turbulent flow by the Reynolds Number (R_N). This number is a value calculated from the flow velocity (V), the size of the flow section (L), the density (ρ) and viscosity (μ) of the fluid. For high values of the Reynold's Number, flow will be turbulent whereas flow will be laminar and dominantly viscous forces for low values of the Reynold's Number which is expressed as:

$$R_N = \frac{\rho V L}{\mu}$$

Mass movements may thus be visualized as a consequence of the strain which results from the interplay of strength of materials and the stress which is applied. Movements may therefore be induced as a result of a reduction of soil strength by long-term swelling, by exceptional ground water conditions when pore-water pressures are increased following heavy rain, or by weathering which reduces the strength of the material and the angle of shearing resistance (Chandler, 1977). Alternatively, increases in stress can be occasioned by increased loading upon a slope due for example to the construction of man-made features or to excavation by rivers undercutting the base of a slope which will induce an increase of shear stress. The shear strength of a mixed soil is indicated by Coulomb's Law which relates the shear strength to the cohesion of the material, the pressure normal to the shear plane, and to the angle of shearing resistance or internal friction as shown in Section 1.2.

Knowledge of materials, stresses and strains can be employed to interpret specific slope situations in the drainage basin. On a given slope it is possible to use laboratory tests of the materials to give an indication of the shear strength of the material and of the density of the material and these can be related to the angle of slope of the ground surface and to the critical height at which failure will occur. It is then possible to employ theoretical relations between these variables to suggest the critical heights to which slopes can exist before failure will occur. A specific material occurring in a particular position and subjected to a definite water regime under the prevailing climate will be able to exist up to a certain limiting or threshold angle of slope. At or above that slope angle value the slope will fail because the downslope gravitational component force will be greater than the strength of the material (e.g. Figure 2.14). Thus studies on Exmoor in England showed (Carson and Petley, 1970) that angles of 26° were extremely common in areas where the underlying rock weathers into debris composed of a mixture of rock rubble and coarse soil particles. The existence of these angular values was attributed to the angle of stability of the debris when subjected to a particular pattern of pore-pressure distribution within the mantle.

An investigation of a more arid area (Carson, 1971) showed that in the Laramie Mountains of Wyoming there were many slopes containing straight sections and that modal groups of slope angles had values of 33, 28–25, and 18 degrees. Slope stability analyses employing theoretical equations when applied to the area suggested that these three groups of slope angles corresponded to angles of limiting stability at different stages of weathering of the mantle. A theoretical analysis of a particular slope situation can be a valuable prerequisite for decisions about the location of engineering structures on a slope. Thus if a road is to be constructed along a slope a stability analysis will indicate, according to the nature of the local materials and conditions, the limiting angle that must not be exceeded when slopes are cut by man.

2.3b Fluids in Channels

A similar approach is possible in the study of fluid flow in channels. The velocity of the water at any one river channel cross section reflects the balance between the forces directed downstream and those resisting movement. The downstream force is affected by the head of water and the kinetic energy possessed by water further upstream and also by the gravitational component which is proportional to the sine of the slope angle. The resisting forces are provided by the shear resistance at the fluid/bed and fluid/banks contact and to a lesser degree at the contact of fluid and air above. Thus in a river channel cross section the lines of equal velocity (isovels) have lowest values near the bed and banks and highest values immediately below the water surface and near the centre of the channel. The discharge *(Q)* of water is a product of velocity *(V)* and cross sectional area *(A)* and the mass rate of flow *(G)* is equal to the product of discharge and density of the fluid *(ρ)*. Thus in a single cross section $Q = VA$ and $G = ρQ$. If density remains constant from one section to another then along a river reach discharge remains the same, if no water is received from tributaries or is received or lost from bed and banks, so that $Q = A_1V_1 = A_2V_2 = A_3V_3$ etc., where $A_1, A_2 A_3$ etc. are the cross sectional areas of successive river channel cross sections and V_1, V_2, V_3 etc. are the corresponding mean velocities. This is a form of the continuity equation which expresses the notion that water discharge is constant in successive reaches.

Velocity thus changes inversely with cross sectional area to maintain continuity and a constant discharge value. A number of equations have been available for many years to relate discharge to channel characteristics at particular channel cross sections. An example is provided by the Manning equation in which velocity *(V)* is related to slope *(S)*, hydraulic radius *(R)* which is the channel cross sectional area divided by the wetted perimeter *(P)*, or A/P, and channel roughness *(n)*. This equation is expressed as:

$$V = \frac{1.009}{n} R^{2/3} S^{1/2}$$

and it can be used to estimate a discharge value at a particular site. If after a large flood it is necessary to calculate the peak discharge; this can be done by measuring the channel cross section occupied by the flood so that a value may be derived for hydraulic radius; by surveying the slope of the former water surface; and by estimating roughness. Roughness is included in the Manning equation to give an index of resistance to fluid flow. Very smooth channels such as plastic or glass have low roughness values of the order of 0.010 whereas mountain streams with rocky beds have a roughness value nearer 0.050. Use of the Manning equation therefore depends upon a good estimate of channel roughness for the particular site under consideration. The Manning equation expresses the fact that velocity will be proportional to the square root of slope. For large wide channels hydraulic radius $A/2d + W$ is approximately equal to A/W which approximates to $W.d/W$ or depth *d*. Thus we can see from the Manning equation that in large shallow channels, if roughness is constant then velocity will be proportional to the product of the slope and depth, i.e. $S^{1/2} d^{2/3}$.

Flow equations can be employed to provide an indication of discharges that would occupy a channel of known dimensions. In the North York Moors of England a glacial drainage channel (p. 272) called Newton Dale has been regarded as the channel responsible for draining water southwards from a proglacial lake and from melting ice to the area of the Vale of Pickering on the south side of the North York Moors. Assuming that the glacial drainage channel, which is up to 60 m in depth, was occupied by water to a depth of 10 m for the highest discharges occurring for two or three weeks in the year during deglaciation, it is possible to estimate the peak discharges likely for this channel during deglaciation conditions according to surveys of channel cross section, channel slope and estimates of channel roughness. These estimates of peak discharge can be employed to show that approximately 10 years could be sufficient to drain all the water stored in the proglacial lakes, ice and snow to the north of the glacial drainage channel (Gregory, 1978).

Movement of sediment is primarily a reflection of sediment availability and, whereas a number of such equations have been proposed to express total bedload and suspended load movement according to flow and channel characteristics, they assume that unlimited amounts of sediment are available. Such a capacity load situation is rarely encountered in nature and this is why there can be great differences between the load calculated by equation and that load measured for a particular river reach. The empirical relation between discharge and sediment or solute concentration at a particular site may be expressed in the form of a rating curve. Thus when discharge values *(Q)* over time are related to values of sediment concentration (SSC) (e.g. Figure 2.15A), it is possible to derive a rating curve with the form $SSC = aQ^b$ and similar curves can be produced for bedload or for solute (Figure 2.15B) concentrations. Because such curves provide a

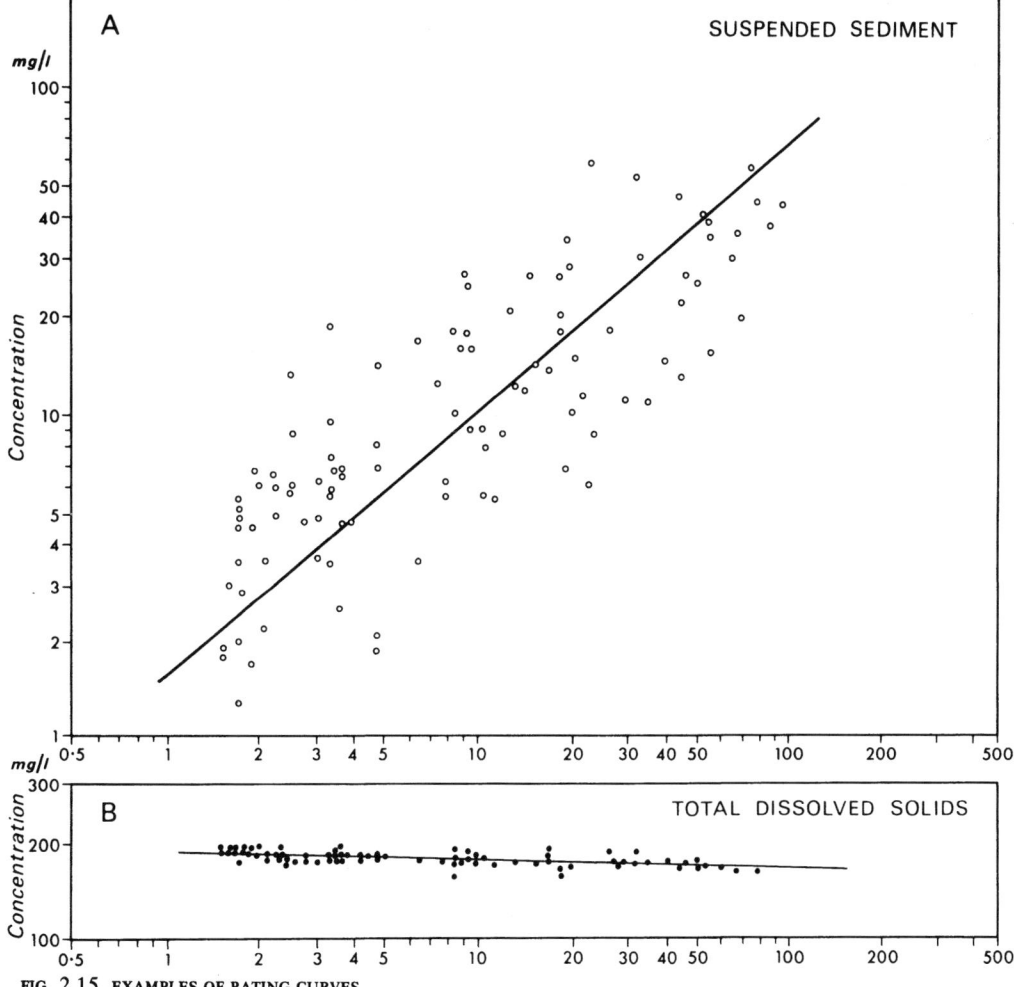

FIG. 2.15. EXAMPLES OF RATING CURVES
(A) shows relation between suspended sediment concentration and discharge, giving suspended sediment rating curve; (B) relates total dissolved solids concentration to discharge, providing rating curve where solute concentration decreases with discharge because of dilution effect.

relation between fluid flow and sediment or solute concentration they can be employed to calculate amounts of sediment yield over time. This is achieved by converting continuous discharge into a continuous sediment or solute trace from the rating curve and then calculating the volume of sediment moved per unit time. Use of rating curves assumes that the relationship between discharge and sediment or solute concentration is identical for rising and falling limits of the hydrograph and that the discharge hydrograph and the sediment or solute hydrographs are synchronous.

In the example illustrated in Figure 2.16, the discharge hydrograph and the suspended

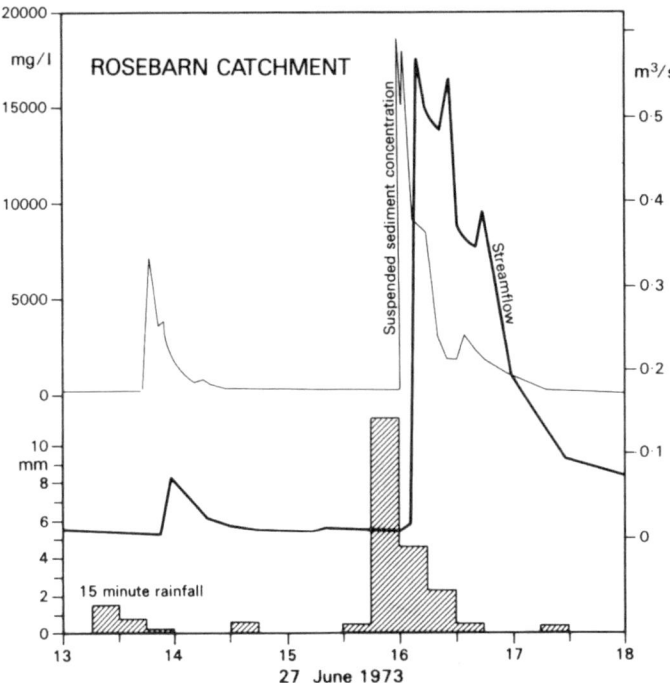

FIG. 2.16. SUSPENDED SEDIMENT HYDROGRAPH AND DISCHARGE IN RELATION TO RAINFALL
The small Rosebarn catchment on margins of Exeter, Devon is sensitive to small amounts of rainfall and lag between peak rainfall intensity and peak discharge can be only 15–20 min (see Table 2.3). Suspended sediment hydrograph obtained from plot of suspended sediment concentrations occurs before streamflow hydrograph, because sediment sources are closer to gauging site than runoff-producing areas. Because the two hydrographs do not coincide there is not a good rating curve relation (e.g. Figure 2.15), and amount of scatter of points is considerable. Where several suspended sediment hydrographs occur in sequence, peaks may become progressively lower because all sediment sources have temporarily been exhausted.

sediment hydrograph do not correspond exactly in time so that there will not be a good rating curve relation between discharge and suspended sediment concentration in this catchment. Depending upon climatic and catchment characteristics the two hydrographs may peak at the same time, the discharge peak may arrive first or it may lag behind the sediment and solute hydrographs.

Whether sediment and solute hydrographs are synchronized with the discharge hydrographs depends upon the sources of sediment, solutes and water and also upon the transformation of a hydrograph as it passes along a river channel. A hydrograph is produced with a particular shape from a specific basin and attempts have been made to derive a generalized or unit hydrograph which reflects the characteristics of a particular basin. As the wave of water which constitutes a hydrograph moves downstream its form will be changed as a result of additional water entering from tributary and other sources, of the pattern and arrangement of these tributaries, and of the size and character of the stream channel. The last factor is significant because of the storage potential of the river channel. If a steep shallow channel leads into a gently sloping wide channel with numerous deep pools then a sharp peaked hydrograph from the steep reach may become more attenuated in the succeeding reach as water is stored in the pools. These changes are the essence of flood routing techniques which are used to model the way in which the flood discharges may be produced for a particular point along a river channel.

The basis of flood routing techniques is that if there are no inputs or losses of water then as water travels along a reach, the discharge at the upstream end (Q_1) may be related to the discharge at the downstream end (Q_2) in relation to the amount of water stored in the reach between the two sites (ΔS). A similar method may be applied to the downstream movement of sediment hydrographs or of chemographs of solutes but the dilution effect of high discharges gives a reduction in solute concentration with higher discharges. Modelling of the way in which flood waves and sediment pulses are routed through the drainage network of a basin remains a complex problem.

The bedload which slides or rolls along the channel bed moves as a result of the shear stress which arises from the fluid flow and from the downslope component of the weight of the material. Movement of material is resisted by the cohesion between particles and by friction with the bed so that a particle in a river channel is subject to forces comparable to those that influence a particle on a slope (Figure 2.14). It is possible to model the movement of bedload and the Du Boys formula related bed-load discharge per unit width of section per unit time (q_b), to bed stress intensity (τ_o), the bed stress required to initiate particle movement (τ_c) and a constant *(K)* in the relation

$$q_b = K \tau_o (\tau_o - \tau_c)$$

This formula therefore derives sediment transport rate from the excess bed stress above a critical stress which depends upon the grain size of the material available for transport. Use of this equation relies upon obtaining values of the constant *(K)* which depends upon the flow parameters such as velocity and turbulence of the water and the sediment contained. Because of the difficulties of evaluating this constant and of determining the critical bed stress, a number of alternative equations have been developed and employed particularly by hydraulic engineers (Statham, 1977).

Many simple equations are developed from the assumptions that river flow is uniform, that is that flow takes place in a rigid, straight open channel, and also steady, in that discharge remains constant over time and that depth is maintained as constant in the downstream direction. In natural river channels flow is not uniform because there are variations along the channel from riffle to pool and there are also variations in planform as the thalweg, or line of the deepest part of the channel, follows a sinuous path. Flow is unsteady because discharge changes and there are associated variations in channel roughness and in sediment in the flow. Modelling of these changes continues to pose a challenge for the mathematical modelling of fluid flow and channel behaviour (Allen, 1977).

Discharge in a channel cross section influences the sediment moved and reference has already been made to the critical erosion velocity or competent velocity which is that minimum current velocity just capable of initiating the movement of grains of a given size or diameter. Such relationships were expressed in a series of curves produced by Hjulstrom in 1935 to relate grain size and velocity to show the fields in which erosion, transport and deposition occurred. These curves have been modified subsequently and a modification by Sundborg (1956) is illustrated in Figure 2.17. Inspection of the relationships

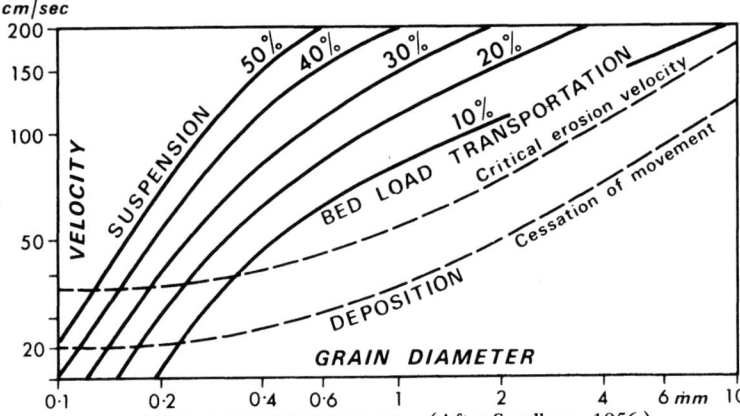

FIG. 2.17. RELATION BETWEEN VELOCITY AND SIZE OF MATERIAL. (After Sundborg, 1956.)

embodied in these curves underlines the time-dependent nature of the sediment in a particular channel cross-section. If a bar exists during low flow, is swept away during higher flows but replaced again during the following low flow, the channel is regarded as stable. If however the channel is constantly changing and banks and bars which do appear are soon swept away again, then the channel is unstable. Because increasing velocity can move greater grain sizes and a larger volume of sediment, a sequence of sedimentary changes can be observed during the passage of a flood wave. The sedimentary structures or bed forms on the bed of a mobile sand bed channel will vary with discharge and a plane bed with no sediment movement is followed by ripples and dunes. There may be a lag or time lapse between the threshold discharge and the resulting bedform, and the amount and rate of change of discharge influence the sequence of forms and the residual bed features.

2.3c Equilibrium Concepts

A concomitant of a theoretical approach to slopes and fluvial processes in drainage basins is that equilibrium concepts should be developed for the components of the basin. A number of equilibrium concepts have been developed for slopes and stream channels and one of the first of these was the concept of grade. This concept has been applied to stream channels and a graded stream was defined by Mackin (1948) as '. . . one in which, over a period of years, slope is delicately adjusted to provide, with available discharge and with prevailing channel characteristics, just the velocity required for the transportation of load supplied from the drainage basin. The graded stream is a stream in equilibrium; its diagnostic characteristic is that any change in one of the controlling factors will cause a displacement of the equilibrium in a direction that will tend to absorb the effects of the change'. Although Dury (1966) recommended that the notion of the graded stream should not continue to be used, the same concept can also be applied to slopes. Carson and Petley (1970) expressed the main thesis of the concept of grade as the fact that most denudational processes operating in the landscape tend to create a threshold angle of slope. Whereas there is rapid denudation at angles greater than this threshold value, producing a slope of lower inclination, slopes will not be produced at angles below the threshold because at lower angles the process can no longer be operative.

A study of straight hillslopes in two areas of strong, well-jointed rock showed that the frequency distribution of angles indicated that angles of 20–21°, 25–27° and 32–34° are characteristic of both areas. Analysis of the shear strength characteristics of the waste mantle on these three slope types suggested that the particular angles are very probably limiting angles of stability for the three types of waste mantle (Carson and Petley, 1970). There is thus a return to the idea that slope form is adjusted to the prevailing conditions operating on the materials available in a specific location.

Angles of slope may be characteristic of a particular area according to the limiting angle of stability for the material. However it must be remembered that the character of materials varies over short periods of time. During a year, seasonal variations in moisture are such that the shear strength may be at a minimum for a few days of the year and the slope angle may be in equilibrium with this value, whereas for the remainder of the twelve months the shear strength is greater. Thus although the slope remains static for the whole year the equilibrium condition actually obtains for a small proportion of the time.

Such a condition of near equilibrium or quasi-equilibrium also applies to sheet flow on slopes. A slope may be fashioned by slope wash for a few days of the year but soil creep or slow mass movement processes may occur typically for the remainder of the time. A similar situation applies to the stream channels within the drainage basin. This is well illustrated by the fact that large boulders or pebbles in the bed of the stream channel may be related to the flows which occur infrequently and which are competent to move the material (e.g. Figure 2.17), whereas for the remainder of the year they are unmoved by lower flows with smaller water velocities.

It is therefore necessary to understand the ways in which fluvial landforms are adjusted to large events and to appreciate the role of large events in influencing landforms (see also Section 2.4). Wolman and Miller (1960) in a commentary on the magnitude and frequency of processes in geomorphology, drew an analogy with the work of a dwarf, a man and a giant in clearing the trees of a forest. The dwarf works for long periods but achieves small results rather like the flows that occur in a river channel for many days of the year or the slow mass movement processes that operate on a slope for the year. The man works less frequently but more effectively and is analogous to the rapid mass movements that occur on slopes each year at certain times or to the flows in stream channels once or several times each year. The giant sleeps for long periods but can effect great damage for short periods and this is similar to the catastrophic slope failures or the great river floods that occur rarely and perhaps only once in several hundred years in a particular location. Wolman and Miller concluded that the most effective processes fashioning the landscape were the ones analogous to the work of the man and they suggested that the processes which occur for short periods perhaps once or several times every two or three years are the dominant processes essential to consideration of equilibrium between the form of slopes and drainage basins and their controlling processes (see Table 2.5).

A river channel cross section may be measured to give an indication of the bankfull capacity of the river channel. Although this is not always clearly defined for many river channels there are many cases where the size of a river channel at a particular location may be determined. The value of this capacity reflects an equilibrium between the river flow, the sediment carried by the river, the local site characteristics especially slope, and the shear strength of the materials comprising the bed and banks. Only when the shear strength of the local materials is exceeded by the velocity or power of the stream can erosion of the channel occur. In unconsolidated materials the velocity required to modify the channel may be quite small so that the channel capacity is adjusted to flows which occur several times each year. In more consolidated materials or in bedrock the capacity may be related to less frequent flows which occur perhaps only once in several years.

Although a number of indices of flow have been related to channel capacity it is apparent that a range of flows affect the size of a river channel. This range of flows begins with the velocity just competent to move the material making up the bed and banks of the channel and it extends at least to the bankfull discharge values. A similar range of discharges has been held to be significant in affecting channel pattern so that the wavelength of meanders has been related to measures of peak discharge in a number of studies. The size of meander arcs increases as discharge increases, and the dimensions of single thread or multithread planforms are related to a range of higher discharges. Although a range of discharges may be significant in influencing the capacity of a river channel and the dimensions of river channel planform, this broad range may include categories of flow which are responsible for particular aspects of channel capacity or planform dimensions. Thus a study of rivers in New South Wales, Australia led Pickup and Warner (1976) to conclude that in the Cumberland Basin the most effective flows which recurred on average two to five times a year were the ones which influenced the sediments in the cross sections and bedload transport, whereas less frequent flows, with recurrence intervals of between 4 and 10 years on the annual series, were significant in determining overall channel dimensions.

The character of a river channel is also influenced by the sediment conveyed through the channel and by the calibre of the sediment in the bed and banks. One way of expressing an aspect of this relationship is derived (Schumm, 1977) by relating the width/depth ratio as a measure of channel shape to the percentage of silt clay in the perimeter of the channel cross section (Figure 2.18).

Equilibrium ideas have been applied to river channels for many years through regime theory. This approach was encouraged by studies of artificial rivers or canals which were constructed for irrigation, in India for example, and where it was necessary to design a channel which would be adjusted so that it would convey the water and sediment supplied to it without either scouring or aggrading. Therefore a channel carrying a particular long-term pattern of water and sediment discharge and subjected to a definite set of constraints, could adjust to the action of the flow and would be expected to acquire a definite regime (Blench, 1972). The regime would be expressed by equations relating water and sediment discharge to some of the variables of width, depth, slope, meander length, meander belt width, grain size of materials in bed and banks, and length of dunes that form on the bed. This approach is particularly useful where flow and sediment transported is steady but is difficult to extend to situations where flow is unsteady. A number of ideas implicit in equilibrium concepts are summarized in Table 2.5.

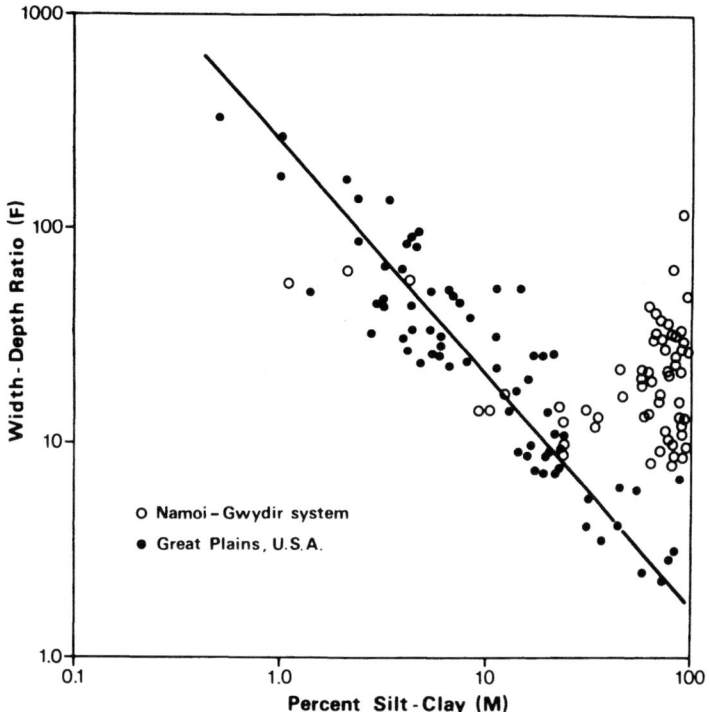

FIG. 2.18. RELATION BETWEEN CHANNEL SHAPE AND SEDIMENT CHARACTER
Channel shape expressed as width/depth ratio (F) and related to weighted mean silt clay percentage (M) in bed and banks of channel for alluvial channels measured in the semi-arid and sub-humid areas of Great Plains, USA and Riverine Plains of New South Wales, Australia (Schumm, 1977). Silt clay was measured as sediment smaller than 0.074 mm diameter. Relation for which the line is plotted was $F = 255 M^{-1.08}$. Data from some areas does not conform to same relationship and for the Namoi-Gwydir river system, Riley (1976) showed that the points (shown as circles) did not support same relationship because F may not be a consistently good index of channel shape, and cohesive sediments can complicate relation between F and M.

It is also possible to relate the drainage network in terms of its extent and density to flows of a particular magnitude. The network expands and contracts according to the preceding moisture conditions, the characteristics of the drainage basin including relief, soil type and vegetation cover, and the input of precipitation from particular storms. Whereas a drainage network may be very extensive during rare flood events the outer channels may be maintained by flows that occur with a frequency of approximately once a year. The channels produced by the rare events may be infilled by slope processes during the many years which elapse between such catastrophic events. This situation may be illustrated for the tropical island of Barbados where the drainage network based upon information from 1:10 000 maps is shown clearly in Figure 2.19. In the north east of the island the underlying rocks are Tertiary clays and the density of water courses is greater than in the remainder of the country which is underlain by Coral Limestone. Although dry valleys occur in many limestone areas of the world, the valleys on limestone in Barbados have ephemeral stream channels along their length and the occurrence of flow along these channels is testified by the culverts which have been constructed where a road or track crosses a stream (Figure 2.19). The stream channels are rarely occupied by water but for

Table 2.5
EQUILIBRIUM CONCEPTS IN THE DRAINAGE BASIN

| Concept | Interpretation | Application to: | |
		Slopes	Fluvial system
Static equilibrium	Balance of forces operating does not occasion movement to maintain equilibrium	Weathered bedrock and material on slopes not displaced because resisting forces greater than stress	Material on bed and in banks of river channel does not move because stress has not reached critical level
Dynamic equilibrium	Movement takes place at a constant velocity and because there is no tendency to accelerate or to decelerate, governing forces are constant and maintain dynamic particle in equilibrium or in steady state	Scree slope in which material is moving continuously could maintain its angle and form when dynamic equilibrium is provided because material supplied from free face above replaces material moved down scree	Movement of particle of sediment or drop of water in fluid moving through river channel. As particle moves from its position it is replaced by another identical one and form of channel or channel bed maintained in dynamic equilibrium
Quasi equilibrium	State of near equilibrium (Langbein and Leopold, 1964). Number of events of specific size maintain morphology	Slope processes may occur intermittently and slope form maintains characteristic form between events	River channel morphology may be determined by flows which occur for several short periods each year. During remaining time channel is in apparent equilibrium
Dynamic metastable equilibrium and metastable equilibrium	Erosional development of landscape proceeds in pulses by episodic erosion (Schumm, 1977)	Slope morphology is transformed by episodes of mass movement	Erosion of channel or change of channel planform proceeds in stages induced by episodic erosion

the channels to be maintained, and not filled in by mass movement processes, flow must occur at least once every few years. The drainage network shown in Figure 2.19 is therefore probably adjusted to moderate frequency events. The size of the river channel cross section, the size of the channel pattern and the extent of the network are all related to flows which may occur several times a year or once in several years.

2.3d Utilization of Models

A theoretical basis is necessary for the development of models (Section 2.1d) for stream and slope processes. Models based upon physical laws include those based upon flood routing methods (see p. 81) and upon assumptions that the rate of operation of mass movement is dependent upon the height above the slope base. Using assumed physical relationships it is possible to build computer models to model slope and basin processes based upon flow diagrams such as Figure 2.5, and by optimizing parameters in the models it is possible, for example, to provide hourly values of streamflow for a particular river according to the characteristics of the basin and of the prevailing and preceding climatic conditions. Alternatives to such physical models are statistical or parametric ones in which

FIG. 2.19. DRAINAGE NETWORK OF BARBADOS
The drainage network is based on 1:10 000 maps and the culverts, indicating likelihood of occasional channel flow, also derived from this map series and from field observation (Gregory, 1979). Density of stream channels greatest in north east of island where there are Tertiary rocks including shales. Elsewhere coral limestone outcrops allow some streams to end at sinks, accounting for lower drainage densities of the south and north.

a large number of values are measured and inter-related by statistical methods. Thus by obtaining measurements from a large number of slopes the rate of mass movement could be related to angle of slope, precipitation characteristics and other factors. Similarly measurements of flow in a number of rivers could be the basis for statistical relationships established between a flow index such as peak discharge and measures of precipitation and characteristics of the basin such as relief, area and density of the drainage network.

A third type of model may be based upon probability considerations. Thus slope processes may be modelled using Markov chains. A probability or stochastic approach has also been applied in the case of the study of river channels. Langbein and Leopold (1964) argued that there are two possible tendencies for a river system. One tendency is for an equal distribution of energy expenditure throughout the river course. An opposing tendency is for the river to flow in such a way that it effects the minimum total work. They conclude that the most probable state along any river will be one which lies between these two limits, and the long profile of a river which is concave upwards, the downstream change of channel geometry and the channel properties at a given cross section can all be interpreted in this way. At any location along a river or stream the stream has a certain amount of available stream power *(P)* which is the time rate of energy expenditure and is necessary to maintain motion at a given rate. This depends upon the specific weight of water (γ), the hydraulic radius $(R = $ cross sectional area divided by perimeter), the water surface slope *(S)* and the mean velocity *(V)*, and may be expressed as $P = \gamma RSV$. Unit stream power is defined as the rate of potential energy expenditure per unit weight of water. An alluvial channel tends to adjust its velocity, slope, roughness and channel geometry to maintain a balance between unit stream power and the rate of sediment transport and this can be used to calculate sediment transport and to model fluvial hydraulics (Chih Ted Yang, 1976).

Once ways of expressing slope and stream processes have been achieved it is possible to envisage how changes will take place over time.

For the reasons outlined above it is not always easy to identify equilibrium relationships or to suggest how changes will occur over periods of time. Difficulties arise because processes are not steady over time so that it is difficult to ascertain the processes that are really responsible for particular forms and threshold values may have to be exceeded before an adjustment takes place. Thus if an equilibrium exists between slope form and slope processes the effect of man removing the vegetation on the slope is to alter the processes operating. However the slope form may not change until an event of a certain magnitude occurs to effect the change of slope form. There may be a delay or lag of several months or years between a change of process and a compensating adjustment of form. The significance of such thresholds was emphasized by Schumm (1973) and he also elaborated the significance of complex responses. A change in the processes affecting two adjacent slopes or drainage basins could produce a distinct effect in each case. Thus two adjacent slopes could be affected by greater rainfall amounts and intensities. Depending upon local conditions and materials the result on one slope could be an increase in surface wash whereas on the other slope rapid mass movement phenomena could be more frequent. Two adjacent basins similarly affected could see an increase in the extent of the drainage network in one basin as gullies were eroded, whereas the adjacent basin could be modified by enlargement of the existing channels without any extension of the channel length.

Analysis of changes during timespans of various length is of particular concern to the geographer and is essential to an understanding of past landform development and of possible future changes.

2.4 PATTERNS OF PROCESS

Processes in the drainage basin vary spatially over the surface of the basin, and temporally throughout each year. A recurrent problem has been to measure processes at sufficient locations to obtain some overall idea of the processes taking place over the entire basin surface and a second problem has been the need to measure processes for sufficiently long periods of time to provide an understanding of long term rates. The essential nature of short term variations will be outlined before illustrating the magnitude of rare events and proceeding to show how measurements of processes have been made.

2.4a Temporal Patterns and Sequences

Mass movement processes on slopes are seldom continuous and slow mass movement processes are probably made up of a large number of small movements unevenly distributed over time whereas rapid mass movements, although occurring in shorter time periods, are similarly made up of a number of discrete movements. A study of mudflows in north eastern Ireland involved continuous measurements (Figure 2.20) of the rate of mudflow movement (Prior and Stephens 1972). One result of this study was to show how surges of mudflow movement (Figure 2.21) at rates up to 2.8 m/hr may reflect the sequence of storm rainfall amounts as shown in Figure 2.20. The discharge hydrograph in the river channel provides an analogous example and in Figure 2.16 the discharge hydrograph from a small experimental basin is represented in relation to the rainfall record. In the same diagram the suspended sediment hydrograph is shown and although this presents a sequence of peaks very similar to the streamflow trace the suspended sediment hydrographs occur slightly in advance of the streamflow hydrographs. Both examples reflect temperate situations and the two illustrations (Figures 2.16 and 2.20) typify the events which occur with moderate frequency. The Rosebarn catchment has been monitored since 1968 (Gregory and Walling, 1973) to indicate the effects of man expressed in building activity and urbanization of a small basin on the margin of Exeter, Devon, U.K. The concentrations of suspended sediment recorded during the storm in June 1973 (Figure 2.16) were higher than any recorded up to 1971 when building operations began to influence the sediment sources exposed in the basin.

The significance of major infrequent events, often of a catastrophic nature, has been increasingly investigated since 1970. Such major events inspire mass movements after periods of high precipitation and in Hong Kong a period of 24 hours in June 1966 saw the receipt of up to 401 mm of precipitation at the end of a period when a trough of low pressure crossed the area five times (So, 1971). The isohyets during one hour during this period are shown in Figure 2.20 and one consequence of the enormous volumes of water produced were mass movements on slopes which included rotational slips, and washouts including debris avalanches, boulder falls and rock slides. Under the tropical conditions of

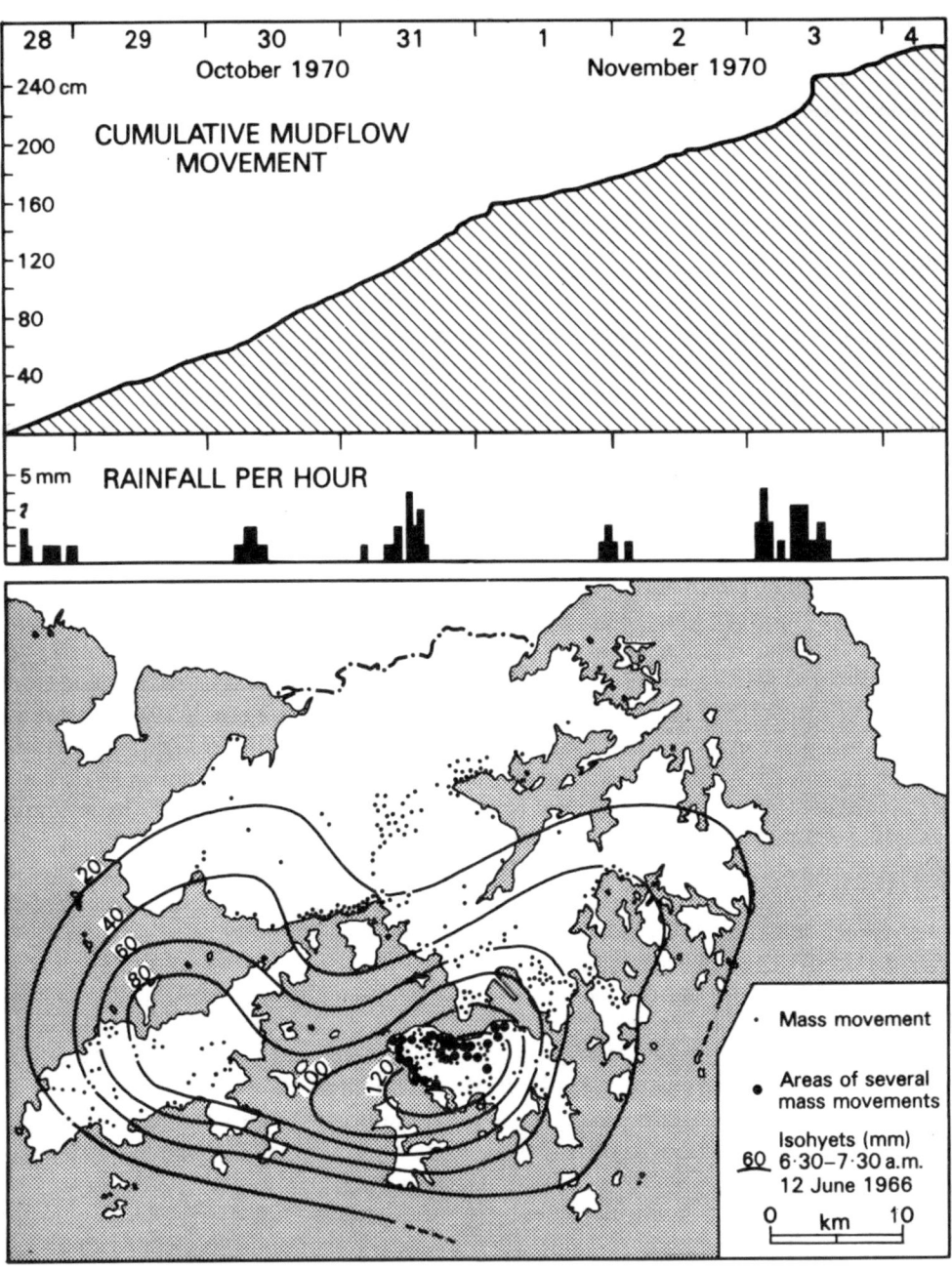

FIG. 2.20. EXAMPLES OF MASS MOVEMENT DURING SPECIFIC EVENTS
Cumulative rate of mudflow movement measured by continuously recording movement of pegs inserted into mudflow tongue and connected to a recorder (Prior and Stephens, 1972). Comparison with rainfall amounts per hour shows how successive storms induce surges of movement. Mass movements over slopes of Hong Kong instigated by severe storm in 1966; isohyets shown for one hour during storm (after So, 1971).

FIG. 2.21. MUDFLOW IN NORTH-EAST IRELAND
People can be seen at end of mudflow lobe, and a similar feature measured by Prior and Stephens (1972) is seen in upper part of Figure 2.20. (Photograph: K. J. Gregory)

Hong Kong a deep regolith of chemically weathered material occurs above bedrock in many areas. Many of the mass movements which occurred arose because the water lubricated the deep weathered material which then had a reduced shear strength so that movements occurred to achieve a lower angle of slope sympathetic with the limiting angle possible for the material under these conditions.

After the storm the landslide scars will gradually be revegetated and covered by the deposition from slower mass movement processes. Under conditions of high precipitation one would normally expect to find the most dramatic effects on deforested land because forest would intercept some of the precipitation and restrain some of the movements over the slopes. In this case the precipitation was so great that the most numerous mass movements occurred in the forested areas. It was in these areas that more water was allowed to infiltrate to lubricate the weathered material whereas in the deforested areas a great amount of the precipitation was removed as surface runoff.

Unusual events inspired by high rainfall amounts not only produce mass movements but also lead to high surface runoff, to peak streamflow discharges and to river flooding. Exceptionally high peak discharges may erode enlarged river channels, induce deposition on the flood plain, cause adjustments of the river channel pattern and lead to the development of new first order channels which increase the density of the drainage network. Processes associated with floods have been analysed for a number of events including the flood of December 1964 on Coffee Creek in California (Stewart and Lamarche, 1967). This small stream draining 300 km² in northern California had a flood in 1964 with a recurrence interval estimated to be 100 years and the peak discharge recorded was five times greater than any previously recorded on the Creek. Determination of the channel changes was accomplished by mapping on post-flood aerial photographs which were then compared with pre-flood photographs (Figure 2.22). Many new channels were eroded and old channels were completely infilled. Erosion in the valley bottom was considerable and flood waters scoured into fluvial deposits and into fan material. Deposits of sand and poorly sorted gravel were deposited over 70 per cent of the flooded area, and in places to an average depth of 0.4 m thick across the valley bottom; natural levées were formed during the flood along the sides of the main flood channels; and in some areas deep scour of pre-existing deposits occurred.

Although flood plain construction is thought to result from lateral migration producing coarse point bar deposits and fine overbank deposition, it was concluded (Stewart & Lamarche, 1967) for Coffee Creek that the coarse deposits in flood plain sediments were produced largely during rare events. The significance of such rare events is accorded increasing attention to establish their importance in shaping slopes and fluvial landforms. A probability approach involves assigning a frequency of occurrence to particular events. If records of rainfall or streamflow are available for periods of at least ten years it is possible to take the highest recorded value from each year to make up a probability series and then to calculate the recurrence interval for events of a particular magnitude (Gregory and Walling, 1973).

2.4b Measuring Process Rates

Measurements are required to demonstrate the types of processes operating, to indicate how effective the processes are, and to give the basis for an understanding of how processes operate. When values of change are available it is possible to calculate rates of erosion although such values should be interpreted with great caution (see p. 96). Three main groups of methods are available for the derivation of process rates. *Empirical methods* involving measurements of processes over short term periods including groups of individual events. A second group of *historical methods* are also empirical or field-based in character but they cover longer periods of time and do not readily provide information on individual events. *Experimental methods* employ laboratory or simulated situations to provide information upon the character of processes and upon the rates at which they operate.

A further distinction may be made between instantaneous measurements and continuous records. *Empirical measurements* of streamflow usually require continuous recording apparatus and there are several methods available to provide records of water stage or depth of water at a gauging station. Measurements of depth may be converted to discharge employing a water stage-discharge relationship which is established by rating the section or control structure (Gregory and Walling, 1973). Samples of water may be collected instantaneously or by pumping samplers and these can then be analysed to give concentrations of suspended sediment or of total dissolved solids or of individual ions. It is also possible to obtain samples of bedload or to collect all the bedload passing a particular

COFFEE CREEK , CALIFORNIA

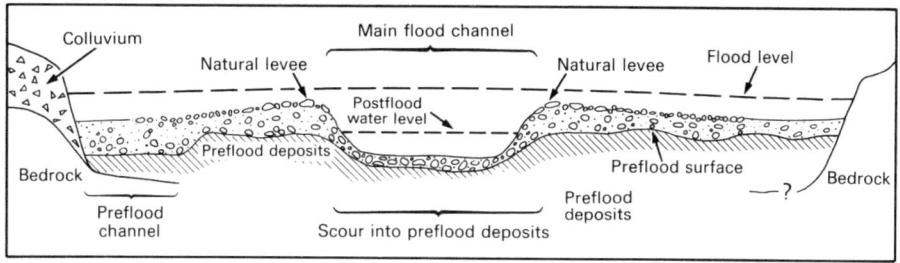

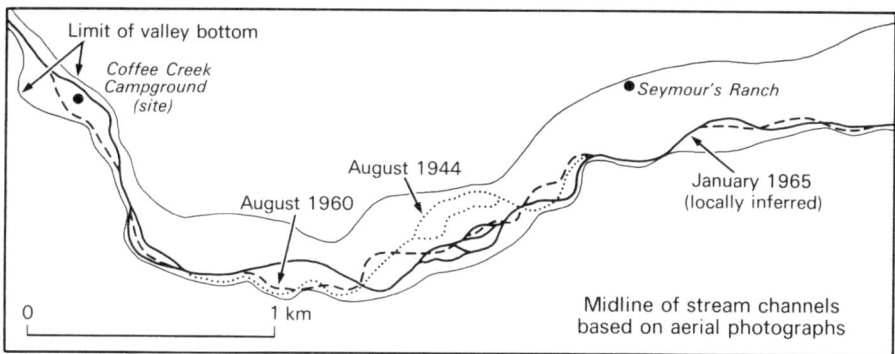

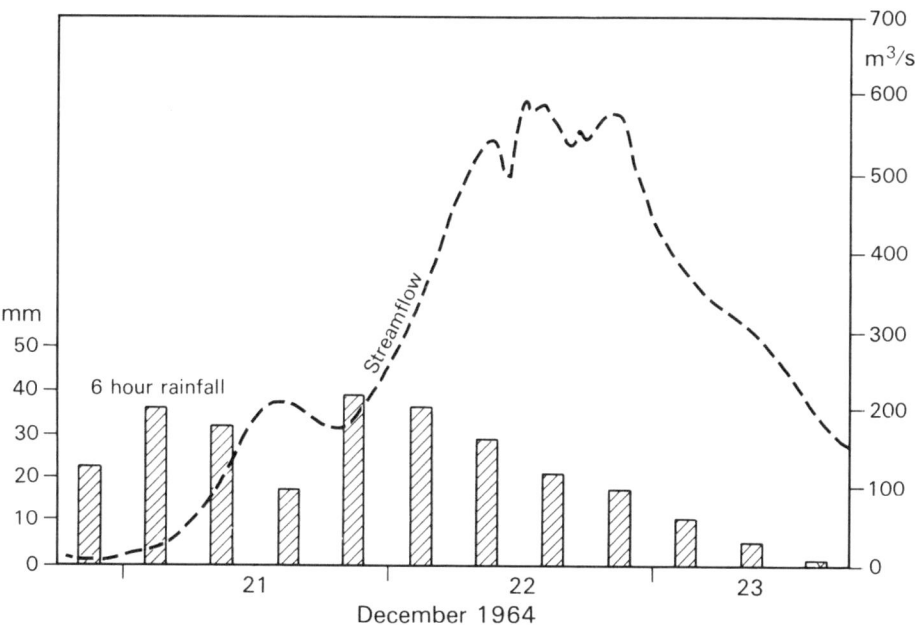

FIG. 2.22. EXAMPLE OF EFFECT OF RARE FLOOD DISCHARGE ON CHANNEL FORM (Based on Stewart and LaMarche, 1967.)

section and there are devices available to monitor suspended sediment by turbidity techniques and to monitor dissolved solids by conductivity measurements.

A variety of ancillary empirical methods are available to indicate the effects of water and sediment discharge in stream channels over short periods (Table 2.6). These include measurements of bedrock erosion with a small micro-erosion meter equipped with a vernier scale, the tracing of marked pebbles or tracer material, and indications of scour and fill provided by chains inserted in vertical holes so that measurable lengths of chain will be exposed by scour and then covered to a certain depth by subsequent alluviation. In river channel cross sections metal pins with washers can reveal the extent of lateral erosion and repeated surveys or photography of specific channel cross sections can demonstrate the amount of short term change. Measurements of the size and shape of bedload at a number of sites along a stream can help to suggest the way in which bedload is changed during transport.

The principal difficulty confronting the interpretation of short term measurements is that the time period may not be representative of longer periods and it may contain too many or too few large events. *Historical methods* have therefore been devised to facilitate calculations of the effects of processes operating over longer periods of historic time (Table 2.6). Frequently employed is the method of comparing present surveys of channel cross sections, channel planform or drainage networks with earlier surveys, aerial photographs, documentary records or maps and plans. This is illustrated in Figure 2.23 which compares the course of the Goulbourn River, Victoria, Australia at several different dates according to a number of topographic surveys. Whereas this method can provide information on change over several centuries it does not indicate how frequently change occurs and in what proportion of the time available. Other intriguing methods have been devised

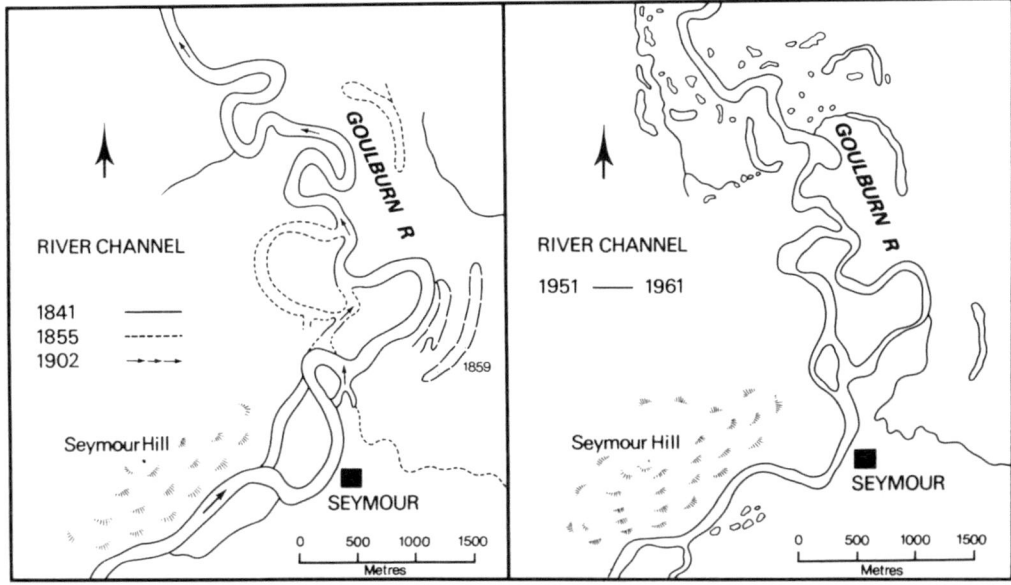

FIG. 2.23. RIVER CHANNEL CHANGE INDICATED FROM TOPOGRAPHIC MAPS
Area near Seymour, Victoria, Australia and maps of scale 1:63360.

(Table 2.6) and include calculation of the volume of sediment accumulated in reservoirs since the date of dam construction, and the volume of sediment accumulated above datable artifacts such as beer bottles and motor vehicle licence plates.

Whereas empirical measurements, both short and long term, relate to empirical situations which are affected by a great number of controlling factors, *experimental studies* of stream channel processes employ more rigorously controlled conditions. Laboratory channels or flumes have been employed by hydraulic engineers for the analysis of river channel processes and these can demonstrate sequences of bedforms, rates of scour, cut and fill, and the consequences of a change of base level. Tablets of limestone suspended in actual streams have been employed to show how much material is lost by solution and/or by mechanical abrasion.

The same three groups of methods have been employed to document slope processes (Table 2.6). Rainsplash erosion and surface wash may be recorded by a variety of techniques which include erosion plot experiments designed so that the precipitation falling on the plot can all be collected in troughs or trays at the downslope margin of the plot. The movement of water below the surface can be monitored by troughs installed in pits to collect the water flowing from designated soil horizons or by small structures designed to document flow through pipes. Other techniques (Table 2.6) to document slow and rapid mass movements may indicate the cumulative movement between readings or they may give continuous records by suitable apparatus (e.g. Figure 2.20). General estimates of slope change over historical periods have been derived for example, by measuring the difference in elevation between the ground surface and the rock surface immediately below perched blocks deposited by a glaciation of approximately known date.

More specific historical interpretations have been derived using the measurements of material accumulated on the upslope side of tree trunks or field boundaries, by measuring the amount of exposure of tree roots, or by interpreting the curvature of tree trunks as creep has occurred. In all these cases the measurements are related to the age of the trees interpreted from a count of the tree rings by dendrochronology. Experimental methods for the study of slopes include those employing a rainfall simulator to demonstrate the nature of processes acting upon an artificial slope or the use of glass-sided troughs to indicate the types of slope movement which occur with specific materials subjected to carefully controlled conditions of slope and moisture content.

Only since 1950 has geomorphology fully begun to realize the opportunities available for the measurements of processes and only with such measurements has it been possible to begin to appreciate the rate at which slopes and stream channels change in drainage basins. Such rates of erosion have to be interpreted with caution because not only are they affected by the techniques employed for measurement and analysis, and by the characteristics of the area studied, but they also reflect the size of the area investigated. Some of the largest rates will be derived for the smallest areas and any comparison of widely separated areas must take the scale factor into consideration. At particular sites along a river a continuous record of river flow may be used in connection with measurements of total sediment and solute load to give estimates of total volume of sediment and solutes moved in a particular time period. Although the rate of sediment and solute delivery will increase downstream the rate per unit area may decrease because of the storage of some of the material in flood plains and because intense storms are usually localized over a part of

Table 2.6

SOME METHODS FOR MEASUREMENT OF SLOPE AND STREAM CHANNEL CHANGES

Subject	Methods		
	Experimental	Empirical	
		Long-term (Historical)	Short-term (Sequences of events; individual events)
Slopes: Surface wash	Rainfall simulator	General: Height of pedestals below perched blocks; height of earthworks above surrounding area; estimates of valley accumulation in relation to surface lowering. Specific: – Accumulation upslope and lowering downslope of field boundaries – tree trunks – age and exposure of trees of known age, e.g. bristlecone pines	Rainsplash erosion plus wash – high speed photography – erosion plots, sediment collected in Gerlach troughs or trays – radioactive tracers to indicate pattern of erosion – erosion pins to indicate lowering of surface – lines of stakes to indicate lowering/accumulation
Creep	Experimental wetting and drying of soil blocks. Blocks of soil in glass-sided troughs, pegs in sides to indicate movement	Curvature of tree trunks in relation to age of trees	Painted stones or surface markers. Plastic tubes or metal stakes along contour } resurveyed to indicate movement. Tilt bars (T-bars) later surveyed to indicate upslope or downslope tilting. Deformation of auger hole filled by coloured beads or sand. Young pits, pegs set in side of pit and movement measured after re-excavation. Buried metal plates connected to potentiometers to measure resistance. Movement of buried aluminium 'creep cones' connected by wires to reference stake
		Deformation of shrinkage cracks	Solifluction – probes attached to electrical resistance strain gauges – linear motion transducers connected to buried metal plates
Landslides	Experimental studies of weathering to indicate rock fall, e.g. frost action	Records of rock falls	Measurement of accumulation from rockfall on sack carpets and wire netting, in box traps. Repeated survey of lines of stakes e.g. on mudflows, earthflows, debris slides. Continuous recording of movements by pegs attached to recorder. Accumulation behind barriers
Spatial patterns		Comparison of air photographs with present surface	Mapping of distribution, character and quantitative significance of movements during short time period (e.g. catastrophic event)

Subject	Methods		
	Experimental	Empirical	
		Long-term (Historical)	Short-term (Sequences of events; individual events)
Stream channels:			
Bedrock erosion	Tablets suspended in stream for measurements of loss in weight due to abrasion		Micro-erosion meter for direct measurement of lowering of rock surfaces
Bedload	Measurement of change in size and shape after rotation in drum; Flume studies of bedload movement	Comparison of size and shape of material in bed with material in deposits, e.g. terraces	Measurements of particle/pebble size and shape either repeated at a cross section or compared for a number of sites along river channel to indicate spatial changes; Particles marked with paint, fluorescent or radioactive tracers and relocation of material used to indicate amount of movement; Collection of bedload by mesh basket or tray samplers, sampling of bedload in slot traps, collecting basins,
Alluvial bedforms	Flume studies on sequence of bedforms with specific flow conditions		Acoustic and pressure-sensitive devices to record movement continuously; Repeated observation at specific cross sections
Channel cross section – erosion and deposition	Artificial flume apparatus to indicate relative erodibility of channel materials	Exposure of tree roots; Growth suppression occasioned by exposure of tree roots; Comparison with early surveys or detailed maps; Accumulation of volume of sediment in reservoirs (construction date known) or behind landslide dams (dated e.g. by C^{14}); Accumulation of sediment above litter which can be dated (e.g. beer cans)	Scour and fill chains inserted vertically to indicate amount of erosion (by length of exposure) and deposition by depth of fill; Monumented cross sections resurveyed at intervals of time possibly incorporating reference stakes or erosion pins in the section; Repeated photographs of monumented cross sections
Spatial pattern of network or reach	Simulated basin to indicate network or channel pattern development; Flume study	Comparison of old large-scale surveys (or air photographs) with present patterns; Dating of rocks (e.g. K.A. date for volcano) reconstruction of surface by generalized contours and computation of erosion	Pins or stakes to indicate limits of channel or network (e.g. gullies) to be resurveyed; Mapping and quantitative estimates of changes occasioned by a single event

large drainage basins. To give some idea of relative rates of erosion it is customary to relate the volume of material removed to the area from which it may have been derived. This can then give rates of lowering which may be expressed in millimetres of lowering per 1000 years–units known as Bubnoff units. Such measurements (Table 2.7) may be based upon measurements for a single year however and so caution is necessary before we assume that the processes of that year are typical of ten centuries!

2.4c Some Controls of Spatial Patterns

Rates of slope and fluvial processes vary according to a number of controls and particularly according to lithology and structure, climate, and endogenetic influences. Processes in the drainage basin involve the movement of water, of sediment and of solutes. Movement may occur vertically or laterally, below, on or above the land surface. Movement may be entirely of water as in the case of clear streamflow, exclusively of sediment as in rockfalls, or more commonly an intermediate mixture. The rate of movement may be slow, as in the case of slow mass movements, or it may be rapid as in the case of fluid flow or rapid mass movements. A fundamental problem therefore is to understand why the rate of movement *(m)* or (dm/dt) varies in different areas of the earth's surface. The controls affect the availability of water, sediment and solutes and this availability is affected by surface or exogenetic controls including type of material and climate, by endogenetic or subsurface controls, and by previous landscape history.

Rock type is a major factor influencing the distribution of the several types of mass movement. Whether a particular form of mass movement (Figure 2.6) occurs will depend upon the lithology of the rock and of its weathering profile and upon the structure which includes the frequency of joints, bedding planes and lines of weakness. Only where there is suitable rock or weathered material can a particular form of mass movement occur if the correct climatic conditions are applied. On the islands of St Lucia and Barbados Prior and Ho (1971) analysed slope instability in relation to the materials locally available. The several types of mass movement which occur including rotational, translational and complex slides and earthflows are the result of heavy rainfall on clay-rich soils. Analysis of the types of clay present in the surface materials of different areas showed that morphological differences between types of slides were related to the mineralogy of the soils. Those soils with the greatest proportion of swelling clay minerals and the highest plasticity were susceptible to the greatest distortion and disintegration of the soil structure so that they exhibited the greatest tendency to soil flowage. Translational slab slides developed largely on non-swelling kaolinite clay soils.

This type of study demonstrates how spatial variations can occur in small areas as a result of mineralogical variations and there are further contrasts in that certain types of mass movement (Figure 2.6) phenomena will be best developed on certain types of lithology. Thus cambering and valley bulging are most effective where an impermeable clay or fine shale occurs in the sequence and is susceptible to squeezing when saturated by the pressure of the rocks above. A similar sequence of contrasted rock types favours the occurrence of rotational slips (Figure 2.6). The frequency and orientation of lines of weakness such as joints and the dip of rocks will influence the potential for slope failure.

A further factor affecting the stability of a hillslope is the presence of erosion or

Table 2.7

RATES OF SLOPE EROSION IN DIFFERENT CLIMATES

Ranges of values are given based upon data from 109 studies collected together by A. Young. The rate of slope retreat. *Progress in Geomorphology* Institute of British Geographers Special Publication No. 7, 1974, pp. 65–78.

Climate	Soil creep		Solifluction		Surface wash Bubnoff units mm lowering/ 1000 years	Slope retreat from landslides Bubnoff units mm lowering/ 1000 years	Rates of cliff retreat Bubnoff units mm lowering/ 1000 years
	cm³/cm/year volumetric movement	mm/year movement of surface or upper 5 cm	cm³/cm/year volumetric movement	mm/year movement of surface or upper 5 cm			
Polar Maritime	–	–	5–25	0–200	–	–	–
Polar Continental	15.0	6	50	0–100	1–10	36	20–500
Cold Temperate Maritime	–	0.04–11	–	–	–	–	–
Temperate Maritime	0.2–3.2	1–7.6	–	–	2–19	11	200–900
Temperate Continental	6.0–8.4	–	–	–	0.03–508[a]	–	650 000
Warm Temperate	1.3–3.2	–	–	–	50–100	–	–
Humid Sub-Tropical	–	10–50	–	–	19	–	–
Mediterranean	650	3–12	–	–	0.4–253	–	–
Semi-arid	4.9	–	–	–	2000–11 700	204	2000–13 000
Arid	–	–	–	–	–	–	0.04–600
Savanna	4.4–7.3	–	–	–	1.6–39	–	–
Rainforest	7.5–12.4	4–5	–	–	260–1500	227	2000–20 000
Montane Zone	–	–	2–337	–	–	109	700–1000

[a] Values can be as high as 230 000 on badlands

– indicates no values available

Note: Values given are the extremes of ranges quoted from different areas. Variations occur with vegetation cover, slope angle and location and details of these can be found in the paper by Young (1974) and in the original papers cited by him.

deposition at the slope base. In an attempt to relate processes to the form of slopes Dalrymple, Conacher and Blong (1967) proposed a hypothetical nine unit landsurface model. This model endeavoured (Figure 2.24) to incorporate the nine slope components that could occur in world slopes. Each component was associated with a particular

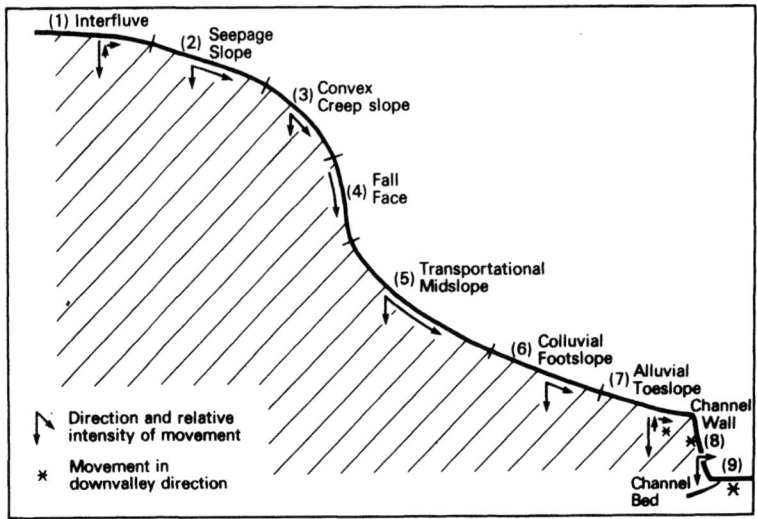

FIG. 2.24. HYPOTHETICAL NINE UNIT LANDSURFACE MODEL
Proposed by Dalrymple, Conacher and Blong (1967). Length of arrows proportional to assumed importance of movement in direction shown.

assemblage of processes and it is therefore feasible to suggest how slopes may appear under contrasted climatic and morphogenetic conditions. A semi-arid slope could be composed of components 1, 4, 5, 6, 7 where surface wash and rock fall are dominant, whereas in humid temperate areas creep would assume a greater significance. In this case the fall face may be a fossil feature or non-existent, and surface wash on the lower slopes would be comparatively rare. The humid temperate hillslope could thus be made up of components 1, 2, 3, 5, 7, 8, 9. Although the model is a hypothetical one it represented a significant advance upon previous models because it embraced both form and process of slope components whereas earlier models had concentrated upon morphology. This model provides a means of analysing major world slopes according to broad climatic patterns.

Variations can also occur within a small area according to differences in meso climate. Such differences can be inspired by aspect and exposure. Differences of aspect and exposure are important in relation to exposure to wind, to differences in precipitation due to windward and lee effects, and to contrasts in the receipt of solar radiation. Differences in incident energy will be found between slopes facing north and south. Thus in middle latitudes the south-facing slopes of the northern hemisphere will receive greater amounts of insolation than the opposing north-facing slopes. These contrasts are such that across a single valley there may be substantial contrasts in plant cover, in soil type and in slope processes. In the central Appalachians Hack and Goodlett (1960) found a distinct contrast between slopes facing northwest and southwest and those facing northeast and southeast (Table 2.8). In the area of California east of the Berkley Hills, Beaty (1956) found that

Table 2.8
DIFFERENCES BETWEEN OPPOSITE SLOPES IN CENTRAL APPALACHIANS
(Hack and Goodlett, 1960)

Characteristic	Northwest or Southwest Facing	Northeast or Southeast Facing
Angle	Gentle	Steep
Moistness	Dry	Wet
Surface mantle of stones	Coarse	Fine
Predominant vegetation	Yellow pine forest unit	Oak forest unit
Density of cover	Dense with many shrubs	Open with few shrubs
Drainage density	High	Low
Postulated most important process	Slope wash and channel erosion	Creep

between 70 and 78 per cent of all landslides occurred on slopes facing northwest, north and eastwards. This tendency was attributed to the fact that north- and east-facing slopes tend to receive less insolation especially in winter; they therefore retain more moisture, and thus reach the critical conditions for the development of landslides before other areas. Such contrasts in process have often been associated with the development of asymmetrical valleys (Kennedy, 1976). Such valleys were described in Siberia in 1927 and thought to result from differences in the active layer above permafrost on opposing slopes. On the Beaufort plain, northwest Banks Island, Canada, French (1971) showed that the southeast facing slope was warmest and that the southwest facing was one of the coldest. At depths greater than 10 cm there were temperature differences between the slopes of 2–3 °C and occasionally of 4 °C. Slopes facing northeast had an active layer thawed to a depth between 0.3 and 1 m and had solifluction and nivation processes. Slopes facing southwest had an active layer up to 0.6–1.0 m thick, soils were well-drained and gully dissection occurred.

Lithology and structure exercise an important influence upon stream processes because they influence the nature of flow, the provision of sediments and solutes for transport, and the potential for erosion of the channel and of the transported material. Many substances in rocks will dissociate into ions on contact with water, the degree of ionization varies from one compound to another and the degree of mobilization also varies. The permeability of the rock, the character of the structure and the degree of jointing or fractures will affect the extent to which material can be detached by fluvial processes.

In addition to these influences however, climate exercises a supreme control in providing the water necessary for stream activity. Climate is reflected in the regime of a river and major world contrasts occurring in river regime are illustrated in Figure 2.25 for five major world rivers. Some rivers show a marked seasonal regime which reflects the seasonal pattern of precipitation. Thus the Guadiana of Portugal (3, Figure 2.25) shows a winter maximum and a summer minimum and the Sanaga of west Africa (5, Figure 2.25) includes low flows during the dry season and much higher flows corresponding to the wet season of the seasonal tropical rainfall regime. Other river regimes may be influenced by temperature regimes and snow melt contributes to the annual hydrograph of the Rhine (2, Figure 2.25), and the melting of permafrost, ground ice and snow affects the discharge of the Usa river, USSR (1, Figure 2.25).

Mean annual discharges do not readily indicate the extent of the seasonal variation and it is often necessary to analyse discharge patterns as shown by the upper diagram in Figure

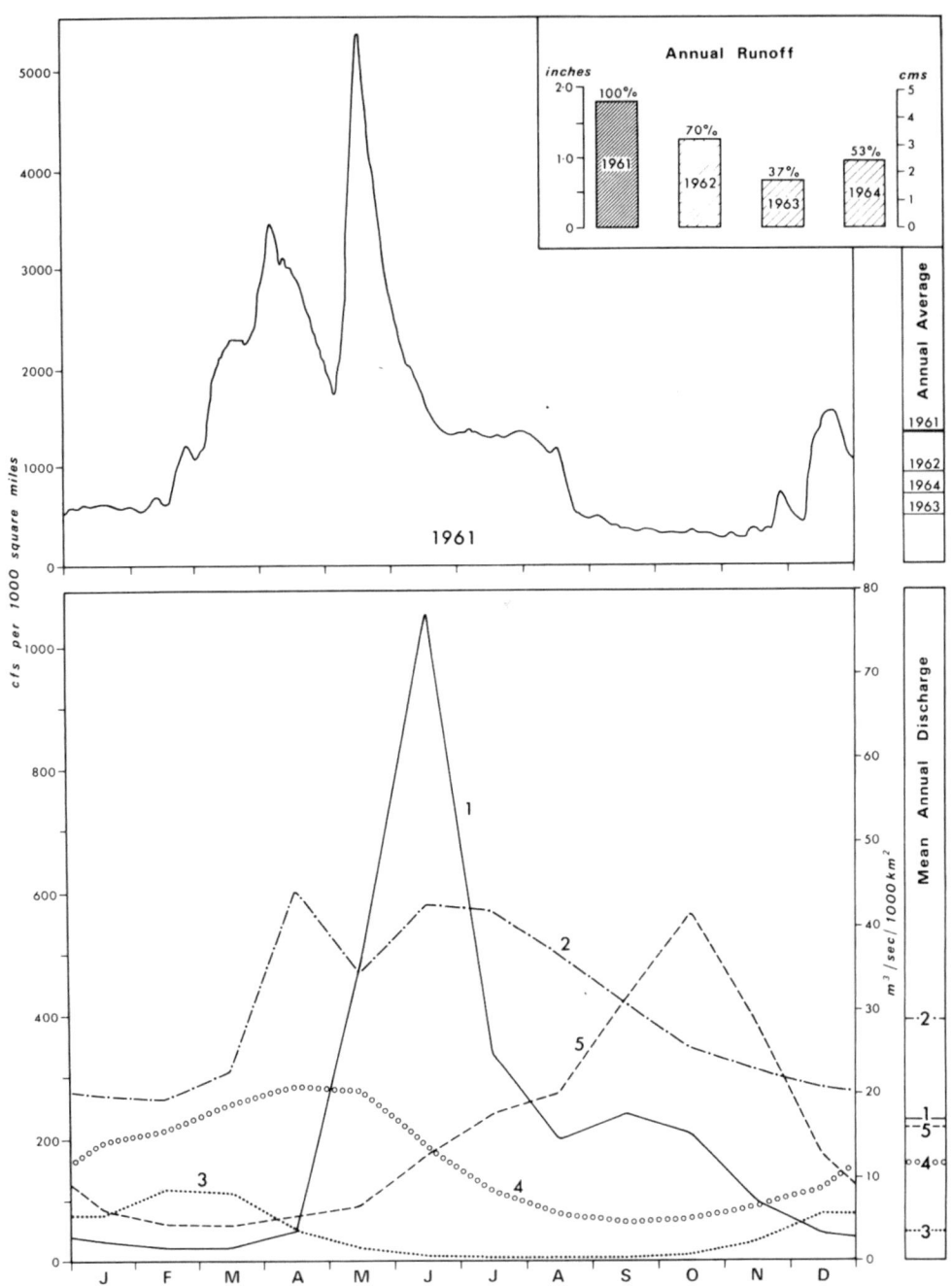

FIG. 2.25. ANNUAL RIVER REGIME
Upper diagram: annual regime for White River, Arkansas, plotted from daily mean discharges in 1961. Lower diagram: average annual river regime shown for 1- Usa, USSR, 2- Rhine, Europe, 3- Guadiana, Portugal, 4- White River, Arkansas, USA, 5- Sanaga, West Africa. To reduce influence of size of drainage area, discharges plotted per 1000 square miles (2580 km^2).

2.25 which is based upon average daily discharge values for the White River in Arkansas whereas in the lower diagram the average hydrograph for the same river is based upon average monthly discharge values. Considerable variations in flow occur from year to year and annual runoff may be calculated as the total volume of water which passes a particular gauging station in a single year. Calculations of annual runoff expressed as a layer from the entire drainage basin can facilitate comparison with the precipitation over the basin and variations over four years for the White River are shown in Figure 2.25. World river regimes have therefore been classified in various ways and one distinction has been made between simple regimes, which reflect seasonal precipitation distribution, and complex regimes which are influenced by additional factors such as snow melt. The occurrence of very high discharges which cause flooding may be a result of large storms with high intensity precipitation, of more moderate precipitation following periods of continuous precipitation so that soil moisture levels are high inducing high percentages of runoff, or by other factors such as the drainage of an ice-dammed lake or dam failure.

River regime reflects and also influences the structure of the drainage network and it is necessary to model the typical forms of water flow downstream from the divide as shown in Figure 2.26. This diagram endeavours to model the types of flow along the valley axis projected to the divide and so is analogous to the hypothetical nine-unit landsurface model depicted in Figure 2.24. Whereas all elements may be present in some areas, in others a

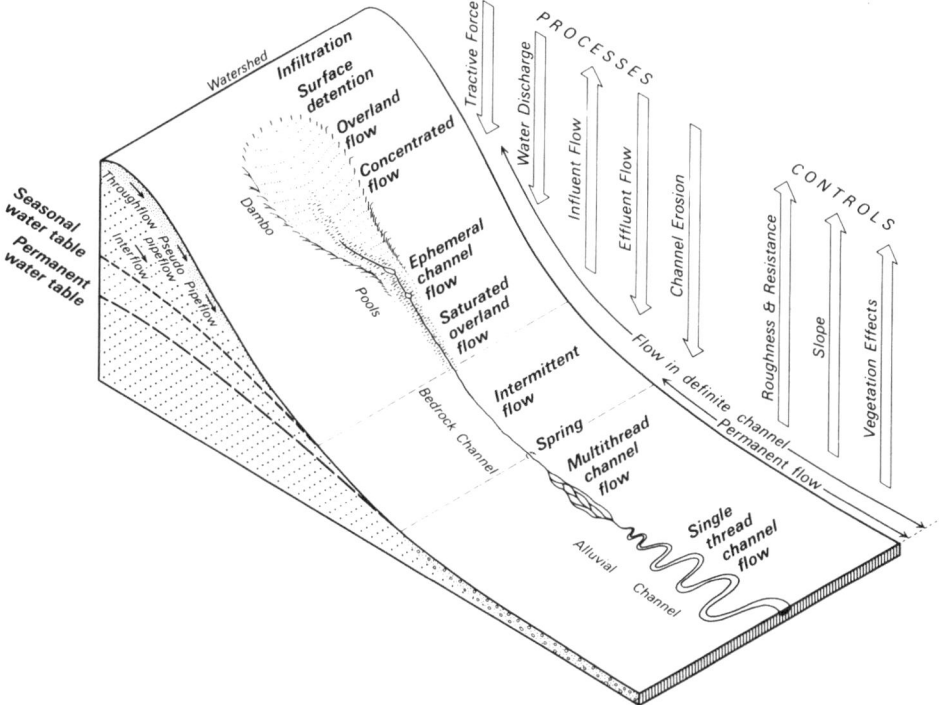

FIG. 2.26. MODEL OF CHANNEL CHARACTER
Way in which channel character varies downstream, portrayed in relation to flow types which may occur, distribution of definite channels, processes involved, and associated major controls. Compare with land surface model based on slopes in Figure 2.24.

few of the components shown will be represented. Thus in humid temperate areas all components except the dambo and multithread channel flow may occur. In arid and semi-arid areas a more restricted number of components will be found, composed probably of overland flow, ephemeral flow and multithread channel flow terminating in an alluvial fan or basin of inland drainage. In seasonal tropical areas a broad depression lacking a definite channel but characterized by concentrated flow along the axis of the depression is often found below the zone of overland flow and above the beginning of an ephemeral channel. The existence of a channel reflects the fact that on at least one occasion a flow has had a velocity sufficient to overcome the shear strength of the vegetation and surface material combined. Where the shear strength of the vegetation and surface material is greater than the force exerted by the flowing water then a channel will not be formed, but flow can still occur over the vegetation surface.

Calculations of erosion rates according to stratigraphic evidence indicate that the sedimentation rate earlier in geological time may have been considerably lower than at present. One estimate suggests that whereas the sedimentation rate from the Carboniferous to the Jurassic was of the order of $0.73-3.43$ $km^3.year^{-1}$ the contemporary rate may be nearer to 12 $km^3.year^{-1}$. A number of reasons may be advanced to explain this disparity. Obviously the influence of man has increased the rate at which processes operate, and soil removal by accelerated sheet and gully erosion is one reason for increased sediment yields. Chemical weathering may have amounted to 10 billion tonnes.$year^{-1}$ before, but 24 billion tonnes.$year^{-1}$ after, the advent of man.

The world land area is larger now than at earlier stages of geological time and lowering of sea level during the last 3 million years of Quaternary time has also facilitated higher sedimentation rates. Although many rates have been calculated for middle latitude areas a further significant factor is introduced by earth movements in the Quaternary. Indeed, Schumm has estimated (1963) that average rates of denudation range between 30 and 90 mm per 1000 years but up to 900 mm per 1000 years in areas of high relief, whereas the rates of land uplift can be as much as ten times the maximum denudation rate. Tectonically active areas are of restricted occurrence on the earth's surface but present rates of uplift have been quoted to be 7.5 m per 1000 years in parts of California, 10.8 m per 1000 years at the head of the Gulf of Bothnia, and 4.5 m per 1000 years in Japan. The role of endogenetic processes is thus of considerable import in a number of world areas and this was intimated by W. M. Davis from California in 1930:

... the scale at which deposition, deformation and denudation have gone on by thousands and thousands of feet in this new-made country is ten or twenty fold greater than that of corresponding processes in my old tramping ground. On shifting residence from one side of the continent to the other, a geologist must learn his alphabet over again in an order appropriate to his new surroundings.

Within the drainage basin such endogenetic forces are expressed in tectonic uplift and in vulcanicity. The areas of the circum-Pacific zone and in the tectonically active belt extending from the European Alps through the Himalayas to south east Asia have been characterized as areas of tectonic relief. In the Sierra Nevada of California total Quaternary uplift may have amounted in some areas to as much as 2700 m and such uplift may be the cumulative result of earthquakes. The Alaska earthquake of 27 March 1964 lasted a mere 3–4 minutes but elevation or depression affected an area of between 200 000 and

250 000 km². Such elevation and depression can create new features on slopes, and earthquake scarplets have been identified within many slopes in Japan, and dammed river courses and river courses modified to follow the lines of fault lines are frequent in New Zealand. In the central mountains of New Guinea Simonett (1968) analysed the distribution of landslide mass movements under tropical rainforest. He showed that slides were much more widespread in the Torricelli mountains than in the West Bewani mountains and this difference was attributed to the fact that the 1935 earthquake had been much more effective in the former area.

Endogenetic movement of the surface may also be stimulated by human influence. In the western USA subsidence has been described due to the drainage of peatlands in the Sacramento valley of California, to the process of hydrocompaction which gives subsidence of the surface after the application of water, and to the withdrawal of fluids. Extraction of oil, gas or water from below the surface may induce surface subsidence and the disposal of fluids underground can also lead to the occurrence of earthquakes. Tectonic movements modify the existing surface form and exogenetic processes, but vulcanicity can completely mask the pre-existing surface and create new forms. Sixty two per cent of all active volcanoes lie around the Pacific Ocean but other areas occur and Mount Etna in Italy has erupted 40 times since 1945. Vulcanicity was classified by Rittman (1962) into persistent diffuse activity including thermal springs and fumaroles; persistent central activity from a central vent; eruptions from central volcanoes; and fissure eruptions.

2.5 PROSPECTS IN THE DRAINAGE BASIN

The prime reason for geomorphological studies of process is to facilitate the understanding of changes in, and evolution of, landforms. This subject is covered in more advanced books (e.g. Carson and Kirkby, 1971; Gregory and Walling, 1973; Young, 1971) and relies upon the modelling of relationships between form and process (e.g. Chorley and Kennedy, 1971). Adequate understanding of slope profile evolution, of drainage network changes and of drainage basin evolution can only be achieved against the background of improved models. Changes of landform are not only time-dependent however, but also reflect man's influence both directly on the earth's surface and indirectly through alterations of processes. The results of urbanization, of changing land use, and of construction activities such as road and dam building will all affect slopes and drainage basins.

In addition there are certain processes or values of processes that are particularly relevant to man's activities. Thus the extremes of river flow associated with floods and droughts are of much greater potential human significance than the intermediate flows. A group of studies has therefore focused upon hazards arising from earth surface processes and, in a context more general than the drainage basin alone, Hewitt and Burton (1971) have listed the ingredients which contribute to the hazardousness of a particular place. Geomorphological processes operating in drainage basins are instrumental in shaping slope and fluvial landforms and the researcher has to identify the magnitude of man's effects and also to express the patterns of process in a way relevant to human activity.

FURTHER READING

More advanced explanation of form and process may be found for hillslopes in: YOUNG, A., 1972, *Slopes* (Oliver and Boyd), for drainage basins in: GREGORY, K. J., and WALLING, D. E., 1973, *Drainage Basin Form and Process* (Edward Arnold) and for rivers in: MORISAWA, M. E., 1968, *Streams: Their dynamics and morphology* (McGraw-Hill). A broad review directed towards environmental management is afforded by: DUNNE, T. and LEOPOLD, L. B., 1978, *Water in Environmental Planning* (W. H. Freeman). A book with many low latitude and Australian examples is: DOUGLAS, I., 1977, *Humid Landforms* (Australian National University Press). Basic principles provide the starting point for the approaches taken in: CARSON, M. A., 1970, *The Mechanics of Erosion* (Pion Monographs) and STATHAM, I., 1977, *Earth Surface Sediment Transport* (Clarendon Press, Oxford). Temporal analysis and change with frequent reference to slopes and rivers is given in: THORNES, J. B. and BRUNSDEN, D., 1976, *Geomorphology and Time* (Methuen). A stimulating approach to the drainage basin and river channels is given by: SCHUMM, S. A., 1977, *The Fluvial System* (Wiley Interscience).

3
Coastal Processes

There has been considerable progress in our understanding of coastal processes in recent years but many important questions remain unanswered. So far, much research has been directed towards theoretical studies and laboratory wave tank investigation of sediment sorting and sand movement by oscillatory waves, with only a limited number of field studies to substantiate the results of such work. The hydrodynamic and sediment variables controlling changes in beach profiles have not been evaluated adequately, either in the field or in hydraulic laboratory studies, and there is still no satisfactory theory of breaking waves, either in deep or shallow water. The almost total lack of reliable field data is probably the greatest single factor inhibiting our understanding of the relationship between littoral sediment transport rate and wave activity, despite available information from model wave basin studies and the knowledge that sediment transport in the surf zone is caused by a combination of shear stresses produced by wave and current action.

There does not appear to be any particular trend in current research projects, most of which are fragmentary and divorced from either short- or long-term management programmes. This may stem from the fact that many coastal problems, particularly those confronting the engineer, are practical rather than academic and consequently require rapid solution. Yet many of these coastal engineering problems have provided a major impetus for research into nearshore sediment dynamics, and the rapid advance in the design and development of equipment for monitoring processes, in such a dynamic environment as the coastal zone, has already made a significant contribution to scientific and engineering interests at large. Competing interests within the coastal zone for residential, recreational, industrial, defence and other facilities are making it increasingly necessary to understand how beaches are regulated by nearshore processes.

In the light of these comments, it is important that geomorphologists should be aware of their contribution to the study of coastal processes, by way of interdisciplinary research. The principal aim of this chapter is to summarize some of the more significant developments and to suggest possible avenues for future research by highlighting the gaps in our current knowledge. Some topics are discussed in less detail than others because they have been evaluated fully elsewhere in basic texts. This slight imbalance in the treatment of subject material is redressed partly by recommending the reader to selected references and publications.

As far as the coastal geomorphologist is concerned, there is a need to study, both quantitatively and qualitatively, those parameters and processes that fundamentally control shoreline equilibrium. The more important of these include the degree to which a particular shoreline is exposed to ocean swell, as well as storm and locally-generated waves; tidal currents; oscillations in mean sea level; variations in the sediment budget; sediment transport by wave action as distinct from tidal and other currents; the morphology of the nearshore zone and the topography of the adjacent continental shelf; the coastline in plan; and the coastal climate. This list is by no means exclusive, but it will introduce students, teachers and researchers to a broad spectrum of disciplines which will now be examined in more detail.

3.1 WAVES

The coastal geomorphologist is concerned basically with three types of wind-generated waves, namely sea, swell and surf which, in association with currents, are responsible for developing and fashioning coastal landforms. Waves are propagated in the direction the wind blows and although they appear to move approximately parallel to the direction of the wind in a storm area they are, in fact, quite irregular with small waves superimposed on larger ones. Such waves in the process of generation within or near a storm centre are called sea but as they leave the centre, or move away from the winds that generate them, they become more regular and resemble sine waves in profile as a result of a process known as wave dispersion. Swell comprises waves, with rounded crests beyond the storm area, which travel for thousands of kilometres across the oceans without losing much energy.

Wind-generated ocean waves may be described as progressive oscillatory waves because the waveform travels through the water and causes an oscillatory water motion. This motion is capable of entraining and transporting sediment because the oscillatory flow associated with the wave motion can reach the seabed and interact with it. Simple oscillatory wave motion, or a regular wave-train, may be described by three parameters: the wave period (T) or the time it takes a succession of wave crests or troughs to pass a stationary point; the wavelength (L), the horizontal distance between successive crests, which is often directly related to the period of the waves; and the wave height (H), the vertical distance between a trough and a succeeding crest, which is independent of the other two wave parameters.

3.1a Wave Propagation Velocity

The periodic motion of a wave in water obeys the equation:

$$C = \sqrt{\frac{gL}{2\pi} \tan h \frac{2\pi d}{L}} \qquad (3.1)$$

where C is the wave velocity in ft/sec,* d is the water depth in ft, g is acceleration due to gravity in ft/sec^2 (980 cm/sec^2) and L is the wavelength in feet as defined above. The wave-phase velocity, or the rate of propagation of the waveform, may then be expressed as $C = L/T$.

Equation (3.1) applies to both deep and shallow water. In deep water where d is larger

*Metric units, in either the c.g.s. or SI systems, have been used in this chapter but, in a few instances, it has been necessary to retain Imperial units where these have been used in cited publications.

than half the wavelength ($d/L > 0.5$) the term tan h $2\pi d/L$ approaches unity. Equation (3.1) then becomes

$$C = \sqrt{\frac{gL}{2\pi}} \qquad (3.1a)$$

Since $L = CT$, the deep water phase velocity can be expressed as a function of the wave period T, where

$$C = \frac{g}{2\pi}T = 5.12T \text{ ft/sec} = 1.56T \text{ m/sec} \qquad (3.1b)$$

and

$$L = 5.12T^2 \qquad (3.1c)$$

In shallow water where the wavelength is large compared with the water depth tan h $2\pi d/L$ approaches $2\pi d/L$ and the velocity expression in (3.1) after cancellation, becomes

$$C = \sqrt{gd} \qquad (3.1d)$$

Waves are somewhat arbitrarily known as shallow water waves when $d < \frac{1}{2}L$ or 0.5. Direct substitution in equations (3.1b) and (3.1d) gives the following expressions for wavelength:

$$L_d = \frac{g}{2\pi}T^2 = 5.12T^2 \text{ ft} = 1.56T^2 \text{ m} \qquad (3.2a)$$

and

$$L_s = T\sqrt{gd} \qquad (3.2b)$$

where d and s refer to deep and shallow water respectively. Thus, the wavelength L is equal to 5.12 times the square of the period* and therefore a one-second wave will be 5.12 ft or 1.56 m long (see also Figure 3.1).

3.1b Wave Types

Theoretical studies of wave motion by Airy, Stokes, Rayleigh, Kelvin and others in the nineteenth and early twentieth centuries were mainly stimulated by the early experiments of Scott-Russell in 1837 and 1844. There are various types of wave which are worth mentioning in the context of wave motion. The simplest type is the Airy wave which is of small amplitude in deep water. It is sinusoidal in profile, and particle orbits are closed circles in deep water and ellipses in shallow water. In contrast, Stokes and Gerstner waves have a trochoidal form and are of finite amplitude in deep, intermediate and shallow water. They are characterized by flatter troughs and steeper crests than sinusoidal waves. Particle orbits are not closed but lead to a slight net transport in the direction of wave propagation. Cnoidal waves are periodic waves of finite amplitude. They can have widely spaced sharp crests separated by wide troughs and therefore are applicable to wave description seaward of the breaker zone. Sinusoidal waves of deep water and solitary waves in shallow water are special limiting cases of cnoidal waves. Solitary waves are isolated crests of finite

---

*Wave frequency, defined as $1/T$, is commonly used instead of the wave period. Thus, frequency (f) is expressed in cycles/kilosecond (c/ks), and for a 10-second period $f = 100$ c/ks; for a 20-second period $f = 50$ c/ks and for a period of 5 seconds f is 200 c/ks.

amplitude moving in shallow water, and separated by relatively flat troughs. A solitary wave has no wave period or wavelength because it consists of a single wave crest.

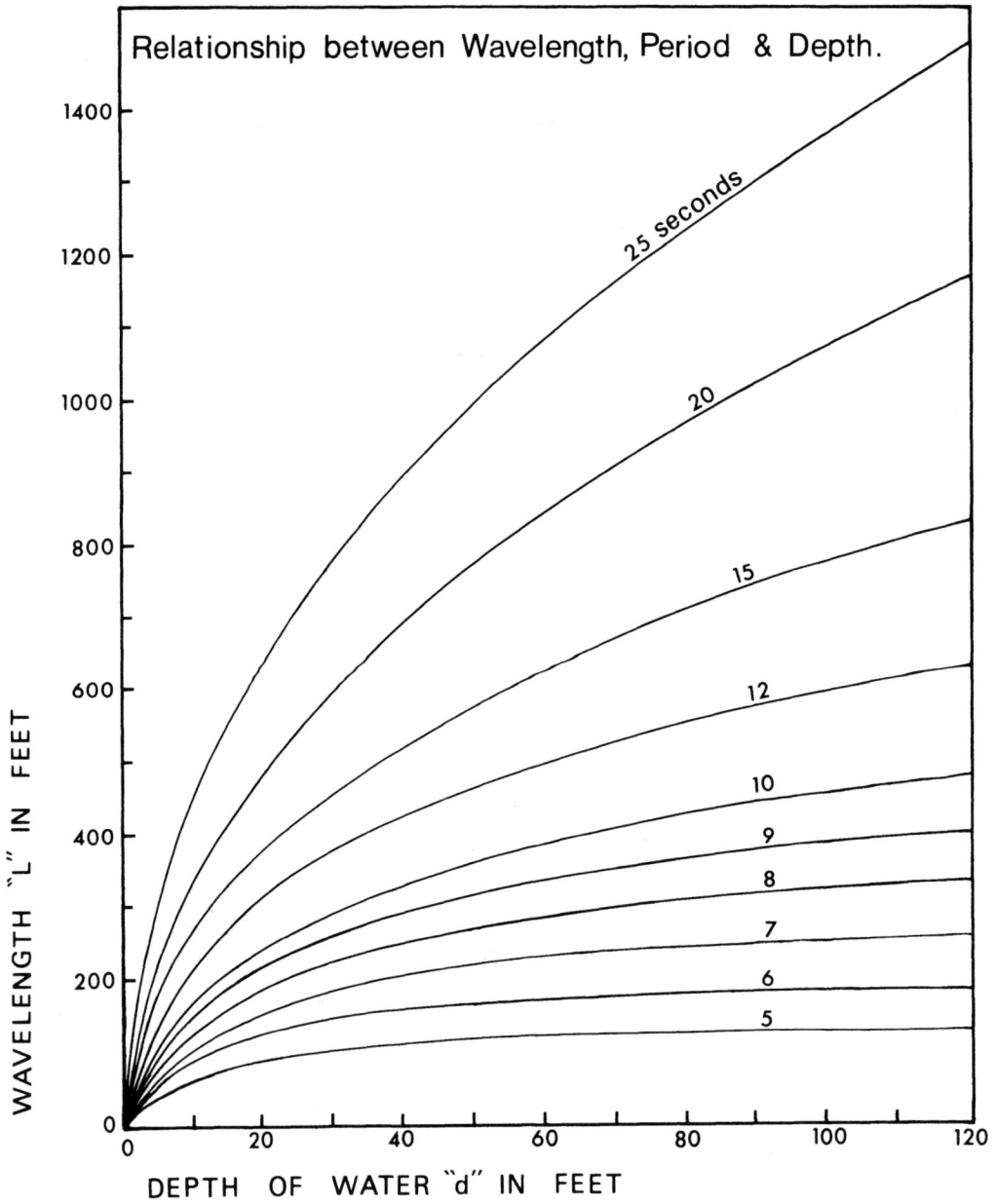

FIG. 3.1. RELATIONSHIP BETWEEN WAVELENGTH, PERIOD AND DEPTH

3.1c Waves Generated by the Wind

In the area of generation, the height of sea waves and their period are functions of the duration that the wind (storm) blows *(D)*, the wind velocity *(U)*, and the distance or fetch *(F)* over which the wind blows. For short fetches the wave depends on the fetch length and wind velocity, and for long fetches on wind velocity and duration. In the introduction to waves it was mentioned that the term sea is used for irregular waves in the process of generation. The basic difference, though, between sea and swell is related to the spectrum changes that waves undergo as they move from the generation area across the open oceans, as swell, into shallow water and finally break. Energy is distributed within a wide spectrum of periods which, for a sea, range from ripples of about 0.1 sec to waves of periods of between 15 and 20 sec. As swell waves approximate a simple sinewave in form, they have a very narrow range of periods. Sea waves are complex because small waves are superimposed on larger waves and any one crest disappears after a short distance.

The length of fetch determines not only the time during which wind energy is transferred to the sea surface, but also the wave height and period. This relationship can be written symbolically as $H, T = f(W, F, D)$. Swell waves usually have periods in the range 6 to 16 sec, but occasionally longer periods may be recorded. Most of the long-period waves are generated in the southern oceans, between latitude 40°S and the Antarctic continent, where there is a large fetch. For example, the most common waves on the coast of New South Wales, Australia, have periods of between 8 and 14 sec, and up to 20 sec during storms. Wave heights over this wider range are known to have reached 9 m (30 ft) and even higher. Isolated extreme waves of 15 m (50 ft) or more appear to be the result of two waves of slightly differing periods coinciding. Summer waves arriving at the coast of southern California are generated in the South Pacific and have travelled more than 8000 km before breaking on the beaches. The prominent part of the spectrum of waves generated in the southern oceans may include wavelengths exceeding 610 m (2000 ft) with periods >20 sec.

3.1d Wave Attenuation: Shoaling Transformation

In deep water, oscillatory waves approximate to a trochoidal form, and water particles rotate in circular paths. This means that the water at the crest of a wave is moving forward, but backwards in the trough. Theoretically there is no net forward movement of water particles but, in fact, there is a very slight forward drift. As waves enter shallow water the circular orbits become elliptical orbits, the ratio of vertical to horizontal axes becoming smaller as the water becomes shallower. There is also a net forward movement of particles which is slight in deep water, but appreciable when a wave is nearly breaking. Also, as waves enter shallow water ($d/L < 0.5$) they 'feel bottom', their velocity and length decrease, and their steepness and height increase until the wave train consists of peaked crests separated by relatively flat troughs. An empirical relationship which gives the depth at which a wave breaks is:

$$H_b = 0.78d \qquad (3.3)$$

where H_b is the breaker height. Short waves in deep water also break when their height-to-length ratio reaches 0.143. It is worth remembering that in shallow water the sinusoidal characteristics of deep water waves disappear, and wave period and wavelength are no longer significant.

3.1e Wave Energy

In deep water, the potential energy is equal to the kinetic energy for an Airy (sinusoidal) wave. Total energy is determined by both wavelength and height and is usually expressed as wave energy per unit crest length. Potential energy results from the wave surface being displaced above the flat, still-water level of the sea, whereas kinetic energy is due to the orbital motion of water particles. The rate at which wave energy is transmitted in the direction of wave propagation is known as the *energy flux*, or *wave power*, and may be expressed by the equation

$$P = EC\eta \tag{3.4}$$

The energy of a sinusoidal wave per unit of surface area, or the energy averaged over the wavelength, is proportional to the square of the wave height, as given by

$$E = \omega L H^2 / 8 \tag{3.5}$$

where ω is the weight of one cubic foot of seawater (64 lb)

$$\text{or } E = \tfrac{1}{8}\rho g H^2 \tag{3.6}$$

where ρ is the density of water and g the acceleration due to gravity. The corresponding expression for the energy of a Stokes (trochoidal) wave per unit of surface area is

$$E = \tfrac{1}{8}\rho g H^2 \left(1 - \tfrac{1}{8}k_d{}^2 H^2\right) \tag{3.7}$$

The wave group velocity, or the rate at which the wave energy and the wave group (individual groups of sine waves that comprise the wave spectrum) travel, in deep water is given by

$$Cg = \tfrac{1}{2}C = \tfrac{1}{2}\left(\frac{g}{2\pi}T\right) \tag{3.8}$$

As a result of dispersion the energy spectrum is narrowed so that a strong energy peak corresponds to the wave period of the swell.

3.1f Breaking Waves

Three main types of breaker can be identified, depending upon the gradient of a beach and wave steepness. *Plunging* breakers result from long, low swell breaking on steep beaches, usually shingle. They are characterized by a hollow, concave upward, front as illustrated in Figure 3.2. On the other hand, *spilling* breakers are associated with short-period wind waves, and the tendency for this breaker to occur is greater if the wind blows in the direction of wave travel. *Surging* breakers do not actually break but surge up beaches with steep nearshore and shore face slopes (Figure 3.2). According to Galvin's (1968) classification, *collapsing* breakers have been identified as a fourth type of breaking wave, and are intermediate between the plunging and surging types.

FIG. 3.2.
(TOP): PLUNGING BREAKER WITH HOLLOW FRONT, PEARL BEACH, NEW SOUTH WALES, AUSTRALIA. (BOTTOM): SURGING BREAKERS, STANWELL PARK, NEW SOUTH WALES, AUSTRALIA (Photographs: J. R. Hails).

3.1g Surf Beat

Variations in the height of waves breaking at somewhat irregular periods, every two to three minutes, on the beach is a phenomenon which has been called *surf beat* by Munk (1949) and Tucker (1950). It is believed to result from the interference of two swell wave trains arriving simultaneously at the shore from two storm centres. Depending upon the wave periods of the two sets of swell, every sixth, seventh or eighth wave is much larger than the others in the breaker train. Thus, the waves have no constant height and tend to increase and decrease in height periodically.

3.1h Long Period Waves

The origin of waves in the range of 30 seconds or longer is still unknown, although it is possible that such waves are part of a spectrum produced by storms. In fact, they may result from fluctuating pressures and wind gusts in the generating areas. Long period waves can cause considerable problems in harbours because ships may be moved violently. However, their effects along exposed shorelines are little understood at present.

3.1i Wave Refraction

Upon entering shallow water, waves are subject to refraction, reflection and diffraction. As their velocity and length decrease, they will tend to alter direction. The part of the wave which enters shallow water first is retarded and the wave front tends to swing towards the shore. An individual wave crest tends to parallel the depth contours as it is refracted and in the case of relatively straight coasts with parallel contours, the crest is approximately parallel to the shoreline.

The refraction of waves is analogous to the bending of light rays and Snell's law of refraction implies that:

$$C_1/C_2 = \sin \alpha_1/\sin \alpha_2 \qquad (3.9)$$

where α_1 and α_2 are the angles between adjacent wave crests and the respective bottom contours; C_1 and C_2 are the successive phase velocities of the two depths.

Refractive focusing of long swell has caused damage at Long Beach in California, and at Botany Bay, New South Wales. Also, offshore banks and shoals, and irregular bottom topography can cause waves to be refracted in such a complex way as to produce variations in the wave height and energy along the coast (see Figures 3.3a and 3.3b). Submarine ridges or headlands and canyons, for example, cause waves to converge and diverge respectively. Wave convergence or divergence owing to wave refraction is an important factor to consider when structures are built along certain sectors of coasts. Komar (1976), for example, draws attention to the Standard Oil Company's oil-loading wharf at El Segundo, California which was constructed before wave refraction computations and techniques came into practice. It is located in an area of strong wave convergence so that under certain wave directions and periods large breaking waves are concentrated at the wharf site which has to be closed for extended periods of time.

Figures 3.3c and 3.3d show how waves are refracted along the New South Wales coast as they enter shallow water in Twofold Bay and Broken Bay respectively. The orthogonals in these diagrams, which would normally be parallel in deep water, are bent and where they diverge the level of wave energy is lower than where the orthogonals converge. Wave crests fit the outline of the beaches particularly well in Twofold Bay. Figure 3.3c shows that the level of wave energy (lowest wave coefficients) in Twofold Bay is lowest at Boyd Town, and at the northern end of Whale Beach immediately south of Mowtree Headland. Wave energy (highest refraction coefficients) is highest along Eden spit which is relatively exposed to southeasterly swell. Lagoon inlets and river mouths are generally located where the level of wave energy is lowest. The rivers on the mid-north coast of New South Wales enter the ocean at a point where the level of wave energy is lowest, generally immediately north of the headlands which define the arcuate bays (Figure 3.3e).

3.1j Wave Refraction Diagrams

Since the early work of Munk and Traylor some thirty years ago, theoretical advances in the application of wave refraction studies to resolve erosion problems have been rather limited. Knowledge of the distribution and dissipation of wave energy along the shoreline is fundamental to an understanding of sediment distribution in the nearshore zone, as will be discussed below. This knowledge, however, is more readily available than in the past because of computer programming and automatic plotting routines. Also, variations in

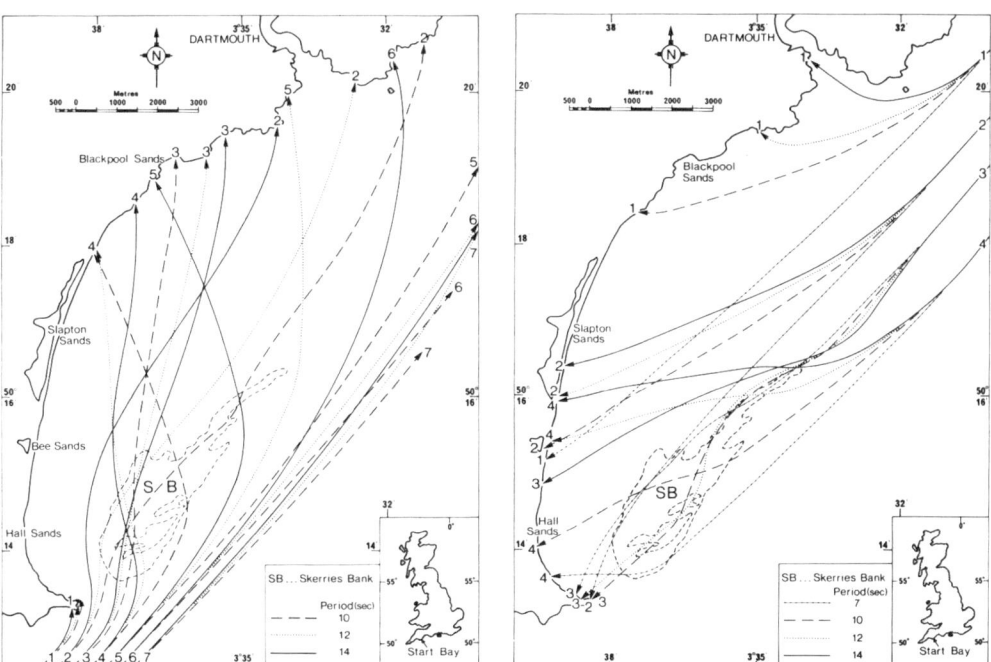

FIG. 3.3a,b. REFRACTION OF SOUTHWESTERLY AND NORTHEASTERLY SWELL
Skerries Bank, Start Bay, Devon, England. Torcross is 1 km north of Bee Sands.

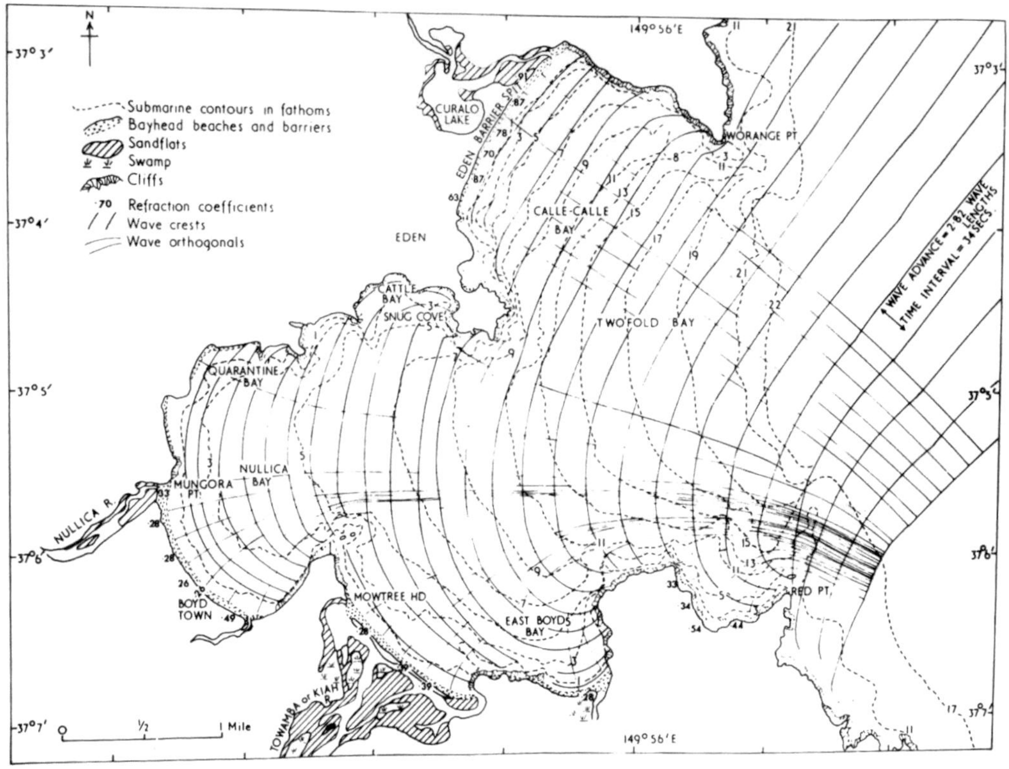

FIG. 3.3c

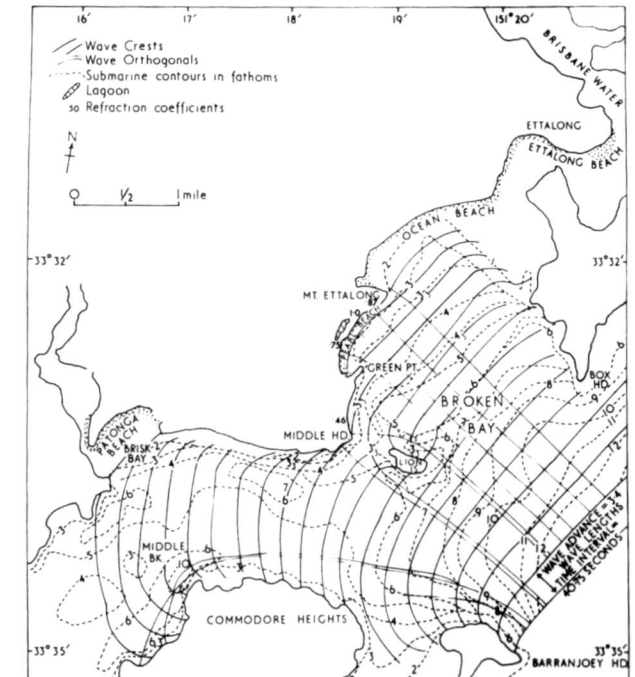

FIG. 3.3d

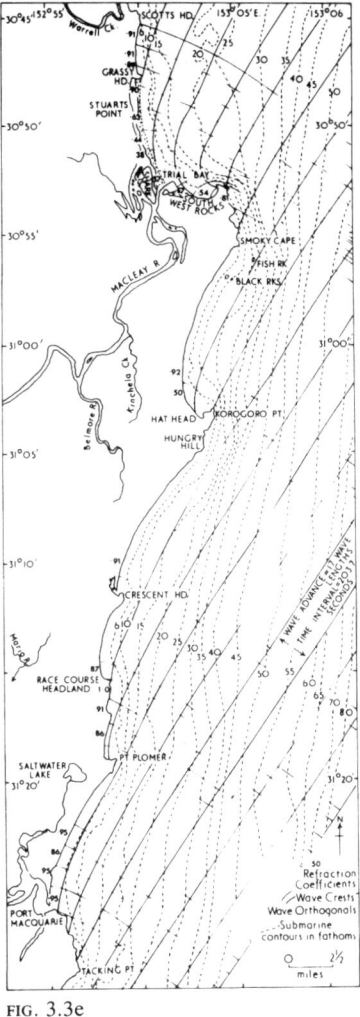

FIG. 3.3e

FIG. 3.3c,d (opposite) and e (above). REFRACTION DIAGRAMS FOR SOUTHEASTERLY SWELL OF 12-SECOND PERIOD Twofold Bay (3.3c), Broken Bay (3.3d), and the mid-north coast of New South Wales (3.3e), Australia. The diagrams have been constructed to show relationship between swell wave pattern and offshore profile.

wave refraction and wave energy relative to theoretical computation can be appraised more readily than previously. Existing programs already take into account such factors as refraction, reflection, diffraction, absorption, shoaling effect and wave breaking, all of which modify the characteristics of deep water waves as they enter shallow water. The primary output of most of these programs is on a computer-controlled graph plotter which is similar to a manually drawn refraction diagram. In addition to the graphical output a full tabulated set of results of each orthogonal is available from the line printer.

Like most theoretical studies supported by minimal field data, conventional wave refraction computations have their limitations. Most coastal researchers, for example, are aware that ocean waves are sinusoidal until they steepen on approaching the shoreline, yet linear wave theory is predicated on the assumption of sinusoidal wave forms. Notwithstanding this, and other examples, the limits of analytical techniques seem to be outweighed by the advantages afforded by wave refraction studies, particularly with regard to a broader understanding of nearshore processes.

Wave refraction diagrams can now provide information on coastal engineering problems. One principal use is to determine the directions from which waves can approach a sector of the coast because this is an important factor in the design of breakwaters, jetties, harbour entrances and coastal protection works, and for analysing littoral drift problems. It is possible to determine the heights of refracted waves at any point on a diagram by comparing the spacing of wave rays. The rate of energy transmission between wave rays remains constant, but as a wave enters shoaling water the velocity of energy movement shoreward alters, of course. Energy is dissipated in shallow water, as already mentioned, owing to the frictional effect of the seabed. On a steeply shelving shoreline wave height may increase because of a rapid decrease in wavelength, whereas on a gently sloping beach the wave height will diminish.

Research undertaken by the writer and co-workers in Start Bay, Devon, has shown that wave energy is focused at the shoreline by offshore banks such as the Skerries shown in Figures 3.3a and 3.3b. It can be seen that the villages of Torcross, Beesands and Hallsands have been built on sites where wave energy is concentrated during northeasterly storms as evidenced on 4 January 1979, when 10 m high waves, accompanied by a Force 9 gale and spring tides, severely damaged Torcross and Beesands. A similar critical combination of storm waves, northeasterly gales and spring tides was responsible for the Hallsands disaster in January 1917, although the legacy of exploiting gravel from the foreshore at Hallsands between 1897 and 1902 was also a contributing factor. Wave energy is seen to be dissipated more quickly along the shoreline when southerly, rather than northeasterly swell enters the Bay.

Refraction diagrams produced on a regional rather than a local scale from the Virginian Sea Wave Climate Model show that there are distinct areas along the east coast of the USA between Cape Henlopen in Delaware and Cape Hatteras, North Carolina, where there is a distinct convergence of wave rays (orthogonals) (Goldsmith, 1976). Goldsmith and his associates have demonstrated that large variations in computed shoreline wave heights and wave energy are reflected by large historical changes in shoreline configuration and, to a lesser extent, by variations in beach grain size. The refraction patterns produced by this model show that wave energy is concentrated downwave from a topographic high such as a linear ridge and then dissipated downwave from a topographic low, such as a submarine canyon.

In the Cape Hatteras region of North Carolina, where barrier islands are frequently eroded by severe storms and noted for their washover fans (see p. 142), wave energy dissipation has been correlated with changes in shoreline configuration, as measured from aerial photographs and volumetric changes recorded in the field. Such correlation indicates that it may be possible in the near future to use refraction models in order to identify areas which are particularly prone to wave attack, although it must be borne in mind that regional wave refraction diagrams cannot be applied widely yet in coastal zone management programmes because they are still in early stages of development. Thus their usefulness remains at the local level at present.

3.1k Wave Reflection

A wave may be reflected upon itself with little loss of energy when it meets a steep rocky cliff rising from deep water. The amount of incident wave energy that is reflected from a coast increases as the slope of the seabed steepens and may approach the entire incident wave energy as is the case of low waves reflected from a vertical wall. Wave reflection from a shoreline depends upon the slope and composition of the beach, water depth immediately in front of the reflecting beach, wavelength and wave height. A short period wave of medium height is nearly always absorbed by a gently sloping sandy beach, while a long period wave of low amplitude is almost completely reflected. For a 5 ft high wave of 10-second period (512 ft long in deep water) the ratio of reflected wave height to deep water height from a uniformly sloping impervious beach varies as follows:

Table 3.1
RATIO OF REFLECTED WAVE HEIGHT/DEEP WATER HEIGHT AT VARIOUS BEACH SLOPES

Slope	Reflection ratio
1:10	0.05
1:5	0.33
1:4	0.60

Waves reflecting from, or breaking on or near, a structure will 'run up' the structure to a greater height than the crest height of the wave. For example, for a wave fully reflected from a vertical seawall, the crest of the standing wave or clapotis will reach a height above still-water level of:

$$H + \frac{\pi H^2}{L} \coth \frac{2\pi d}{L} \qquad (3.10)$$

Partially reflected and breaking waves will 'run up' differently according to the shape and roughness of the structure, the depth of water in front of it, the configuration of the bottom, and wave characteristics.

3.1l Wave Defraction

Wave energy may be transmitted into the lee of an obstacle, such as a breakwater, as a result of wave diffraction. This process can be interpreted as producing a lateral spread of wave energy along the wave crest from where the height is large to where it is lowest, and arises from the refraction process. Where the seabed is regular the wave fronts will take the approximate form of successive arcs of circles with their centre at the end of the obstruction.

3.1m Wave Direction

Radar has been introduced recently in some coastal research programmes particularly to indicate wave direction in certain conditions of sea state. In Start Bay, Devon, shore-based

Decca Type 919 river radar has proved a reasonably useful and reliable tool in the routine measurement of wave direction and shoreline equilibrium studies, (Hails 1975, Holmes, 1975). The radar, installed on a cliff-edge site 16 m above sea level, has been used on a 2–5 km range. It is operated automatically every three hours and a photograph is taken of one complete sweep of the plan position display.

The radar has been used in conjunction with two pressure type wave recorders operating about 200 m from the shoreline on either side of the radar, the closer of the two being located within 1.2 km of the equipment. Wave records of ten minutes' duration are taken every three hours. In this way, the wave records and radar photographs have been analysed to determine the relationship that exists between them. The advantages and limitations of this technique are still under review. Suffice it to say that wave direction can usually be measured to an accuracy of ±5 degrees, but the radar detects only high wave crests. This means that there is not a good correlation between the wave period calculated from the radar picture and that observed on the wave record.

3.1n Edge Waves

According to the theoretical and experimental wave basin work of Bowen and Inman ordinary incident swell waves may generate standing edge waves. These are surface waves, trapped against a shoaling beach by refraction, with crests normal to the shoreline and with wavelengths from crest to crest parallel to the shoreline. In other words, edge waves are distributed along the length of a beach with an orientation opposite to that of the approaching swell.

It is thought that the interaction of incoming swell and standing edge waves, which have the same period, produces alternately high and low breakers along the shoreline, resulting in a regular pattern of circulation cells with regularly spaced rip currents. Because of this factor the positions where the two sets of waves are in phase, or for that matter out of phase, remains stable. Also, it is believed that the velocity fields associated with standing edge waves of relatively long periods of the order of 30 to 60 sec, account for various sedimentary features exhibiting rhythmical longshore patterns, such as crescentic sand bars and cuspate shorelines, and also that edge waves are the most probable cause of the rhythmic spacing of beach cusps (see p. 148).

However, the existence of edge waves on natural beaches and associated cell circulation patterns not produced by wave refraction has been challenged by some coastal workers. Despite such counter-claims, Huntley and Bowen (1975a, b) are reported to have measured edge waves in the field. They used two-component electromagnetic flowmeters, mounted on tripods standing freely 15 cm above the seabed on a steep shingle beach (Slapton Beach in Start Bay, Devon), to measure the horizontal velocity field or components of flow in the nearshore zone. The meters, which have no moving parts but a relatively fast response time, were connected by cable to recording systems on shore. The tripods were deployed at low water mark so that the flowmeters were covered by the tide for four hours spanning high water, and measurements were made over a distance of 25 m from the shoreline as the tide rose.

So far, this technique has proved both flexible and reliable on steep shingle (slope $\simeq 0.13$) and shallow sand (slope $\simeq 0.014$) beaches under varying wave conditions, includ-

FIG. 3.4
(TOP): REFRACTION OF 14-SECOND PERIOD SWELL WAVES AROUND SHORE PLATFORM, BUNGAN HEAD, NEW SOUTH WALES
(BOTTOM) LONG-PERIOD (14-SECOND) SWELL ENTERING BROKEN BAY, NEW SOUTH WALES (Photographs: J. R. Hails).

ing storm waves on beaches of intermediate slope ($\simeq$0.04) although experiments have not been conducted on exceptionally steep beaches during storms.

It will be realized that the hydrodynamics of steep and shallow beaches differ markedly, both qualitatively and quantitatively. In particular, plunging breakers steepen rapidly on steep beaches and dissipate their energy over a narrow, but turbulent, surf zone (Figure 3.2, top). On the other hand, changes in breaker form on shallow beaches occur slowly because the waves traverse a relatively wide surf zone (Figure 3.2, bottom).

At present, the effects of waves on beach sediment are better understood than the influence of beach slope on nearshore velocity fields (see also the later discussion on standing edge waves and the formation of each cusps, p. 148). Thus, following these preliminary studies, further measurements of edge wave spectra at wavelengths and frequencies suitable for nearshore processes should produce interesting results. The significance of edge waves in the movement of sediment nearshore has been discussed by Bowen and Inman (1969), Komar (1971b, 1972) and Huntley and Bowen (1975).

3.2 TIDES AND VARIATIONS IN MEAN SEA LEVEL

Tides, with periods between 45×10^3 and 90×10^3 sec, are very long-period waves like *tsunamis* and *storm surges* that have periods between 1000 and 2000 sec. Reference has already been made to waves in the range of 30 sec or larger which are known as *surf beats*; often these have periods from 5 to 300 sec. *Tsunami* is the Japanese term for a wave or a series of waves produced by earthquakes. Sometimes these are erroneously called 'tidal waves' because, in fact, they do not generate or propagate tides. In the mid-Pacific where the water depth can reach 5000 m, the corresponding tsunami speeds exceed 700 km/hr. As tsunamis cross the continental shelves and slow down they transfer their kinetic energy to potential energy, a process which can cause extensive damage if it is concentrated over a limited length of shoreline. Storm surges are generated within low-pressure tropical and extra tropical cyclonic centres both by a 'suction' effect over the ocean within the cyclone and by stress exerted on the water surface by the accompanying wind.

The attraction of the sun and moon on the water of the oceans causes tides. As the earth revolves around the sun, and the moon revolves around the earth, the forces of gravity and centrifugal force are exactly balanced at the centre of gravity of the earth. However, the water nearest the sun or moon causes a slight excess of gravitational pull over centrifugal force with the result that the water level rises. This effect produces the tide which has a wavelength of approximately half the earth's circumference and a period of about 12 hours. The velocity or celerity of the tidal wave across the oceans is governed by depth *(d)*. Its wavelength and period are sufficiently long for the tide to be a shallow-water wave even in the deepest ocean; its velocity of propagation is given by

$$C_s = \sqrt{gd} \tag{3.11}$$

where g is the acceleration due to gravity (980 cm/sec^2) and d is the depth of water. For example, the Atlantic Ocean has an average depth of about 4000 m so that the wave velocity $\sqrt{gd}$ is 200 m/sec or 383 knots. Compare this with the earth's rotational speed at the equator of 450 m/sec and 225 m/sec at 60° latitude.

The water farthest from the sun or moon causes a corresponding decrease in the gravitational pull which, in turn, causes a net outward force on the water. Thus, there is a tide similar to that on the opposite side of the world. The water rises under the moon and produces a high tide, but flows from other parts of the earth's surface to give low water at positions 90° away.

When the sun, moon and earth are aligned at the time of a new and full moon as shown in Figure 3.5A, the tidal bulges of the moon and sun are additive and produce extreme tides, called *spring tides*. These occur every two weeks throughout the year or, more precisely, every 14¾ days. On the other hand, there are *neap tides* when the sun and moon are at right angles, or opposed, to each other and produce counter-bulges at the first and third quarter of the moon (Figure 3.5B).

The moon's elliptical orbit around the earth also produces another significant variation in the height of the tides. At its nearest point, *perigee*, in its orbit the moon is about 357 000 km from earth, and at *apogee*, the farthest distance, it is 407 000 km. This change in distance and attractive force causes tides that are respectively 20 per cent higher and lower on average once a month, or every 27.55 days. Many coasts have semidiurnal tides whereas some places have only one high and one low tide each day. Most of the Indian and Pacific Ocean coasts have mixed tides of high and low waters of unequal height.

Davies (1964) has classified tides by their spring tidal range into three categories. These are microtidal (<2 m), mesotidal (2 to 4 m) and macrotidal (>4 m). The first two are usually located on exposed coasts and also in virtually landlocked seas such as the Black, Mediterranean and Red Seas. Well-known macrotidal coasts include the Bay of Saint Michel (Brittany, France) and the Bay of Fundy in Canada. These two examples accentuate the control of shoreline configuration on combined convergence and resonant effects, because as the tidal front approaches an embayment or gulf, relatively narrow indentations on a coast, its movement is constricted by the enveloping shores so that the height of the tidal wave increases. Similar funelling effects occur in the English Channel and along parts of the north coast of Australia where they upset the semidiurnal tides.

Chesapeake Bay, USA is one of the few areas where the sine wave pattern of the tide can be observed directly. The high and low tides advance up the Bay as a series of 'progressive' waves which travel slowly at the 'square root of *gd*' velocity of a long wave, as mentioned previously.

Large tidal ranges are associated with wide continental shelves, and accretion rather than erosion. The main effect of large fluctuations in water level is to extend wave action over a greater width of the beach or shore profile, and to dissipate energy over an enlarged surf zone so that waves break farther from the shoreline, the implications of which are discussed later.

The full mathematical analysis of tides, which is a separate subject outside the scope of this chapter, is available in several text books (see, for example, Clancy, 1969, Lisitzin, 1974). Nevertheless, it is pertinent to mention here that the analysis of tidal variations is similar to that of other waves at the shoreline whereby spectral analysis is used to differentiate between individual swell wave trains. However, a much longer record is needed for tidal analysis because, as already stated, the periods of the component tidal waves are far longer than those of swell waves.

Sea level is the height that the sea surface attains if it is undisturbed by waves, tides and

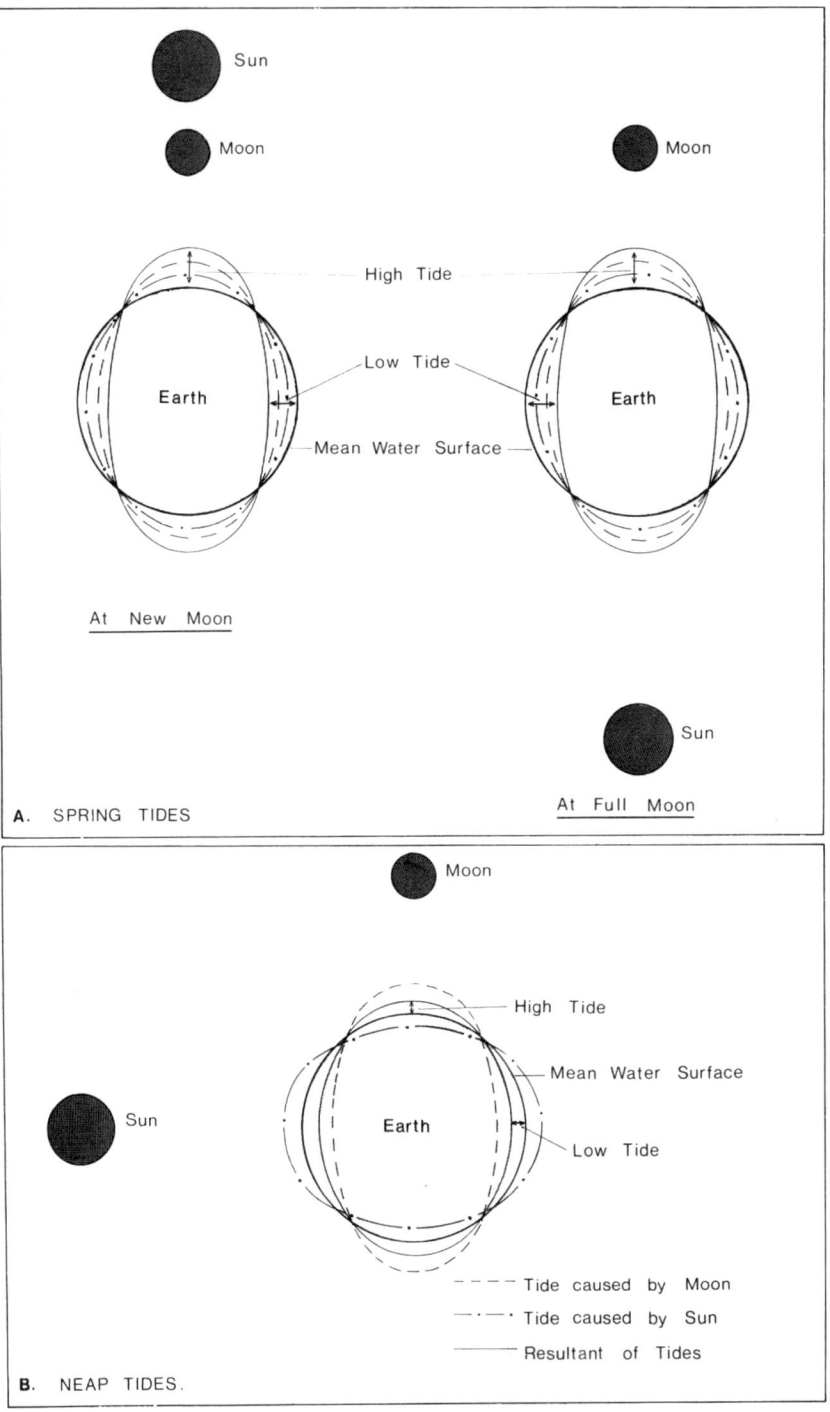

FIG. 3.5. SCHEMATIC CROSS-SECTION THROUGH EQUATOR SHOWING EFFECTS OF SUN AND MOON IN CAUSING TIDES
(A) Spring tides produced by sun, moon and earth being aligned. (B) Neap tides produced by sun and moon acting in opposition, producing counter-bulges.

winds. However, as these disturbances exist, mean sea level has been adopted as a datum plane from which heights of tides and depths of water are measured. Fluctuations in sea level at the coast are of a daily, seasonal, annual and, more importantly, long-term nature. It is, in fact, the response of sea level to climatic changes, particularly during the Pleistocene epoch, that has been of prime interest to coastal geomorphologists. However, the short-term changes that cause coastal erosion and modify depositional coasts are significant in the study of coastal processes.

The effects of variations in sea level on beach profiles are little understood because of the changes caused by waves and tides, and therefore future studies are likely to be conducted in the laboratory. Notwithstanding these facts, Bruun (1962) developed a conceptual model in order to ascertain the response of beaches to a rise in sea level. The essential argument in his theory is that material eroded from the upper foreshore and deposited offshore will keep pace with the rising sea level so that the water depth remains constant.

Radok and Raupach (1977) have shown in recent studies, for example, that weather systems raise the mean sea level against the south coast of Australia through barometric pressure and wind stress operating over large areas of the Southern Ocean. In turn, tides and waves are superimposed on these mean sea levels. Because variations in mean sea level are relatively slow they are able to penetrate into confined water bodies like the Gulf of St Vincent and Spencer Gulf in South Australia and may consequently induce notable water exchange and sediment movement. Coastal erosion along the southern margin of Australia can be linked to these variations, and studies in progress at present may show the futility of engineering solutions to local erosion problems.

Coastal oceanographers and engineers, and to a lesser degree geomorphologists, rely upon tide gauge records to determine variations in mean sea level. These records, though, should be reviewed cautiously because sites for tide gauges must be chosen carefully with respect to extreme weather conditions, wave action, impounding (a gauge will not record the full tidal range owing to the effect of impounding), variations in salinity, the strength and direction of tidal streams (a strong stream will cause a change in pressure at the orifice of the well) and sometimes, the availability of power supplies. Although the stilling well is reported to smooth out wave action, its efficiency is suspect when wave action is strong. There is still some doubt about the reliability and frequency response of some tide gauges and therefore the extent to which records from these instruments can be used to study, for example high order tidal components and seiches must be critically reviewed. Following these comments, it is reasonable to say that interpretations of tide gauge records made by geomorphologists in the recent past may well, in fact, be spurious.

Further investigations of overall shoreline changes and beach processes, regardless of whether they are of academic interest or applied in context, should attempt to correlate variations in mean sea level with local weather parameters. Such a correlation involves comparing daily mean sea levels with corresponding barometric pressures and wind components. Furthermore, continuous records will provide a basis for forecasting the occurrence and magnitude of surges because tidal fluctuations spread the energy of waves over a wide section of the beach. As far as possible, tidal predictions should be supported by empirical work which should continue on a yearly basis, otherwise the significance of differences in mean sea level may be missed.

3.3 NEARSHORE SEDIMENT BUDGET

Empirical correlations between sediment supply, drainage basin area and effective annual precipitation, which are fundamental to an evaluation of the littoral budget, have been discussed in Chapter 2. In coastal studies, little attention has been paid to either the rate at which sediment supplied to a beach by cliff retreat is subsequently dispersed by waves and currents, or to the role of coastal streams in regulating the offshore sediment supply. Measurements of river discharge within individual catchments provide not only an estimate of contemporary erosion rates as controlled by such multivariate parameters as topography, lithology, vegetation, climate and man, but also some idea of the amount of material that reaches the coast to be dispersed subsequently by waves and nearshore currents. The response of the shoreline to variations in sediment supply are unknown in quantitative terms, and unless catchments are instrumented and monitored continuously it will be extremely difficult, perhaps impossible, to forecast what short-term changes may occur in the adjacent offshore profile.

Many activities upstream from the coast such as the construction of dams, diversion of river channels, flood mitigation schemes and land use, have significant and far-reaching consequences on shoreline equilibrium. The amount of sediment reaching the coast of southern California, for example, has decreased significantly since dams have been constructed on several rivers in the Los Angeles area as part of a flood mitigation programme. The marked reduction in sediment supplied naturally to the coast in the past has indirectly promoted a net loss of material from beaches because nearshore currents transport beach sand towards neighbouring submarine canyons. Such a net loss of sediment can pose problems along a coast like that of southern California between Santa Barbara and San Diego which can be divided into four discrete sedimentation cells.

According to the work of D. L. Inman, T. K. Chamberlain, and J. D. Frautschy, each cell is a system in which rivers supply the beaches with sand which, in turn, moves south as littoral drift to be trapped within submarine canyons and lost offshore to deep water. The canyons are in fact sinks which dissect the continental shelf and intercept sand as it is transported along the beach. Man's land use practices within the catchments of coastal rivers also reduce sediment discharge to the coast and cause coastal erosion, as exemplified by the beach erosion that has occurred several kilometres south of the St John's River mouth in Florida (see Hails, 1977).

Langbein and Schumm (1958) have demonstrated that the sediment yield from a drainage basin varies as a function of climate, and that the maximum yield occurs for an effective precipitation of about 300 mm. The effective precipitation – defined by Langbein and Schumm as the amount of precipitation that would be required at a mean temperature of 50 °F to produce the actual annual runoff from the basin – is obtained from actual precipitation by correcting for evapotranspiration losses to a standard temperature of 50 °F. In the context of a fluctuating sediment supply controlled by climatic change the type of climatico-geomorphic process model originally described by Carr and Hails in 1972 is worth mentioning here because, indirectly, it illustrates how coast protection schemes and management plans may ignore the subtle interaction between sediment supply and transport, wave climate and minor oscillations in mean sea level – an interaction actually dependent upon the paramount control effected by climatic change. It can be deduced from Figure 3.6 that over geological time, climatic change influences rates of weathering, river discharge and ultimately sediment yield to the nearshore and offshore zones.

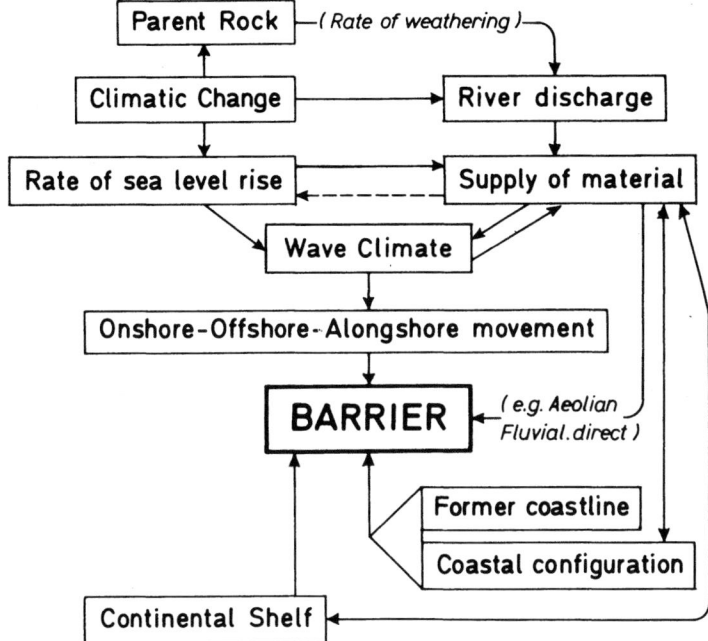

FIG. 3.6. CLIMATICO-GEOMORPHIC PROCESS MODEL
(After Carr and Hails, 1972).

3.3a Fluvial Sediment Discharge

Fluvial sediment discharge to the oceans from the landmass of the United States has been summarized by Curtis *et al.* (1973). These workers have estimated that the annual amount of material carried to the Atlantic Ocean is approximately 14×10^6 short tons; to the Gulf of Mexico 378×10^6 tons, and to the Pacific Ocean 99×10^6 tons. Data were analysed from 27 drainage areas and may be used to extrapolate part of the world sediment yield to the marine environment. The average annual suspended-sediment concentrations for the 27 drainage basins were computed by dividing the daily average suspended-sediment discharge in tons, by the daily average water discharge in cubic feet per second (ft^3/s) and by using appropriate conversion factors. As an example, for an average annual sediment discharge of 460 000 short tons (417 000 tonnes) and a daily average water discharge of 23 500 ft^3/s (665 cubic metres per second, m^3/s), the conversion factor cited by Curtis *et al.* is 0.0027, and the average annual daily sediment concentration, in milligrams per litre (mg/l) is computed as follows:

$$460\ 000/365 \times 23\ 500 \times 0.0027 = 20 \text{ mg/l.}$$

For the metric system the conversion factor is 0.0864:

$$417\ 000/365 \times 665 \times 0.0864 = 20 \text{ mg/l.}$$

The daily amount of sediment discharged into the Atlantic and Pacific Oceans is 38 915 tons and 271 400 tons respectively, while about 10^6 tons are carried to the Gulf of Mexico. Another way to express these figures is to say that a train with 778 wagons (each one carrying 50 tons) would be required to move the sediment load each day to the Atlantic; 5428 wagons to transport the Pacific coast sediment, and 20 740 wagons to carry suspended sediment to the Gulf of Mexico.

3.3b Estuaries

The role of estuaries in regulating the supply of sediment to the shoreline is a major topic and, therefore, the following comments are only a brief overview of the salient factors which should interest those engaged in coastal zone research programmes. Any analysis of the nearshore sediment budget must evaluate the efficiency of estuaries as sediment traps because some are particularly effective in preventing material from reaching neighbouring beaches. The complexity of estuarine sedimentation is controlled by the pattern of circulation of estuarine waters – the interaction and transition between marine and fluviatile agencies which themselves produce the characteristic estuarine conditions controlling erosion, transportation and sedimentation. Frequently, though, it is difficult to differentiate between bedload and suspended load within an estuary because sediment that is bedload at one stage of the tidal cycle may be suspended at another time.

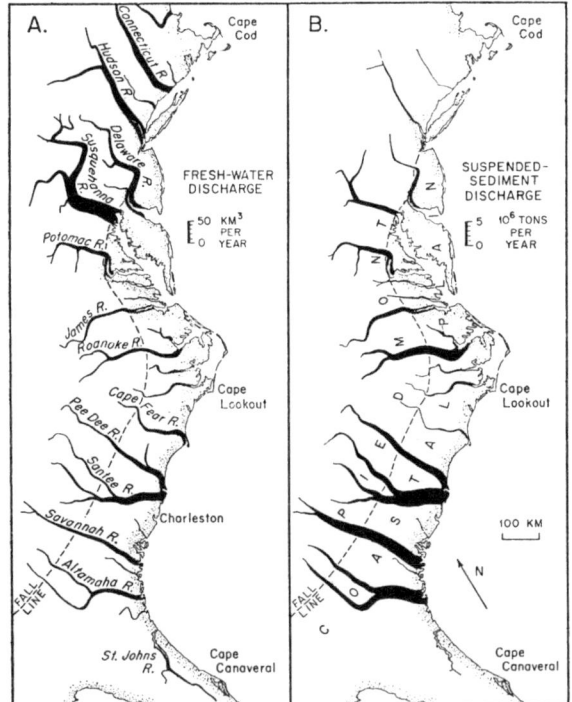

FIG. 3.7. WATER AND SEDIMENT DISCHARGED BY RIVERS OF ATLANTIC DRAINAGE BETWEEN CAPE COD AND CAPE CANAVERAL, USA
(After Meade, 1969).

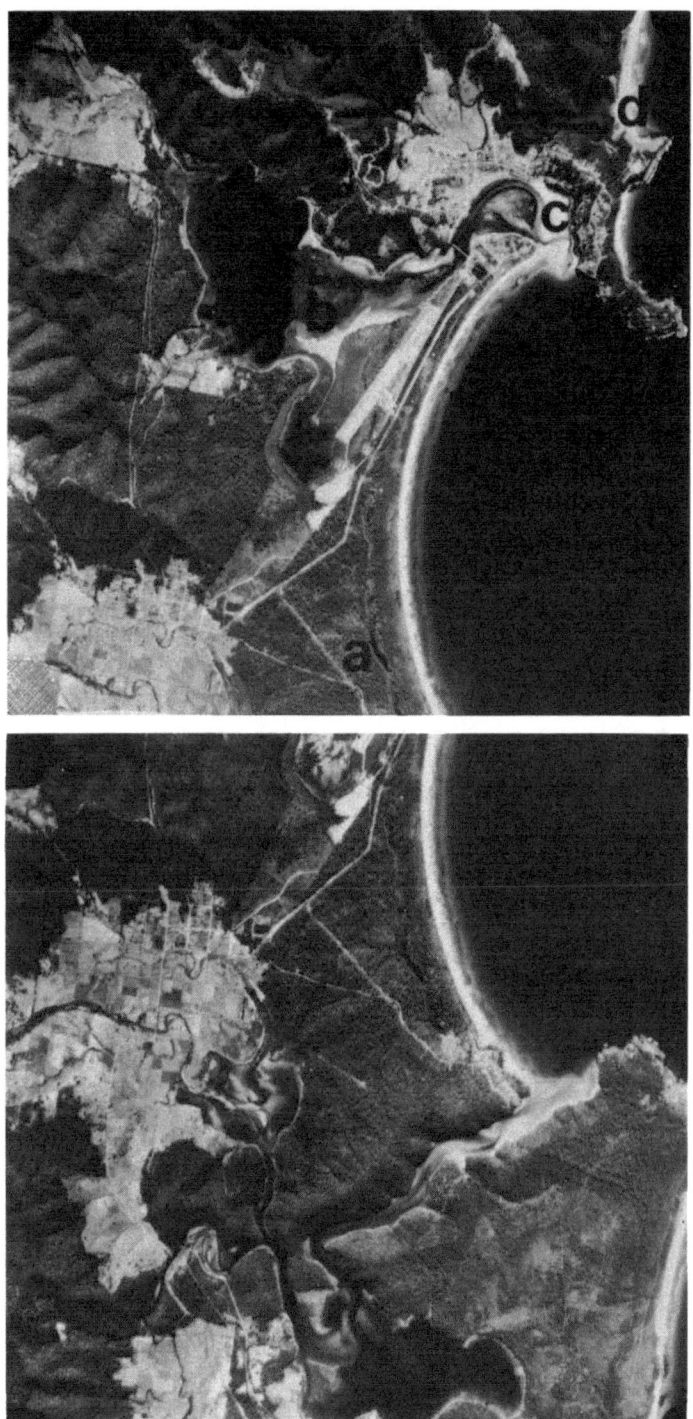

FIG. 3.8. VERTICAL AERIAL PHOTOGRAPH OF PAMBULA SPIT AND ADJACENT AREAS, NEW SOUTH WALES (NSW 1156–5079).

It has now been established from studies of diagnostic mineral suites and the textural properties of sands on different coasts that material from adjacent beaches and/or offshore sources is transported into the mouths of estuaries (Meade, 1969). Beach sands move towards and into the mouths of some estuaries at rates of several hundred thousand cubic metres per annum. For instance, of the average 5.5×10^6 m³ of sediment that was deposited in the navigation channel of the Savannah estuary annually between 1953 and 1957, only about 1.5×10^6 m³ is believed to have been transported into the estuary by the Savannah River, according to the records of the US Army Engineer District, Savannah, Georgia, USA. Measurements of the sediment flux (suspended-sediment concentration and water velocity measured at depth intervals through tidal cycles) show that the sediment is moved progressively landward by estuarine bottom waters which also trap river-borne sediment in estuaries.

According to Meade, on the Atlantic Coastal Plain of the USA the large rivers north of Cape Lookout, such as the Susquehanna and Delaware, transport disproportionately small sediment loads that have not yet infilled the large submerged valleys which were eroded during glacial times, whereas the smaller estuaries to the south, like those of the Savannah and Altamaha Rivers, are mostly filled with sediment because the rivers carry larger sediment loads relative to the size of their valleys (Figure 3.7). Man has caused a significant increase in river-borne sediments by converting forests to farms, a practice which has produced about a tenfold increase in the sediment yield of the land; by mining coal and by encouraging urbanization (see Meade, 1969 for references).

There is substantial evidence to indicate that sand is being transported into coastal lagoons, and up some coastal rivers for about 5 km, under the action of flood tides on the New South Wales coast, Australia. Figure 3.8 is a vertical aerial photograph of Pambula barrier (a), the landward margin of which is separated from bedrock by a lagoonal swamp. Sand transported through the entrance (c) of the lagoon, known as Merimbula Lake, has formed a *flood-tidal* delta (b). Similar features are quite common in other lagoons along the coast.

Many questions still remain unanswered about sediment transport and sedimentation in estuaries. How much material is trapped permanently? How much is carried offshore? Over the long term, is more sediment entering or leaving the mouths of estuaries? Some estuaries, of course, are in equilibrium with present sea level in that sediment inflow and outflow are reasonably well balanced over periods of several years.

3.4 HYDRODYNAMICS OF THE NEARSHORE ZONE

The nearshore zone includes that part of the beach face which is covered by the uprush of water following wave breaking – a film of water <6 mm deep at its upper feather edge known as the *swash* – to beyond the area in which waves break. As shown in Figure 3.9 it embraces three zones, but should not be used synonymously with the term *littoral zone* which includes the offshore area beyond the breakers. It is, in fact, a zone in which waves dissipate their energy, nearshore currents are generated and sediment is moved along-shore. Many important questions still remain unanswered about the mechanics of wave motion and how energy is transformed and expended in and near the surf zone. Our

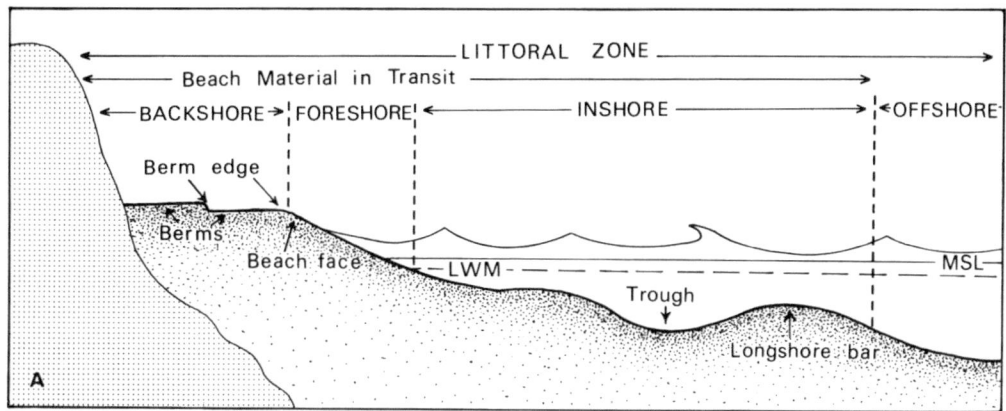

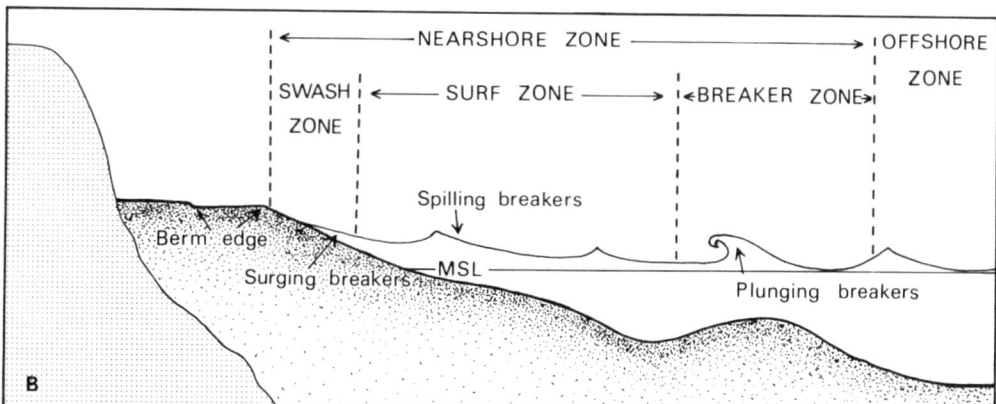

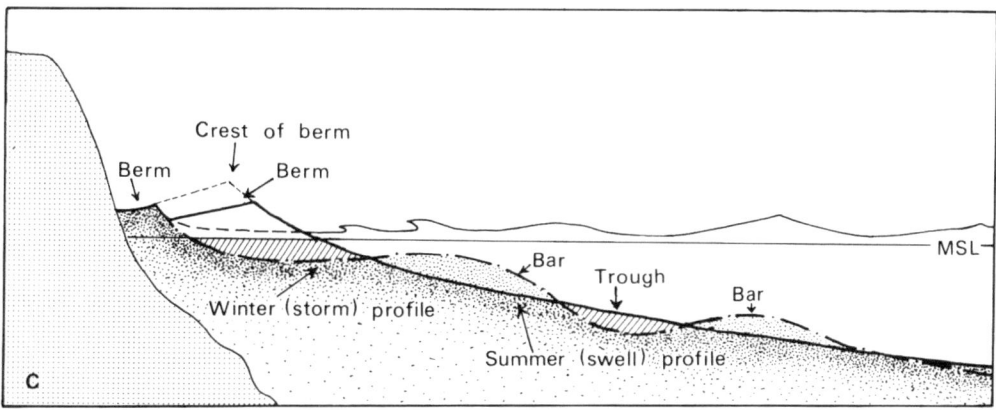

FIG. 3.9. SCHEMATIC DIAGRAMS SHOWING ZONES USED TO DESCRIBE (A) BEACH PROFILES, (B) WAVE AND CURRENT ACTION, (C) PROFILE CHANGES DUE TO STORM AND SWELL WAVES

understanding of nearshore currents is such that two systems can be identified and these will now be discussed briefly in turn.

3.4a Longshore Currents

Longshore currents are generated by waves breaking at an angle to the shoreline. Such currents, which are mainly confined to the surf zone, interact with the wave surf and produce sand transport parallel to the shoreline. Until a decade ago, most theoretical models on longshore currents and the mechanics of sediment transport on beaches considered the continuity of the water mass, the energy flux and the momentum flux (Galvin, 1967). Since then, considerable progress has been made towards an understanding of nearshore processes as a result of studies that have evaluated the generation of longshore currents in terms of the longshore momentum flux (radiation stress) of the waves (see papers by Bowen and Longuet-Higgins).

Bowen has indicated that longshore currents resulting from an oblique wave approach may be generated by the longshore component of the *radiation stress* according to the equation:

$$S_{xy} = E\eta \sin \alpha \cos \alpha \qquad (3.11)$$

where the x-axis is normal and the y-axis parallel to the shoreline; thus, S_{xy} is the onshore flux of momentum which is directed alongshore. The *radiation stress* has been defined by Longuet-Higgins and Stewart (1962, 1963, 1964) as 'the excess flow of momentum due to the presence of the waves'. Besides being used in the study of littoral processes to evaluate the generation of longshore currents, it has also been applied to forecast changes in the mean water level (*setdown* and *setup*) in the nearshore zone. The *setdown* is the lowering of water level below still-water level when no waves are present outside the breaker zone whereas the *setup* is a rise of mean water level above still-water level within the surf zone. Thus on a beach with wave crests arriving parallel to the shoreline there will be a shoreward flux of momentum S_{xx}, given by the equation

$$S_{xx} = E \left[\frac{2Kh}{\sin h\,(2Kh)} + \frac{1}{2} \right] = E\,(2\eta - \frac{1}{2}) \qquad (3.12)$$

where $K = 2/\pi L$, E is the wave energy density, L is the wavelength, and h is the water depth.

In shallow water the x-coordinate, which is positive normal to the shoreline in the onshore direction, is

$$S_{xx} = \frac{3}{2} E = \frac{3}{16} \rho g H^2 \qquad (3.13)$$

because $\eta = 1$, and $E = \rho g H^2/8$ for wave energy (see p. 112).

3.4b Rip Currents: Nearshore Cell Circulation

Figure 3.10 is a schematic diagram illustrating that the nearshore cell circulation consists of feeder longshore currents, rip currents, and a mass transport of water returning to the surf

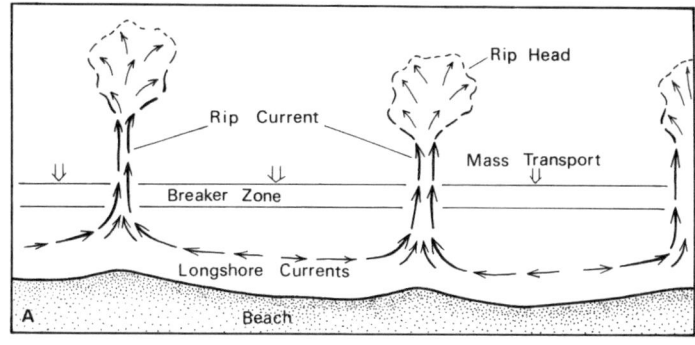

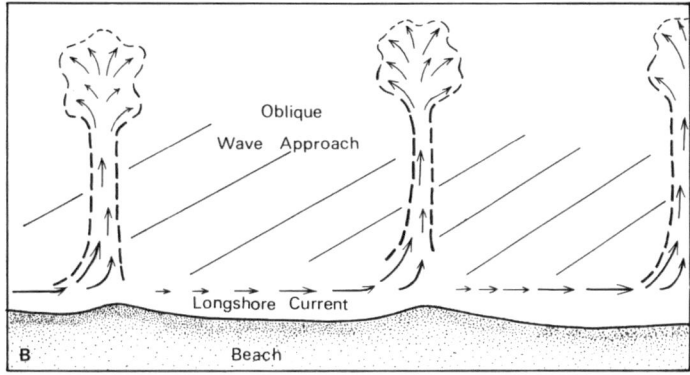

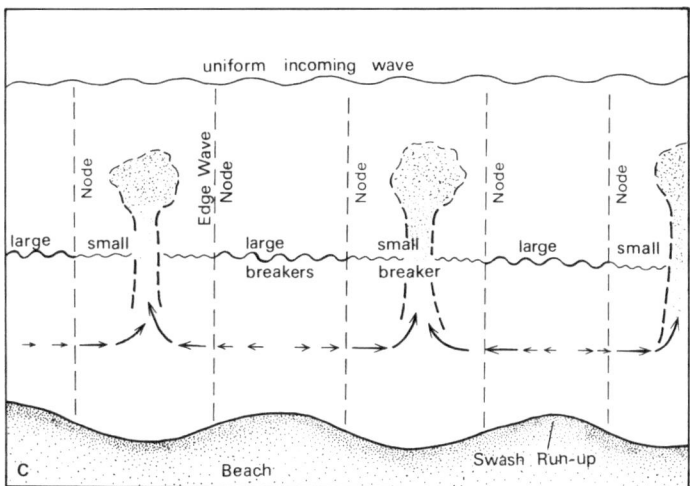

FIG. 3.10. NEARSHORE CELL CIRCULATION

(A) Feeder longshore currents, rip currents and mass transport returning water to surf zone (after Shepard and Inman, 1950). (B) Longshore current pattern in nearshore zone resulting from oblique wave approach (after Komar and Inman, 1970). (C) Position of rip currents with edge waves and incoming waves 180° out of phase (after Komar, 1976).

zone. Rip currents are narrow, strong bodies of water that flow seaward from the surf zone. They were first observed by Shepard *et al.* (1941) who related their velocity, and the distance they flowed seaward, to the height of incoming waves. Measurements in the field were obtained by Shepard and Inman about a decade later at Scripps Beach, La Jolla in California. They recognized that circulation cells with rip currents could exist on long, relatively straight shorelines with regular bottom relief, although offshore topography and its effect on wave refraction probably controlled the position of rips along a beach.

Bowen (1969a, b) and Bowen and Inman (1969) have shown both theoretically and experimentally since then that variations in wave setup may provide the necessary long-shore head of water to drive the feeder longshore currents and produce the rip currents. Any variation in the wave setup is related to longshore variations in the incoming wave height. It follows that the wave setup is greater if the waves are higher. Thus, the feeder currents appear to flow from the area of highest breaker height and become rips where wave height is lowest. It is still a matter of conjecture as to whether the variations in wave height can be attributed to the effects of wave refraction or the interaction between incoming waves and edge waves trapped within the nearshore zone.

Debate about the origin of rip currents still continues, as the following examples show. The work of Harris (1967), and more recently that of Guza and Inman (1975), shows that the dominant longshore spacing of rips on non-erodible dissipative beaches (where waves break and nearshore circulation cells are important) exceeds the maximum possible longshore wavelength of synchronous edge waves. On the other hand, Hino (1972, 1973) has suggested that the setup sea surface associated with breaking waves is unstable and that such instability is a rip current. Le Blond and Tang (1974), in their study on energy coupling between waves and rip currents, have put forward a criterion for the spacing of such currents based upon the minimization of relative energy dissipation rates. However, their results have not been supported by empirical work.

In concluding this brief account of nearshore currents it can be said that the development and application of radiation stress concepts for water waves have been major advances in theoretical studies of longshore current generation, and that sediment transport in the surf zone is caused by a combination of shear stresses produced by wave and current action.

3.4c Longshore Sediment Transport

It was mentioned in the previous section that wave-induced longshore currents are responsible for moving sediment alongshore. However, the rate at which sediment moves, or the littoral transport rate, is difficult to measure and still remains a problem to be resolved. Estimates of the amount of longshore transport can be made where groynes, breakwaters and other man-made structures intercept longshore transport. The following examination only summarizes the main points of more detailed reports on this complex subject which have been published in some of the advanced texts listed at the end of this chapter.

According to Komar, attempts at evaluating the sand transport rate have relied mainly upon empirical correlations with

$$P_1 \, \vee \, (EC\eta)_b \sin \alpha_b \cos \alpha_b \qquad (3.14)$$

where $(EC\eta)_b$ is the wave energy flux evaluated at the breaker zone, and α_b is the breaker angle. P_1 is variously known as the longshore component of the wave energy flux or the longshore component of wave power, but some workers have objected to the use of these terms. Inman and Bagnold (1963) have stressed that the littoral transport rate should be expressed as an immersed weight transport rate I_1, rather than as a volume transport rate S_1, because of the problem of sediment transport from the viewpoint of energetics. These two rates are, in fact, related by

$$I_1 = (\rho_s - \rho)ga' S_1 \tag{3.15}$$

where ρ_s and ρ are the respective sand and water densities, and a' is a correction factor for the pore space of the beach sand ($c.$ 0.6 for most beach sands).

I_1 is related to P_1 through the equation

$$I_1 = K(EC\eta)_b \sin \alpha_b \cos\alpha_b = KP_1 \tag{3.16}$$

where K is a dimensionless proportionality coefficient. As shown in equation (3.15) the main advantage of using the immersed weight transport rate is that the density of the sediment grains can be considered. It follows, then, that there is a reasonably good correlation between I_1 and P_1 for beaches predominantly composed of quartz and coral sand (see also next section on tracers).

Material may be transported alongshore as suspended load or bedload. The former is moved within the water mass and is kept in suspension by turbulent eddies, whereas the bedload is transported along or near the seabed. It is difficult to equate quantitatively data on suspended sediment obtained from laboratory experiments with those recorded in the field because of the inherent problem of devising reliable techniques for measuring waves and a general lack of knowledge about the characteristics of turbulence of the wave boundary layer. Consequently, little is known about the mechanics of suspended sediment resulting from wave activity, and quantitative data on the transport of fine sediment are lacking for water depths exceeding a few metres.

Following the wide use of time integrated samplers, *in situ* techniques are now being adopted in an attempt to measure suspended sediment concentration in water. The factors affecting synoptic surf zone sedimentation patterns at Point Mugu, California, for example, have been studied by emplacing three almometers (each one consisting of 64 photo-electric cells, 1 to 2 cm apart, and a high intensity fluorescent lamp encased separately in watertight acrylic cylinders) on the beach. The cylinders are anchored in a vertical position about 1 m apart so that sediment-laden water can flow between them. In this way it is possible to determine, by Eulerian means (see next section), the movement of sand normal to the shoreline and across the beach face under the influence of breakers, surf and swash. Measurements have shown that suspended sediment within the nearshore zone occurs in pulses of short duration, known as sand fountains, which occur mainly at the still-water level; within the breaker zone sand is rarely suspended 15 cm above the bottom; sand moves predominantly as bedload outside the breaker zone; in the outer surf fountains are almost non-existent.

Twice as many suspension clouds have been recorded on a rising tide as on an ebb tide owing to the lag of the groundwater table behind the tidal elevation. The tidal level is not significant in influencing the height the sand fountains attain. At Point Mugu they reached an average height of 24 cm and lasted 10 seconds. The amount of sand thrown into suspension 15 cm above the bottom for a strip 30 cm wide, each tidal cycle, was between 150 to 300 m^3, or 105 to 210×10^3 m^3 per annum. This is between two and five times that moved by longshore currents. For more information on this latest technique, the reader is referred to the work of Brenninkmeyer.

This and similar studies discount the claim by engineers that the suspended load is more important than bedload transport in the nearshore zone. The removal of fine-grained material from the nearshore zone by simple offshore diffusion and by rip currents seems to account for the prevalence of bedload material on beaches.

The distribution of heavy minerals (specific gravity >2.90) in beach sands can often indicate the direction of littoral drift along a coast, as exemplified by Trask's (1952) classic study at Santa Barbara in California, USA. Santa Barbara, about 144 km northwest of Los Angeles was the first southern Californian area to sustain serious beach erosion after the construction of a breakwater in 1929. Contrary to the assumption that littoral drift could not be interrupted, the movement of sand along the harbour beach was arrested because the breakwater caused the stilling of wave action inside the harbour. The average rate of accretion at the breakwater was estimated at around 206 550 m^3 per annum before remedial measures were taken to resolve the problem (for a more detailed account, see Hails, 1977). Trask, by using augite – a black ferromagnesian silicate commonly derived from basic igneous rocks – as a heavy mineral tracer, was able to show that much of the sand entering Santa Barbara harbour had been transported more than 160 km alongshore (see also p. 126).

However, the use of heavy minerals to determine the source of beach sediments is a technique that should be employed cautiously because it is often difficult to differentiate between locally derived minerals and those that have survived several phases of rock formation and weathering. Preferably, a diagnostic mineral that can be identified in neighbouring bedrock, should be used in littoral transport studies. The writer, for example, has identified a medium-grade metamorphic suite of heavy minerals (andalusite, staurolite, garnet, kyanite, zoisite and epidote) in Pleistocene and Holocene beach sands on the mid-north coast of New South Wales, Australia (Hails, 1969). Yet there is no trace of metamorphic rocks in the area which is drained by coastal rivers.

Therefore, it is possible, firstly, that the minerals were derived from a source which has been entirely removed by erosion; secondly, from other regions by being transported alongshore; and thirdly, from offshore. If the minerals were eroded from cliffs or delivered to the coast by rivers outside the region, the longshore movement of material around headlands and over considerable distances would have to be invoked. However, the presence of picotite – a diagnostic heavy mineral of very restricted occurrence derived locally from serpentine rock – in the modern beach sands on the mid-north coast, suggests that heavy minerals reaching the coast are not transported around headlands by longshore currents, although nearshore experiments, using radioactive or fluorescent tracer sand, are required to verify this preliminary evidence.

As just mentioned, measuring rates of accretion or bypassing of sand at a breakwater and/or other man-made structures is another indirect way of evaluating littoral drift. Again, the reader is referred to another example in California, namely Port Hueneme harbour which was constructed in 1940. Approximately 920 000 m^3 of material were eroded annually from the shoreline downcoast of the harbour after its completion, and the west jetty of the harbour diverted material moving alongshore into the Hueneme canyon (Herron and Harris, 1966; Hails, 1977). Shoreline configuration in the vicinity of coastal structures, man-made and natural, often affords some evidence of the direction of sediment movement alongshore, although observations are needed at regular intervals over a number of years in order to determine long-term trends in sediment movement. Littoral drift can also be computed from statistical wave data obtained from wave refraction diagrams, and this technique may prove the most reliable, until direct field measurements can be made of net sediment transport.

The pattern of transport of beach material along the east coast of England, for example, has been discussed by Kidson (1961) and summarized by King (1972). Continuous and temporary erosion is particularly evident at Holderness, and in parts of Lincolnshire, Norfolk and Suffolk. Yet, on other sectors of the coast there has been rapid accretion, particularly in north and south Lincolnshire and along the north Norfolk coast (see also, Hails, 1977). However, the coastal geomorphologist must be aware of man as an agent of erosion because a detailed inspection by the writer of the cliffs along the coast of Holderness has revealed that mass-wasting and not direct wave attack has been the trigger mechanism for cliff recession. Tile drains, a metre or so below ground level in the cliff face, are actually responsible for slumping because after heavy rains water flows from them down the cliff face. Large plumes of turbid water can be seen drifting along the coast as suspended material is moved by wave and current action. One may well enquire about man's activities elsewhere which have passed relatively unnoticed.

3.5 ESTIMATES OF SEDIMENT TRANSPORT FROM TRACER DISPERSAL STUDIES

Fluorescent tracers and radioactive isotopes have been widely used in an effort to measure rates of littoral drift along natural sand and, to a lesser degree, shingle beaches. However, most published papers discuss qualitative rather than quantitative techniques and the methods used so far neglect the nearshore flow regime, wave climate and seabed topography. By summarizing a few recently published studies a comparison is given here of fluorescent, radioactive (both natural and artificial) and activatable tracers.

At present, there is an obvious need for concurrent measurements of bulk transport, bulk dispersion and overall sediment transport paths. In the case of quantitative work, the accuracy of the measured transport rate depends upon the tracer budget, or the amount of the original tracer that can be recovered in a sampling programme. Therefore, if for some reason a large proportion of an injected tracer cannot be recovered, the data derived from sampling that tracer is suspect. It is also essential to measure the thickness of movement of the sand layer which is in transit in order to calculate the littoral drift rate of sand along the beach. Furthermore, caution must always be exercised when one evaluates quantitative data simply because, as published research so admirably illustrates, field studies differ significantly with respect to injection and sampling methods, not to mention the interpretation of results.

Such different methods serve to amplify the complex interaction, in different geographical areas, between parameters such as the topography and composition of the beach, tidal range, offshore bathymetry and wave conditions. Even so, it must be borne in mind that the purpose of a tracer experiment, whether it is of short/long duration or attempts to examine what relationship(s) exists between the various parameters just mentioned, is to determine the mode of injection and sampling methods, Langrangean or Eulerian, to be used. By using the former the movement of the tracer is followed and at a given time the space distribution of the tracer is examined, whereas with the Eulerian method the observer is at a fixed point and examines the time variation of the tracer at that position.

Considerable effort has been expended by researchers at Scripps Institution of Oceanography, for example, to obtain synoptic measurements of the direction and flux of wave energy and the transport rate of sand in the surf zone with an array of digital wave sensors. Quantitative measurements of the longshore transport rate have been obtained from the time history of the position of the centre of gravity of sand tracers injected on to the beach. Output from each wave sensor has been recorded on magnetic tape and subsequently programmed on a computer to obtain the wave spectra and cross-spectral analysis for the various members of the array. So far, the field data indicate that the longshore transport of sand is directly proportional to the longshore component of wave power, regardless of the grain size of the beach material.

The Hydraulics Research Station at Wallingford, in England, has used radioactive tracers to plot the movement of fine sand in Liverpool and Morecambe Bays, as well as the Ribble and Severn estuaries. Radioactive tracers, used in studies of sand movement on bars at the confluence of the Dordogne and Garonne rivers, have shown that injected material is displaced upstream during flood tides and that there is also a net sediment drift upstream during periods of large tidal amplitude. However, this trend is reversed during periods of decreasing amplitude proportionally with effects from fluvial discharges, thus illustrating that the minimum time required for tracer work is often more critical than the actual duration of the experiment.

A radio-isotopic sand tracer (RIST) system, developed by the Coastal Engineering Research Centre, USA, has now overcome some of the technical difficulties involved in tagging, injecting and sensing the movement of radio-isotopic sand tracers in the nearshore zone (see summary in Hails, 1974). Although this system provides some insight into the mechanics and pattern of sediment movement, and partly identifies zonal differential transport rate, it has not yet provided a solution to the problem of zonal volume rate of sand transport, a factor of practical value to both the engineer and scientific researcher. This is because the differential movement of sand in various environmental zones and different lateral transport rates introduce complexities into the realistic determination of volume sediment movement in the nearshore zone.

Most of the shingle tracer experiments reported in the literature have been conducted around the coast of the British Isles. Attempts have been made to measure the littoral drift of beach shingle along parts of the south coast of England by making multiple injections of artificially-prepared fluorescent shingle daily over a period of one month, and to relate the longshore movement of introduced material to the incident wave energy on shingle beaches. However, these and several other types of shingle tracer experiments have severe limitations because they are only feasible in certain localities and require quantities of material that can resist abrasion or fragmentation.

Price (1968) has argued that many fluorescent tracer experiments conducted on beaches can produce misleading results, particularly with respect to the movement of tracers in an offshore-onshore direction. He has evaluated with the aid of an arithmetical model, variable dispersion and its effects on the movement of tracers on beaches, and tentatively suggests that it would be useful to measure in the field the quantity of material travelling as littoral drift at various levels on the beach rather than *in toto*. De Vries has subsequently evaluated the quantitative interpretation of tracer measurements and has indicated that available *overall* and *dispersion* models have only restricted validity because they are more or less implicitly based upon the assumption that the transport conditions are homogeneous in the area under consideration. Obviously, errors are introduced if the assumption is invalid. De Vries refers to Price's argument that the average movement of the tracers will be directed from places with low dispersion to places with high dispersion and suggests that this leads to the wrong conclusion about a net transport being present. It is apparent from these comments that more research is needed on the application of tracers to heterogeneous transport conditions, with or without net transport.

A few remarks about the significance of tracer techniques are apposite at this juncture in order to put their usefulness in perspective. Although fluorescent tracers have been widely employed in studies of sediment dispersal patterns they can only be observed during fair-weather conditions. During storms, when the most significant events are probably taking place in the nearshore zone, tracers are invariably removed from injection points as beaches are eroded and material is moved offshore. Their value, then, diminishes considerably in the offshore zone because measurements become more difficult in deeper water. In the case of radioactive tracers it must be apparent to the reader that the large expenditure of money and manpower involved can only be borne by large government research establishments. As well, there is the inherent difficulty of detecting a sufficiently high proportion of the injected tracer material to undertake a quantitative interpretation of the measured spatial distributions (Crickmore *et al.*, 1972). Because radioactive tracers indicate the direction of net sediment movement only, the actual evaluation of transport by this method is still questionable. From the environmental viewpoint, precautions must be taken to ensure that wave conditions are suitable during injection and that radioactive particles do not pollute the study area and adjacent beaches.

3.6 BEACH PROCESSES AND MONITORING TECHNIQUES

There are several definitions of the word *beach* which, for the purpose of this chapter, is defined as the zone which extends from the extreme upper limit of wave action (excluding catastrophic waves such as *tidal waves, tsunamis,* and *storm tides* that accompany hurricanes and cyclones) to offshore beyond low-water mark where bottom material is moved by wave and current action.

The composition and textural properties (particle size, shape and density) of beaches vary considerably from locality to locality. According to their geographical distribution, beaches may be composed entirely of calcareous, siliceous or volcanic material. Quartz is the most abundant mineral in sands along the continental margins because of its resistance to weathering and abrasion during transportation. It is usually associated with ferromagnesian and other constituents of igneous rocks which have survived the disintegration of igneous rocks over geologic time. In tropical and subtropical regions, where there is pronounced biological activity, shell and coral fragments as well as spherical sand-size carbonate grains called *oolites* are common. In contrast, the beaches on volcanic islands may consist almost entirely of minerals derived from acidic and basaltic lavas.

There is now substantial geological evidence to indicate that most of the terrigenous material deposited on the exposed continental shelves during the last glacial maximum, when sea level stood between 100 and 140 m below its present level, has now been transported shoreward and incorporated in modern beaches. The seafloor, though, is still the direct source of beach sands even if there is virtually no offshore supply to be moved onshore in some areas, except for small quantities resulting from seasonal variations in the overall sediment budget.

3.6a Beach Profiles

Natural beaches, which may exist in dynamic equilibrium, are essentially buffers against wave attack. The concept of the equilibrium beach, although not explicitly defined, was introduced by Tanner in 1958. Such a beach may be described simply as one which undergoes very little change in plan view and profile because there is little or no additional material being supplied to the shoreline by littoral drift. In other words, the morphological components of a nearshore-beach-dune system and the beach outline in plan are an open system which is in dynamic equilibrium. It can be mentioned in passing that groynes, breakwaters, jetties and other man-made structures interrupt the natural equilibrium between the sources of sand supply and littoral drift so that a shoreline changes its configuration in order to attain a new equilibrium.

Beach profiles change in response to the distribution and dissipation of wave energy along the shoreline during storms, the onshore-offshore movement of material, the effects of tides and oscillations in mean sea level, and are related to the composition, size and shape of the beach material which, in turn, determines the slope or gradient of the beach face. This can vary from about 1:10 to 1:30 for coarse and fine sand respectively, and 1:4 for beach gravel – composed mainly of granules, pebbles and small cobbles which collectively are often termed *shingle*. On Chesil Beach, in Dorset, slopes of between 1:2 and 1:3 have been measured which is about the maximum beach gradient that waves can construct. Steep shingle and coarse sand beaches are more mobile than their finer-grained counterparts because wave energy is dissipated within a relatively narrow zone. King (1972) for instance reports that one shingle beach in southern England was lowered 1.5 m within two hours.

Variations in beach profiles can be observed daily, weekly, monthly or from season to season. However, such short-term changes do not necessarily provide reliable data for predicting long-term trends. Seasonal variations can be irregular, as reported by Shepard (1973) who observed for many years that the beach south of Scripps Institution of Oceanography was eroded (cut) during winter to expose the underlying gravel. Yet, except for one occasion, the sand cover has persisted for the past thirty years.

The seasonal occurrence of swell and storm waves is manifest in summer 'fill' and winter 'cut' which can be monitored by regular surveys of beach profiles. In summer, constructive swell waves produce beach profiles which are characterized by a relatively wide berm. This feature is an almost horizontal platform formed by the deposition of material at the limit of the swash – the rapid flow or uprush of water onto the beach face, following a breaking wave, which gradually loses velocity because it is opposed by gravity and friction, and water loss through percolation. Some berms measured by the writer on medium- to fine-grained sand beaches in New South Wales, Australia, had landward slopes of between 1° and 3°. The seaward limit of the berm is characterized by a marked break of slope known as the berm crest or edge which usually occurs above the level of mean high water mark. Bagnold demonstrated in wave-tank experiments that the height of the berm crest is about 1.3 times the height of the deep water waves that form it. However, it is difficult to confirm this relationship in the field owing to variations in mean sea level and the effects of wave refraction. The offshore beach profile beyond low water is usually smooth in summer and submarine bars are generally absent except in relatively deep water.

During winter, the berm is cut back or completely eroded by storm waves. Material moved offshore from the beach is incorporated in a series of bars, with intervening troughs, which trend parallel to the shoreline. If, by chance, berms remain after storms or intense wave activity they are characteristically high and narrow in marked contrast to their summer counterparts. Measurements of beach profiles at some fixed point on the coast show that the volume of sand involved in the seasonal exchange of material between berms and bars remains fairly constant. Although this is a standard technique, relatively few workers outside government research establishments have been able to relate their results to instrumented wave data.

Wave energy indicators, like significant wave height and period, can be evaluated from data obtained from wave recorders located a short distance offshore. Pressure-type wave recorders are reasonably reliable for monitoring nearshore wave conditions, while shore-based radar is now able to provide continuous wave climate data including two-dimensional wave spectra. It is anticipated that storms may eventually be instrumented in order to afford a comprehensive and unifying study of the hydrodynamics of beaches and the nearshore zone. The onshore-offshore movement of material associated with changes from swell to storm conditions can be correlated with wave steepness which has already been defined as the ratio of the deep-water wave height, H_w, to the deep-water wavelength L_w (H_w/L_w). The reader will recall that this relationship, in turn, is related to the wave period, T, by $L_w - (g/2\pi)T^2$.

Despite numerous field observations over a number of years, no satisfactory wave measurements have been obtained to explain satisfactorily how the critical wave steepness regulates the shift from summer to winter beach profiles. This may be attributed partly to the irregularity of profile changes on beaches, although it is now established that an increase in wave height during storms causes sand to move offshore and the formation of the winter beach profile, while almost flat waves produce a summer profile.

Model wave-tank experiments on beach profiles show that storm profiles are associated with a bar at the break-point of waves, which is invariably called a 'break-point' bar. Outside this bar there is landward movement of material, the amount of which is determined by the wave height, and the wave period for that height. Thus, owing to increasing wave energy, a greater volume of material is transported as the waves become higher and their period lengthens. In contrast, the direction of transport inside the break-point bar depends upon the steepness of the waves, but is usually seaward. It follows, then, that material must accumulate at the break-point to form a bar which is characteristic of the storm profile.

Different model experiments have shown, however, that the value of the critical wave steepness may vary. Johnson (1949) and Scott (1954), for example, discovered that a storm profile and offshore bar formed with a wave steepness >0.03; a swell profile persisted with a steepness <0.025. On the other hand, both Rector (1954) and Watts (1954) found that a summer swell profile is produced by waves flatter than 0.016. These variations in the value of critical wave steepness may relate to the size of the material used in the experiments. The relationship between grain size and beach gradient has been mentioned and will be discussed in more detail in the next section. Suffice it to say that the height and position of a break-point bar is independent of the beach gradient.

Nevertheless, for a particular size of material, the gradient determines the critical wave steepness at which the bar will form. The steepness must exceed 0.034 for a gradient of 1:5 in order to form a break-point bar, for example. Iwagaki and Noda (1962) have demonstrated the dependence of critical wave steepness on the wave height and grain size in determining changes in beach profiles. More recently, Dean (1973) has devised heuristic models of sand transport in the surf zone which consider the trajectory of a suspended sand particle during its fall to the bottom while it is acted upon by the horizontal water particle velocity of the wave. Depending upon the time the particle takes to fall relative to the wave period, sand grains will move onshore (high fall velocity) or offshore (low fall velocity).

Cycles of beach erosion and deposition associated with the passage of cyclonic storms and the effects of hurricanes on shoreline recession, particularly on the Gulf Coast of the USA, are well documented (see Hayes, 1967, for a detailed account of the effects of the hurricanes 'Carla' in 1961 and 'Cindy' in 1963). Beach profiles change significantly in response not only to varying degrees of physical energy imparted by waves and currents on beaches as hurricanes and cyclones cross coasts, but also to the high energy conditions that invariably continue for several days after the passage of these storms. Rapid and extensive beach erosion is usually followed, however, by gradual accretion as swash bars – constructed by relatively flat waves in front of the break-point, and often built above the still-water level by the swash and backwash – migrate shoreward and sediment is transported onshore. Some sectors of a shoreline rarely recover fully, particularly if storms occur frequently. Along the relatively low-energy barrier island beaches of central Texas, hurricanes account for most of the long-term erosion because the large volume of material removed from the shoreline is rarely returned. After hurricane 'Carla' in 1961 sands were deposited up to 50 km offshore.

Erosion and profile changes, resulting from large waves and 3 to 4 m high storm tides generated by slow moving storms along barrier beaches, are manifest in such features as *washovers* or *washover fans* which have been described by several researchers (see references in Coates, 1973). Sometimes referred to as *overwashes* in the literature, these features can be distinguished by their sedimentary structures which often show high angle cross lamination facing landward, indicating their formation by strong washover currents towards the land.

The shear stress of the sea breeze has not received the attention it properly deserves considering the engineering problem of arresting the loss of beach sand to the backshore and dune system of some coastlines. This is probably because the methods needed to provide the shear stress coefficient under the condition of sea breezes are still at a preliminary stage, but their potential should not be disregarded.

3.6b Effects of Swash Percolation and Groundwater Flow on Beach Profiles

Surprisingly little attention has been paid to the effects of swash percolation and groundwater flow on beach profiles although the relationship between these two processes and beach dynamics was first reported by Bagnold in 1940. Since his laboratory investigations, field studies have shown that fluctuations in the level of the water table near the beach face occur over a wide range of frequencies, and such fluctuations during the tidal cycle influence the movement of material as the swash-backwash zone migrates across the beach. Also, the role played by foreshore permeability is significant in controlling wave energy dissipation, beach slope and the phase difference between swash duration and wave period.

Duncan (1964) and Strahler (1966) have investigated the readjustment of beach profiles resulting from variations in water level during tidal cycles. Strahler, working at Sandy Hook in New Jersey, noticed that deposition occurred at any point on the beach as the tide rose but this was followed, firstly, by erosion as the particular site came under intense swash and, secondly, by deposition as the breaker zone approached it. A reverse process was recorded during an ebb tide. Duncan found that during a flood tide, water from the swash is lost by infiltration into the beach and sand is moved onshore because the seaward edge of the water table slopes shoreward as a result of the water level rising faster than the water table. During the ebb tide the slope of the water table is reversed, so that water is added to the beach face and material is transported offshore by the backwash (Figure 3.11).

Some attempt to quantify changes in the water table and foreshore sand volume has been made by Harrison (1969, 1972) who used multi-regression analysis on a 30-day-long series of observations of variables in the beach-ocean-groundwater system at Virginia Beach, Virginia, USA. He found that a change in ocean still-water level is the single most important variable influencing changes in the quantity of foreshore sand over intervals of half-tidal cycle to tidal cycle length. Harrison's method is reported fully in the literature but the reader should be aware of the fact that it has a few serious limitations because it is difficult to quantify the relationship between ocean still-water level and the water table outcrop, owing to the problems related to well maintenance in the swash-backwash zone during storms.

3.6c Beach Gradients

The relationship between beach gradients, the length and steepness of waves, and the size of beach material has been studied extensively, both in the laboratory and the field, since Bagnold published his benchmark paper on beach formation by waves in 1940. Bagnold's main series of model experiments on beach profiles was conducted in a wave tank with rounded beach pebbles of diameters ranging from 0.5 to 0.9 cm and other material of diameters ranging from 0.3 cm to 0.05 cm, in order to determine how far his results were modified by changes in particle size. The beach angle was found to be entirely independent of both the wave amplitude and water depth over the whole of the experimental range, and

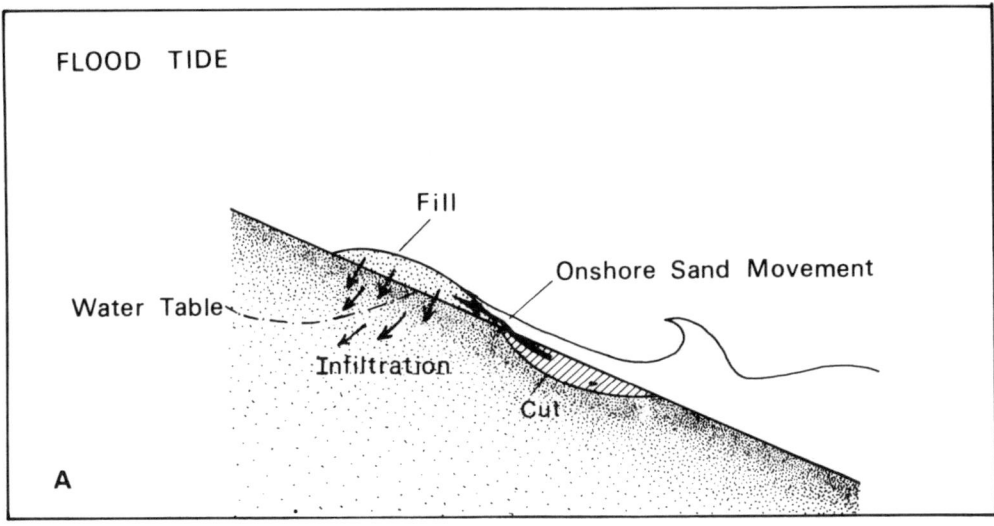

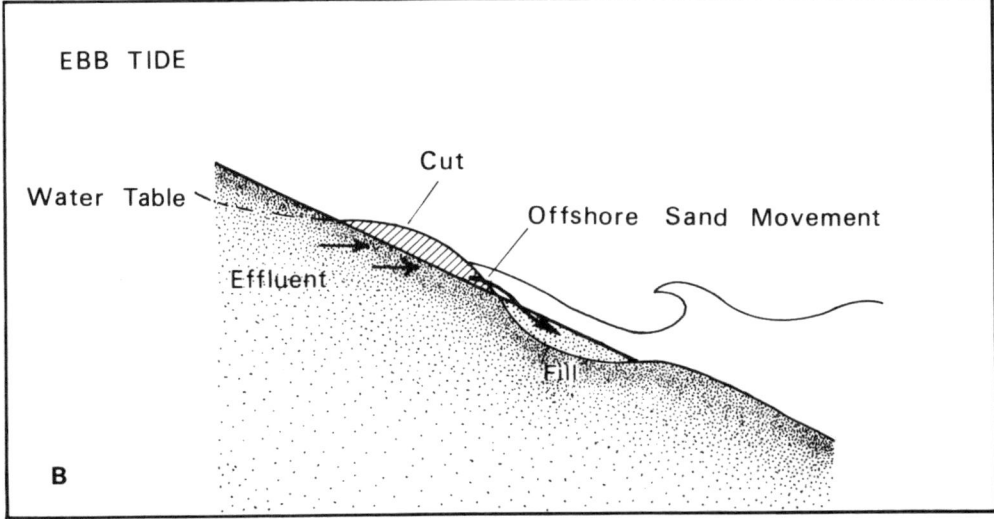

FIG. 3.11. EFFECTS OF WATER TABLE ON EROSION AND ACCRETION OF BEACH PROFILES DURING FLOOD AND EBB TIDES (After Duncan, 1964).

appeared to depend only upon the absolute size of the grains composing the beach. For example, the gradient was 22° for a mean grain diameter of 0.7 cm; 19.5° for 0.3 cm and 14° for 0.05 cm.

The relationship between beach-face (foreshore) slope and sand size at the mid-tide level on exposed beaches has been examined by Bascom (1951) and Wiegel (1964) whose work has shown a noticeable difference in the level of wave energy between the Pacific (average high) and Atlantic (average low in the case of New Jersey, North Carolina and Florida) coast beaches of the USA. The part of the foreshore subject to wave action at mid-tide level has been termed a *reference point* by Bascom. He sampled four beach profiles along the beach at Half Moon Bay, California, which is protected from prevailing northwesterly swell by Pillar Point. Behind this headland in the sheltered part of the bay the beach is relatively flat and composed of fine sand, whereas toward the south the material is coarser and the gradient correspondingly higher, thus illustrating how beaches adjust to the wave climate.

The shape and grain size of beach material are related to sediment sorting which indirectly influences the rate of percolation on the beach face. Krumbein and Graybill (1965) discovered that well-sorted coarse sand beaches have steeper slopes than poorly-sorted ones. On composite beaches, with a mixture of different grain sizes or modes (see p. 140) the influence of the degree of sediment sorting on the beach face slope is particularly marked. Usually when modes are mixed, there is poorer sorting and percolation and beach slopes are reduced. In contrast, shingle beaches are very permeable and much of the swash infiltrates the beach.

The effects of wavelength and wave steepness on the gradient of the swash slope have been studied both in the laboratory and in the field (see references in King, 1972). Model experiments show that the slope at a particular point on the beach face is related to wave steepness and any one size of material. Equations have been derived to show that as the wave steepness increases the beach gradient decreases. For example, the swash slope of Chesil Beach has a gradient of 1:4 or even steeper under calm weather conditions, but this can be reduced to 1:9 during erosion by steep storm waves.

As mentioned in the section on edge waves, experiments have been conducted to measure the velocity fields on beaches of widely different slopes, under both plunging and spilling breakers, by using a two-component electromagnetic flowmeter mounted 15 cm above the seabed and orientated so as to measure the horizontal velocity field in turbulent nearshore water. The influence of beach slope on nearshore hydrodynamics can be determined by this instrument and eventually it should be possible to monitor the interaction between water motion and beach profiles.

The coastal geomorphologist is now able to benefit from aerial remote sensing and satellite photographs for monitoring beach profiles. The Earth Satellite Corporation in the USA has already employed these techniques by co-ordinating aerial and orbital imagery to calculate rates of beach change along parts of the North Carolina and New Jersey coasts. The reader is referred to Verstappen's recent publication on remote sensing for a detailed account of the merits of this technique.

3.6d Ridge and Runnel Profiles

Often, the smooth profile of some sand beaches is made irregular by a system of ridges and runnels trending parallel to the shoreline. These features are smaller than offshore bars and generally develop on sandy beaches with a large tidal range, low wave energy and limited fetch. Wide low-tide terraces with a very gentle offshore gradient are characteristic of such beaches. The ridges, which are exposed on the beach at low water, are the natural equivalent of the swash bar and are therefore formed by constructive waves. Ridge and runnel beaches are fairly common along the shores of the Irish Sea, North Sea (in Druridge Bay in Northumberland, along parts of the coast of Lincolnshire and East Anglia, and between north Holland and Cherbourg), and the Firth of Clyde.

On the Lancashire coast of England, for example, ridges are constructed during calm weather accompanied by moderate swell and are reduced in height during storms. As King (1972) states in summarizing the earlier work of Gresswell, the ridges represent an attempt by the waves to build their equilibrium gradient on a beach that is flatter than the equilibrium gradient. It follows then, that on beaches sheltered from ocean fetch wavelengths will be shorter and the gradient steeper, while a relatively larger sediment supply results in an almost flat overall gradient and the possible development of ridges and runnels. During an ebb tide channels are cut across the ridges at their lowest point by water escaping seaward from the runnels. These gaps allow the water of the next flood tide to enter the intervening runnels while the waves break on the seaward face of the ridge.

Intertidal sand transport on a multibarred (ridge and runnel) foreshore at Formby Point, southwest Lancashire, has been studied by Parker (1975) in an attempt to relate sand mobility to coastal erosion (Figure 3.12). Even if the interrelationships between energy

FIG. 3.12. VIEW SOUTH OVER FORMBY POINT, TOWARDS TAYLOR'S BANK (ARROWED) FROM FISHERMAN'S PATH
Ridges, runnels and rip channels, wet areas on the upper foreshore (W), and transition from cliffed facet (1) to uncliffed though eroding coastline (2) and dune systems. D – Dole Slack Gutter, V – Victoria Road and Northwest Mark, L – Lifeboat Road. (Oblique from 140 m: courtesy John Mills Ltd).
(Reproduced by kind permission of John Wiley & Sons)

and transport are reasonably well established, as already mentioned, it is difficult to prove mobility and continuity of transport from supposed source areas to particular sites, especially along foreshores with complex patterns of sand transport. Parker reports that on the foreshore at Formby Point the ridges dry rapidly and lend themselves to aeolian sand supply. However, the runnel between the farthermost inshore ridge and the upper foreshore remains perpetually wet and completely blocks landward aeolian sand movement from the foreshore ridge and runnel zone. Also, during both the flood and ebb tides the most intensive sand transport occurs when swash water sweeps sand landward, over the crest of a ridge, into the next runnel. When this does not happen, sediment transport is either alongshore or seawards into the runnel seaward of the ridge.

The main conclusion to be drawn from Parker's study is that the stability of an alluvial coast relies on the dynamic equilibrium between the supply of sediment to the multibarred foreshore and its subsequent removal therefrom. The presence of ridges and runnels, in this context, has a dominant influence on mobility and, combined with the transport patterns in the intertidal zone, on the stability of the coast.

The beach at Blackpool, which is composed of fine sand and has a tidal range of 7.64 m at spring tide, is another good example of a ridged tidal beach. At spring low-water mark the width of the beach is 1220 m with an overall foreshore gradient of 1:150. However, the slope of seaward faces of the ridges is much steeper with a range from 1:32 near the upper foreshore to about 1:60 near low-water mark. Also, there is a noticeable increase in the depth of the runnels seaward because the ridges tend to increase in size towards low-tide level. Field observations have shown that the ridges do not move systematically towards the shoreline because they trend parallel to the coast, which is aligned normal to the direction from which the dominant ridge-building waves travel across the Irish Sea.

Depending upon the direction of approach of the constructive ridge-building waves, ridges may occur at an angle to the shoreline because they tend to orientate themselves to face the direction from which waves travel. This is the case on the coast of England in south Lincolnshire, and between Liverpool and Southport in Lancashire.

It will be seen in the next section that it is often difficult to distinguish between ridge and runnel systems, longshore bars and troughs, and bars generated by tidal currents. Therefore, the reader should be cautious about the use of these terms in the literature. Strictly speaking, a bar is an elongate, submerged sand body which may be exposed at low tide.

3.6e Summary Comments

A case might be argued here that it is virtually impossible to forecast accurately rates of shoreline erosion and accretion unless beach processes have been monitored for some time in order to analyze *extreme* rather than average conditions. One realizes, though, that this is often impracticable because of events such as a 50- or 100-year storm. As mentioned already, many researchers have resorted to theoretical studies of beach processes which have also involved laboratory wave-tank investigation of sediment transport and sorting by oscillatory waves, but with only a limited number of field studies to verify the results of such work. Moreover, severe limitations are imposed on *in situ* measurements by heavy swell and wave turbulence in the breaker zone, and increasing public use of beaches is restricting the type of instruments that can be installed and left safely for long-term data collection.

It is imperative to collect relevant data so that they can be used meaningfully, otherwise both monitoring and the compilation of statistics are uneconomical exercises. This consideration, in turn, necessitates some systematic and standardized method of data collection. What is more fundamental, however, in the case of beach processes, is the type of monitoring that may be required from one locality to another along a coast. In other words, what should be monitored, and for how long? Sometimes such questions cannot be answered until crude data are available. Furthermore, the two questions are even more difficult to evaluate when existing monitoring programmes in the coastal zone are often unrelated to clearly defined aims and objectives. It is the responsibility of geomorphologists and others researching coastal processes to ensure that adequately prepared programmes are implemented.

3.7 RHYTHMIC SHORELINE FEATURES

Beaches are rarely straight, or smooth in curvature. Generally, they are characterized by a variety of rhythmic shoreline features, the most common of which are cusps and crescentic bars (Figure 3.13). There is still disagreement about the origin of cusps and, so far, no satisfactory theory can account for the predominant and regular spacing of a given series of cusps and how such spacing relates to the wave parameters and the composition of beach material. Cusps have been observed in all types of beach material from fine sand to gravel, but usually they are best developed in a mixture of sand and gravel, or shingle. The notable difference in grain size between cusp ridges (horns) and the intervening embayments is related to differences in permeability.

At one time the development of beach cusps was thought to be associated with tides, but this theory cannot be entirely substantiated because these features occur along lake shores, in tideless seas and in laboratory wave basins. However, the morphology of beach cusps can be influenced by the presence or absence of tides, as exemplified by those cusps in areas with a marked tidal range where they form a series of simple ridges down the beach face.

Field observations support the contention that cusp formation is most likely when wave crests parallel the shoreline. Measurements obtained by Longuet-Higgins and Parkin on Chesil Beach and at Seaford, for example, showed that beach cusp spacing is related to wave height, swash length and wave period; in fact, their observations agreed with those made in 1910 by Johnson (1919), and later by Kuenen (1948) and Otvos (1964), that a doubling of wave height approximately doubles the spacing of the beach cusps. Longuet-Higgins and Parkin also showed from their results that a better correlation exists between cusp spacing and the *swash distance*, the width of the beach between the break point and the highest area reached by the swash.

Johnson also proposed that a regular succession of swash flows on a smooth beach will start to erode any slight depression and the eroded material will be removed by the backwash, to build deltas opposite the newly-formed hollows. Gradually, if erosion continues, the hollows are enlarged into embayments. Bagnold (1940) was probably the first person to describe wave swash motions around cusps and within embayments (Figure 3.14a). Observations made by the writer on Pearl Beach, in New South Wales, which is essentially composed of coarse sand and granules, are also shown diagrammatically in Figure 3.14b, as well as being illustrated in Figure 3.14c. The backwash within the embayments moves sediment offshore as described by Johnson. However, as demonstrated by Longuet-Higgins and Parkin with the aid of dyed pebbles, an equilibrium is eventually reached in which no additional sediment is deposited on the cusp ridges or horns by swash action, and in which the sediment is in a continuous cyclical movement.

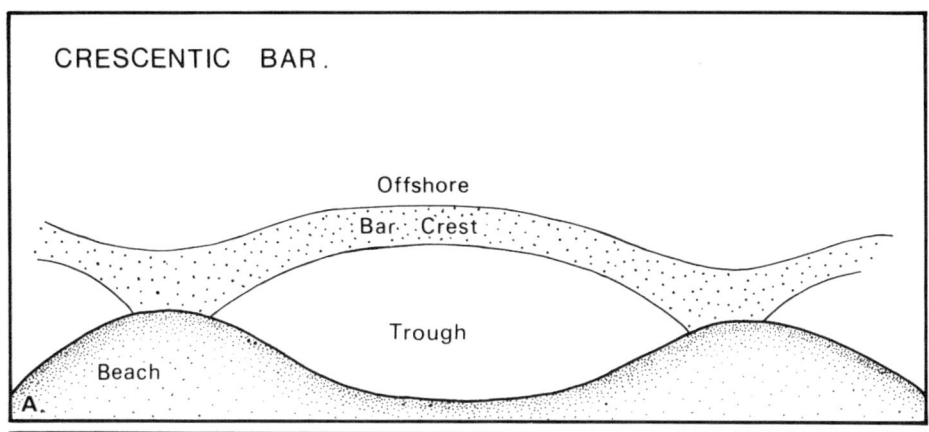

CRESCENTIC BAR.

Offshore

Bar Crest

Trough

Beach

A.

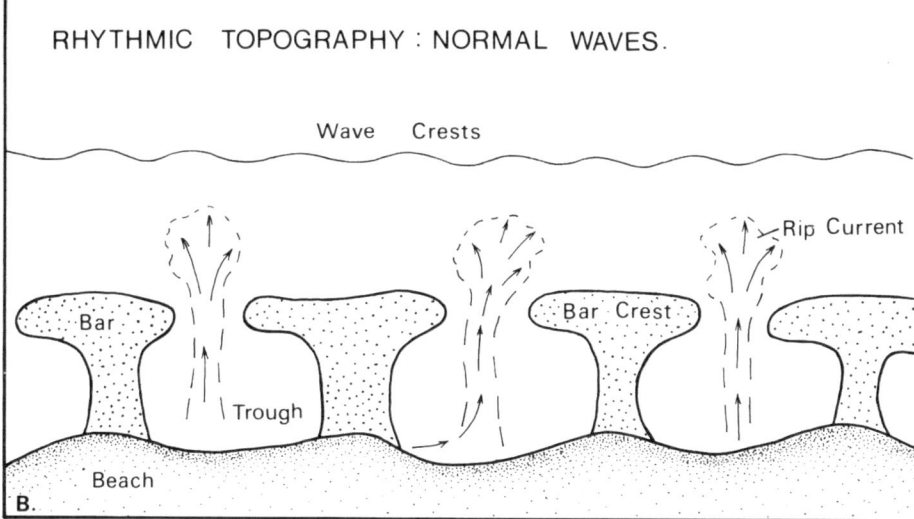

RHYTHMIC TOPOGRAPHY : NORMAL WAVES.

Wave Crests

Rip Current

Bar

Bar Crest

Trough

Beach

B.

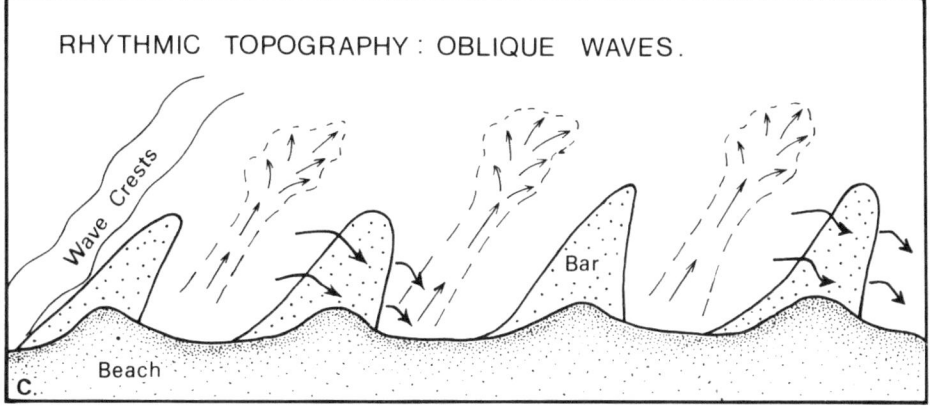

RHYTHMIC TOPOGRAPHY : OBLIQUE WAVES.

Wave Crests

Bar

Beach

C.

FIG. 3.13. RHYTHMIC SHORELINE FEATURES
(A) Crescentic bar and associated cusps along shoreline (after Hom-ma and Sonu, 1963), (B) Rhythmic topography produced by normal waves, (C) Oblique wave approach realigning bars. (Based on Sonu (1973), and Komar (1976).

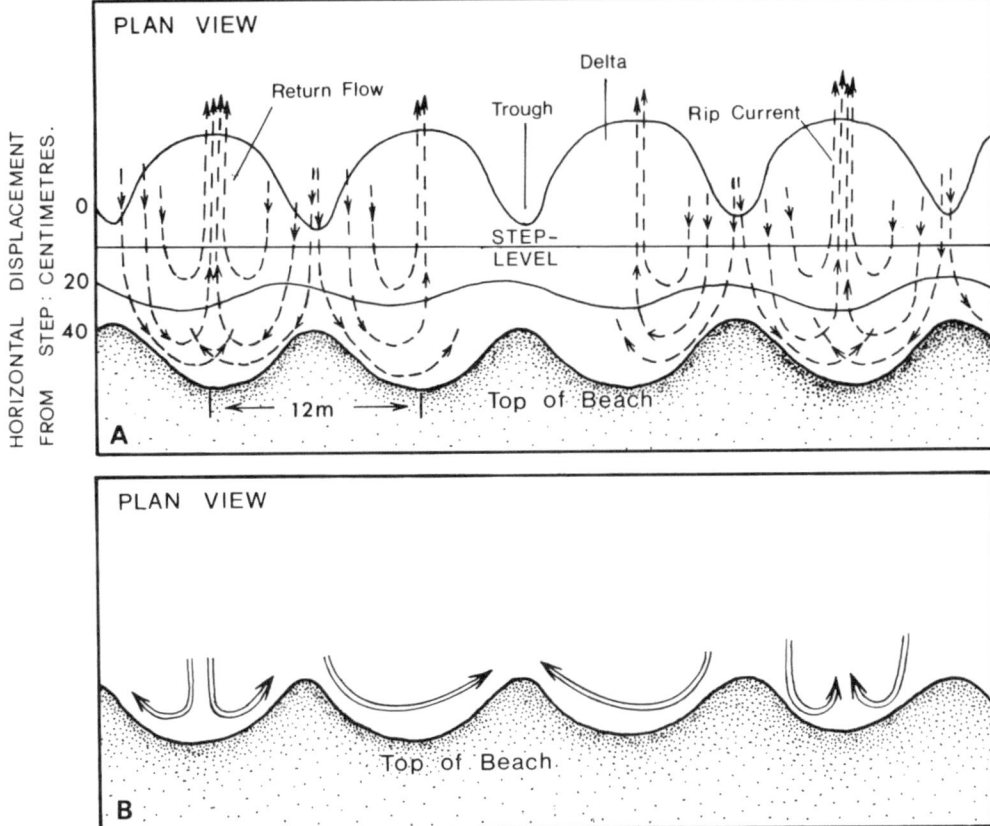

FIG. 3.14. (A) SWASH PATHS AROUND CUSPS AND WITHIN EMBAYMENTS ON SAND BEACH NEAR MERSA MATRUH, EGYPTIAN COAST WEST OF ALEXANDRIA (AFTER BAGNOLD, 1940). (B) SWASH AND BACKWASH PATHS OBSERVED ON STEEP, COARSE-GRAINED BEACH, PEARL BEACH IN BROKEN BAY, NEW SOUTH WALES, AUSTRALIA. (C) CUSPS AT PEARL BEACH, BROKEN BAY (Photograph: J. R. Hails).

Whether beach cusps form during the time the beach face is prograding or eroding is not known.

By experimenting in a wave basin, Escher (1937) attributed the formation of beach cusps to standing waves in the surf zone that are at right angles to the incoming waves and have the same period as the incoming waves. As mentioned earlier in this chapter, recent work by Bowen and Inman, and Huntley and Bowen, suggests that standing edge waves may be responsible for the formation of beach cusps. Huntley and Bowen have demonstrated that on a steep beach, at least, short period edge waves may be responsible for relatively small-scale periodic features like beach cusps while on shallow beaches large-scale features, extending farther offshore, are likely to form.

The dynamics of water movement on the beach face respond rapidly according to whether or not the incident wave is strongly reflected or breaks and propagates onshore as a dissipative bore. An important distinction has been made recently between *reflective* and *dissipative* systems by Guza and Inman (1975). In their view, beach cusps may be genetically linked with these two systems. According to Guza and Inman, the spacing of some cusps formed under reflective wave conditions, both in the laboratory and in certain selected natural situations, are shown to be consistent with models hypothesizing formation by either subharmonic edge waves (with a period twice that of the incident wave) of zero mode number or synchronous (period equal to that of incident waves) edge waves of low mode. Although there is some argument against the importance of edge waves, they are the most probable cause of the rhythmic spacing of beach cusps. Even so, it is obvious from this brief account of cusps that more research is needed into their possible origin(s).

The problem of demarcation between the offshore and swash-surf zones is exemplified by the origin of bars. Despite attempts by Shepard (1952), Zenkovich (1967) and King (1972) to classify intertidal and subtidal bottom features as offshore bars, longshore bars and sand bars, there is still no standard nomenclature on this particular subject. A large number of theories have been advanced to explain the origin of offshore bars whether they are straight, crescentic or transverse in form. The origin of crescentic (lunate) bars is still somewhat obscure, and more information is required about the interrelationships between their growth, sediment movement and wave and current action. Such features may be described, together with beach and giant cusps, as types of rhythmic morphology. The term is also often applied to a pattern of longshore bars separated by rip current troughs.

Although crescentic bars have been reported off long, straight beaches and embayed coasts, they are restricted to areas of small tidal range and are characterized by longshore wavelengths that can vary between 20 m and 1000 m, but generally of the order of 200 m to 300 m. Perhaps it is pertinent to stress here that most rhythmic shoreline features are submerged, apart from associated beach cusps. Crescentic bars are known to migrate slowly in the longshore direction, as indicated by the shifting positions of beach cusps. A migration rate of 15 to 32 m within one day has been observed by E. M. Egorov in the Black Sea, while P. Bruun has reported an average annual displacement of 1000 m on the Danish North Sea coast.

Following the theoretical work of Eckart (1951), A. J. Bowen and D. L. Inman have associated standing edge waves with the formation of crescentic bars in regions of small tidal range, the bars having a longshore wavelength of one-half that of edge waves. It is also claimed by them that, in the absence of large incoming surface waves, the edge waves may also form cuspate features on the foreshore, with the points of the cusps directly opposite the horns of the crescentic bars. Evidence to support this contention has been observed in the field by Hom-ma and Sonu. Bowen and Inman have found that the wave periods of edge waves associated with crescentic bars are relatively long, of the order of 30 sec, yet the drift velocities associated with these waves tend to explain the observed formation of large sedimentary features. One of the main conclusions of their work is that since crescentic bars are expected to be located at a fixed distance from the shoreline, they are unlikely to exist in regions of large tidal range where the location of the shoreline changes substantially during the tidal cycle.

3.8 TEXTURAL PARAMETERS OF SHORELINE SEDIMENTS

It has already been mentioned that the composition and textural properties, such as particle size, shape and density, of beaches vary markedly from locality to locality. Although sand is the dominant constituent of beaches, other material varying from clay to boulders is also common. The mean grain size of beach sediments is related mainly to three variables, namely the sediment source, the level of wave energy and the offshore gradient on which the beach has been built. Before the grain size distribution and sorting normal to the shoreline across the beach face, and in the longshore direction, are examined it is pertinent to mention some facts about sampling and the statistical methods used in sediment analysis. Sampling in the field depends upon a number of factors which can range from the size of a study area, the duration of a research programme, and the frequency of observations. In coastal studies, though, random stratified sampling is a widely used method for examining beach material.

The mechanical analysis of sand particles and gravel by sieving is a conventional technique even if sedimentologists and geomorphologists argue that the behaviour of sedimentary particles during transportation and deposition is more closely represented by their settling velocities. Many researchers still claim that the accuracy of the settling tube method is restrictive, although it is a quicker way of examining particles. Krumbein (1936) was the first person to introduce the concept that the size frequency distribution of sand samples tends to be log-normally distributed, but few workers have attempted to study the relationship between depositional processes and grain-size distributions. Folk (1966), by referring to more than 150 papers, has reviewed many of the graphical and mathematical techniques that have been proposed for the statistical summary of grain-size data.

Yet a cursory glance at the recent literature on grain-size statistics and environmental interpretation shows unequivocally that there is still disagreement about which mathematical technique to use in routine sediment analysis. Although several publications refer to the defects of both moment and graphical methods very few geomorphologists have evaluated in any detail the lack of standard sampling procedures in the field, the different analytical techniques used in the laboratory, the problem of random operator error and, to a lesser degree, operator bias. As this writer has mentioned previously, the relationship that exists between the accuracy of results and the sieve interval used merits more attention than it has received so far (Hails *et al.* 1973).

Cumulative curves, which should always be drawn on percentage probability graph paper, are used in most statistical work although grain-size parameters can be derived directly from data by computer or by hand without drawing such a curve. Many graphical measures of average grain-size and several descriptive scales for sorting have been proposed hitherto. These have been reviewed by Folk who has also evaluated the significance of skewness (or asymmetry) and kurtosis (or peakedness), two parameters that are widely used to measure the non-normality of a distribution.

Undoubtedly, the method of moments is the best way of obtaining parameters of a frequency distribution because the entire frequency distribution is used in the determination instead of a few selected percentiles. Also, because the method includes the entire distribution, it is necessary to make some arbitrary assumption about the grain size of the 'fines' before a computation is made, if the fine-grained fraction (silt and clay) is not analysed. Researchers who have used statistical parameters, though, are readily aware that such estimators are limited by truncation, bias and the grouping process involved in sieve analysis (see, for example, references in Folk, 1966; Jones, 1970). The computed values of the higher moments, skewness and kurtosis, reflect the composition of the tails of a distribution and for this reason all sample data should ideally be complete and not open-ended. Higher moments are very sensitive to small changes and/or inaccuracies in the grain-size distribution data and, therefore, the values of these moments must be treated with caution unless some numerical limit can be placed on their accuracy by virtue of a careful evaluation of both laboratory and sampling techniques. The writer has discovered that both diagenetic and pedogenetic processes may cause aberrations in the third moment measure, skewness (Hails and Hoyt, 1969).

Bearing these limitations in mind, let us now turn our attention to the grain-size distribution across the beach face and in the nearshore zone. Most of the quartz beach sands on the east coast of Australia are *polygenetic* because they have been derived from various sources and their origin is complex in relation to both time and place (Hails 1969). They are very well sorted, showing that the range of grain sizes is small and are negatively skewed because fine grained sand is winnowed from the beach face and moved offshore. The resulting concentration of coarse grains is a characteristic feature of terrigenous beach sands as shown by truncated grain-size distribution curves (see references in Hails, 1972). On the other hand, the presence of shell fragments, granules, or small pebbles within beach sands can result in negative skewness, as demonstrated by Mason and Folk (1958) with samples from Mustang Island, Texas.

Skewness is probably the most significant moment measure of a grain-size distribution to distinguish between dune, beach and river sands and, consequently, it has been used in many studies (see references in Friedman, 1961, 1967; Hails, 1967). Friedman has used two-dimensional plots of standard deviation (sorting) versus skewness in an attempt to differentiate between beach and sands from other coastal environments. Generally, beach sands are not so well sorted as dune sands but better sorted than fluviatile sediments. Greenwood (1969) has applied multivariate statistics in his analysis of sands from Barnstaple Bay, Devon, in order to distinguish dune, and beach foreshore/backshore sands.

3.8a Changes Normal to the Shoreline

Changes in grain-size distribution and degree of sorting normal to the shoreline across the beach face have been reported in several publications, but the processes responsible for them are not well documented or understood. The bottom topography and turbulence associated with wave energy dissipation control, to some measure, the variation in sediment size between the plunge point of the breaking waves and the surf-swash zones. Attempts to correlate swash and backwash energies with sediment sorting have been made by a few workers.

Dolan and Ferm (1966) have measured simultaneously wave height, swash velocity and beach changes in order to establish the relationship between breaking waves (swash-forces), beach configuration and textural properties along the Outer Banks of North Carolina. Their results showed that measurements of swash velocity, based on a system which records the uprush deceleration relative to an array of ground control markers, give a closer relationship to variations in beach slope than conventional inshore wave measurements. Dolan and Ferm also established that correlations between swash velocity and beach slope are better when waves are smaller with longer periods, and that there appears to be a close relationship between swash velocity, beach configuration and particle-size distribution over the beach face. These tentative conclusions need to be assessed more fully because Dolan and Ferm, in a broader context, did not refer to the wave spectra and wave dissipation.

Experiments with sand tracers of different sizes and colours have indicated that there is a general tendency for coarser grains to move seaward from the surf zone towards the breakers, whilst finer grains are mainly confined to the surf zone. It appears that such onshore-offshore sorting and grain-size distributions are associated with zones in which beach material is in equilibrium. A detailed analysis of tracing the onshore-offshore movement of sand in the surf zone is given by Ingle (1966).

3.8b Changes Parallel to the Shoreline

Variations in grain size in the longshore direction or parallel to the shoreline have already been mentioned in this chapter, particularly with respect to Halfmoon Bay in California and to shingle beaches. The relationship between material size and other variables has been investigated extensively by Carr on Chesil Beach, which is virtually a closed system with no net littoral transport and aligned to face dominant storms and high energy waves from the southwest. This shingle beach extends more than 18 km from Chesilton in the east, where it terminates against the cliffs of the so-called Isle of Portland, to an essentially arbitrary limit in the west. Opposite a shallow lagoon known as the Fleet, Chesil Beach is between 150 and 200 m wide, but is narrower at both ends. The shingle consists predominantly (98.5 per cent) of flint and chert, the remaining material (1.5 per cent) being almost entirely quartzite pebbles. Pebble size above low water mark, like the height of the beach crest, increases towards the southeast. Interestingly, though, the pattern of pebble-size variation has remained exceptionally constant for more than fifty years, according to the measurements of Vaughan Cornish in 1897 and of A. P. Carr in 1965.

This close agreement suggests that beach alignment is unaffected by infilling with pebbles of different sizes, otherwise variation in pebble size could be expected if the beach alignment were changing. The nature of the alignment prevents any marked longshore movement which, together with the relict nature of its material, accounts for the grading of Chesil Beach. Thus, it appears in this instance that good sorting parallel to the shoreline occurs when longshore movement is restricted, and the material only moves to that part of the beach where it is in equilibrium with the prevailing nearshore wave conditions, especially wave energy. Of course, this process will also apply to sorting normal to the shore. It can therefore be concluded that the longshore distribution of energy in the swash-surf zone controls the sorting of material parallel to the shoreline.

Longshore variations in grain size may reflect selective transport, whereby fine material is transported farther alongshore than coarse sediments which remain near the original source area, and the progressive movement of certain grain sizes either to the offshore zone or to the backshore of the beach. Such selective removal by waves and wind respectively leaves a lag deposit of coarse grained material. Considerable attention has also been paid to the selective transport of sediment according to shape or degree of particle roundness since the pioneer studies of Wadell, Zingg and others (see references in standard texts cited previously in this chapter). This writer has reservations about the claim made by Shepard and Young (1961) that beach and dune sands can be distinguished on the basis of their roundness, particularly if the sand has a polygenetic origin. Obviously, the wind is able to move rounded grains more easily because angular ones interlock with each other. However, very much depends upon the range of roundness which in the case of the New South Wales samples, for example, was extremely small. Pebbles have been studied widely because roundness and sphericity measurements are relatively easy to undertake compared with sand grains. Three publications, among many, that the reader should consult about beach gravels are those written by Landon (1930), Bluck (1967) and Dobkins and Folk (1970).

3.9 SHORELINE CONFIGURATION: BEACHES IN PLAN

It has been shown that the plan form of the beach is, in part, dependent upon the nature of the beach material. Plan form may be described as the shape of the beach contour between the breaker and swash zones, as viewed from above. A number of theories on the processes which determine the orientation of beaches have been reviewed by King (1972). However, a few pertinent points are summarized here.

Lewis in 1931 and 1938 elaborated upon the work of de la Beche conducted a century earlier by considering that marine depositional features, such as barrier beaches, tend to align themselves normal to the approach of the dominant waves. These, Lewis assumed, are the largest storm waves so that if they arrive at an angle to the shoreline they will induce longshore movement in the opposite direction. Lewis pointed out that the forelands in north Cardigan Bay are aligned to face the direction of maximum fetch, over which the largest waves travel from the Atlantic Ocean. Thus material transported to a beach by prevailing waves will be incorporated into a ridge orientated normal to the approach of the dominant waves.

On the other hand, Axel Schou, working on the east coast of Jutland, advanced the theory that both the direction of maximum fetch and the resultant of winds from every direction determine the alignment of a depositional coast. The wind direction is important in the case of locally generated storm waves, which build shingle beaches, and where long period swell cannot reach beaches and therefore is less important than dominant waves formed by local winds.

J. L. Davies has demonstrated that exposed beaches in Tasmania orientate themselves parallel to the refracted crests of long period (14 seconds) swell waves and develop the same curvature. In this way, beaches fit the waves but are not necessarily straight. The offshore topography, however, is an important variable on which the alignment of the beach depends because it controls the degree to which the swell is refracted. The hypothesis of beach orientation accounts for curved beaches which are far more common than straight ones on sandy coasts. Yet many different wave trains reach a beach, so one may ask what happens at the shoreline as waves arrive from slightly different directions. Is a single prevailing wave train responsible for the curvature of the beach? Certainly, as the curvature of the beach is controlled by wave refraction, which in turn responds to the existing offshore morphology, the beach shape cannot readily assume some prescribed geometric form.

Long straight beaches on the seaward side of offshore barriers occur in the Gulf of Mexico where they extend for 1000 km or so along the coasts of Texas and Louisiana, and for shorter distances east of the Mississippi delta in Alabama and Florida (Russell, 1958). Such features are also characteristic of the barrier islands along the Atlantic coast of the USA. Those near Cape Hatteras are more than 60 km from the mainland. Cape Hatteras, Romain, Fear and Lookout are cuspate features which, collectively, are known as the Carolina Capes. These form part of a system which extends along the Georgia coast to Cape Canaveral in Florida. There are several interesting theories on the origin of the Carolina Capes, none of which satisfactorily explains their configuration. An early concept was that eddy currents, or a series of secondary rotational cells, from the Gulf Stream produced the capes but this was disproved in 1955 after Bumpus investigated the water circulation on the continental shelf in the Cape Hatteras region and established that such eddies were absent. According to Shepard (1952) the Carolina Capes are examples of cuspate forelands the shape of which is controlled by offshore shoals or islands, but several other ideas have been advanced. The processes leading to cape development along other coasts have been reviewed by Steers, Zenkovich, King and Shepard.

Computer simulation models are more versatile and practical to use in shoreline configuration studies because time variations can be introduced and wave refraction and diffraction effects can be considered. Thus, for any sector of a coast, equations for sand transport along a beach, changing wave conditions, and variations in sediment supply to the shoreline through time, can be analyzed.

A two-dimensional mathematical model has been developed by Willis and Price (in Hails and Carr, 1975) which is capable of predicting changes in the plan shape of a beach following the construction of sea defences or an alteration in the wave climate. This model only uses bulk flow sediment transport equations and it must be borne in mind that the laws of sediment transport with respect to waves and currents must be established before a three-dimensional model can be constructed for the entire nearshore seabed. Until the movement of sand can be determined at all points on the beach it will be impossible to build the effects of tides into models. Therefore, research is needed into the transport of sediment in the onshore-offshore direction and the response of beach profiles to wave action. In principle the existing model calculates breaking-wave conditions resulting from refraction over the inshore seabed; the rates of alongshore sand transport on the beach from the breaking-wave conditions; amounts of accretion and erosion from the rates of transport; accretion and erosion distributed over the inshore seabed to a given depth, and recalculates wave refraction.

King and McCullagh (1971) developed a computer simulation (probabilistic) model in order to predict the growth of Hurst Castle Spit in southern England. By selecting a series of random numbers, which determine the order of events in the growth of the spit, it is possible to determine the effects of different waves arriving at the shoreline from different directions. Thus, the proportion of random numbers allocated to the different wave directions can be adjusted until the simulated spit pattern agrees with the natural morphology of the spit. In this way, the most important variables controlling the growth and shape of the spit (whether they are changes of wave pattern, fluctuations in water depth and sediment supply, or the offshore relief) can be assessed.

Despite the advantages of simulation and other models, such as the speed of obtaining results and the different combinations of variables that can be tested, the researcher should answer two basic questions, with corroborative field evidence if possible, before using the computer. What man-made and natural changes have occurred along the shoreline, and over what period of time?

3.10 PLANNING AND COAST PROTECTION

The increasing use of the coastal zone as a recreational resource and as a location for certain types of industrial development and urban expansion has underlined the need for planning coastal land use and for the establishment of techniques to protect the coast from accelerated erosion.

3.10a Beach Nourishment

Most beaches afford some protection for the backshore or contiguous frontal dune system along a coast because they dissipate wave energy and, therefore, they serve the same purpose as man-made structures. However, the Coastal Engineering Research Centre (CERC) in the USA and the Hydraulics Research Station, Wallingford, are now attempting to solve coastal erosion problems by beach nourishment rather than by conventional methods. The artificial supply of sand to restore a deficiency caused by erosion is about the only coast protection measure that does not adversely affect adjacent areas. Nevertheless, more research into littoral processes and a better understanding of the wave climate are required before the behaviour and residence time of beach fills can be predicted accurately. Also, because of the remarks that have been made about variations in beach profiles and the grain size and sorting of beach material, coastal engineers require a thorough understanding of the hydrodynamics of the nearshore zone before beach nourishment schemes are implemented.

It is for these reasons that there is still debate about whether or not this method is the complete answer, or only a short-term solution, to coastal erosion problems. At present, it is not always possible to predict how frequently beaches need to be renourished. Probably the most important factor, though, is to evaluate the respective transport rates for emplaced *borrow* and indigenous beach material. Ideally, the grain size of the restored sand should approximate that of the natural beach sand. Depending upon its source, the material used is either smaller or larger in diameter. Following artificial replenishment, it is possible to evaluate the magnitude and location of changes in the beach profile. This has been done at Atlantic City, New Jersey, for example, where beach replenishment undertaken in 1963 remained effective for six years. Sometimes, though, if a relatively short length of coast is nourished, the material can be redistributed along the shoreline.

Some examples of successful beach nourishment programmes have been mentioned by King (1972), US Army Corps of Engineers (1973), and Hails (1977). It has been estimated that 80 000 km of the United States coast are vulnerable to erosion, particularly the shorelines of southern California, Florida, New Jersey and Texas. In areas such as these, beach nourishment is both practical and economic. Therefore, the study of coastal processes should be an integral part of the design of protective beaches. The role of the geomorphologist is to provide data on the direction and volume of littoral transport; the temporal and spatial distribution of indigenous material within a given area; changes in beach berm elevation and width and variations in foreshore slope.

3.10b Coast Protection Schemes

This writer has recently reviewed, with reference to selected case histories, some of the limitations of coast protection schemes using man-made structures such as groynes, jetties, breakwaters and harbour works (Hails, 1977). Groynes, which are invariably constructed at right angles to the shoreline, are probably the most practical means of intercepting and accumulating littoral material between the seaward limit of breaking waves and the limit of swash, depending upon their height, length and permeability. But groynes are not necessarily an ideal solution to coastal erosion problems, whether they are built individually or in a series backed by a seawall, because they arrest the movement of a certain amount of sand that would otherwise travel along the shoreline. Despite this fact, very few beach restoration programmes have combined beach nourishment with the construction of groynes. Theoretically, of course, sediment will reach the downcoast side of a groyne after the water-line has extended to the toe of the groyne. It can be easily deduced that these structures merely transfer the erosion problem from one part of the coast to another and do not necessarily prevent erosion, particularly during storms. In fact, there are many examples to illustrate that groynes cause considerable damage and expense over the long term if they are built along a coastline without a thorough understanding of the local hydrodynamic conditions and the offshore sediment budget.

Inlets, either natural or dredged, also tend to interrupt littoral transport along the coast. In some areas, jetties flanking inlet channels have been built to serve as training walls in order to increase the velocity of tidal currents so that sediments are flushed from the channels. These jetties also prevent material from entering the channels and cause beach erosion. For example, along the Texas shoreline of the Gulf of Mexico sediments trapped by jetties at the mouths of rivers and inlets are no longer nourishing the beaches. It has been estimated that during the past century the Texas shoreline has lost four times as much coastal land as it has gained, to give a net loss of about 7288 hectares, because the sediment loads of all the rivers discharging directly into the Gulf have decreased markedly in recent years. This situation has been aggravated somewhat by the incidence of hurricanes. The combined effects of a reduced sediment supply and man's mismanagement have been accentuated by a relative rise in sea level caused by subsidence which, collectively, have interrupted shoreline equilibrium.

3.10c Dredging offshore banks

Attention has been focused lately on the exploitation of shingle deposits from British coastal waters because of an increasing demand for sand and gravel at a time when supplies on land in some areas of England are almost exhausted or severely restricted on account of economic and environmental factors, including the somewhat prohibitive cost of transport from source to destination.

Hydrographic surveys around the coast of England have shown that some offshore banks have remained relatively stable, except for some notable changes in their general shape, for more than a century while during the same period others have migrated slowly both parallel and normal to the shoreline. Little is known about the rate, direction and amount of sediment transport across and/or around these banks, and how such movement regulates shoreline equilibrium, if at all. Relatively few combined field and model studies have been undertaken, either to obtain predictive data on shingle movement resulting from wave action or to examine the effects of gravel extraction on the incidence of coastal erosion and seabed topography.

The saltation effects and sediment mass movement along the edges of offshore banks appear to be of importance in shallow water, and more work in this field might lead to an understanding of why offshore sediment movement is often in the direction of dominant tidal current direction when wave action is obviously effective. The coast of East Anglia is an interesting area for basic and applied research into this subject because some offshore banks are relatively close inshore, simple in geometric form, and therefore suitable for model studies.

There is a need, then, to develop a model for water movement around offshore banks which can eventually include estimates of sediment transport. The basic factors to be considered include the tidal current circulation in at least two dimensions, variable water depth and seabed topography, the effects of bottom friction and wind stress on nearshore currents, bedload and suspended sediment transport, and wave energy. The difficulties and cost of investigating and predicting coastal erosion, as well as the effects of changes in seabed morphology, may be reduced considerably if sediment transport models can be developed. One other important area of theoretical research is to establish the relationship, if any, between offshore banks and shoreline equilibrium. It has been mentioned previously in this chapter that apart from determining longshore drift more research is needed into the effect of waves on sediment transport. Thus, the reader can appreciate the complexities of developing and testing models at present.

Quantitative data have been obtained by Crickmore *et al.* (1972) on the mobility of shingle under wave action during winter in water depths of about 10–20 m off Worthing on the south coast of England, by using the radio-isotope Silver 110 m (half-life, 253 days). A short length (3 mm) of activated silver wire (20 μ Ci Silver 110 m) was sealed into the hole of each flint pebble with an epoxy resin. Four selected sites with mean water depths of 9, 12, 15 and 18 m were each seeded with 1000 tracer flint pebbles, laid by divers on the surface of the seabed, over a rectangular area of 30 m by 60 m in order to provide a reasonably well-ordered initial distribution. Subsequent movement of these pebbles was monitored by a wide-sweep detector, developed at the Hydraulics Research Station. Divers measured the depth of tracer material with a portable detector and recorded maximum radiation readings directly above individual tracer pebbles. There were severe storms during the period of the experiment, but no shingle movement was recorded at a depth of 18 m. It might therefore be inferred that it is safe to dredge sites in water depths greater than 18 m where the wave climate is similar to that at Worthing.

It is now possible to predict the effect of offshore dredging by linking wave refraction programmes with beach mathematical models. In fact this has been successfully carried out by the Hydraulics Research Station.

3.11 CONCLUDING REMARKS AND SUGGESTIONS FOR FUTURE RESEARCH

The case has been made in this chapter for better theoretical studies and laboratory experiments, despite the fact that semi-empirical theory combined with carefully designed experiment has produced the best results on the basic principles of coastal processes. In the field, there is a need to know more about the magnitude of energy dissipation as a result of bottom friction and fluid turbulence within the surf zone, the budget of water flowing in the nearshore circulation system, and the relative importance of the sediment transport mechanisms seaward of the breaker zone. It is also evident that a greater knowledge is required of the bulk density of marine sediments, the effect of permeability in various deposits, and the hydrodynamic interactions of sediment and fluid flow. Of course, the major problem is simply the collection of data for a sufficient time to gain a representative picture of the changes taking place in the coastal zone.

Also, as already discussed, the relationship between water movement and sediment transport clearly shows that the former is one of the most difficult parameters to measure accurately. Owing to waves and tides, oscillatory water movements, accompanied by correlated depth changes, are superimposed on the net movement of water particles. In addition, the water velocity at any point can vary with depth from the water surface or height above the seabed because of bottom friction, with the character of wave motion and with other associated effects. It is known, for example, that currents can move in opposite directions simultaneously at different places in a channel between sandbanks because of the horizontal variability in water movement.

Future integrated research projects, then, will provide some insight into the practical problems related to sediment movement, including those arising from the effects of dredging offshore and constructing sea defences. Obviously, specific lines of investigation will be required to determine, for example, the sorting effects of transport on sediments of mixed size, shape and density; the effects of a fluctuating sediment supply offshore; and the interactions between oscillatory and steady flow.

Research into the physics of sediment movement should have a top priority in future coastal geomorphological projects despite the fact that empirical work has already provided some insight into the processes operating on the beach and in the nearshore zone. Further progress will very much depend upon the development of instruments for measuring waves and tides near the beach. The geomorphologist will inevitably play an important role in interdisciplinary programmes if processes are to be quantified; basic research will be required to support applied projects but qualitative work, which has been so paramount in geographical departments until now, will gradually disappear. No doubt the sophisticated techniques currently being developed in other disciplines will become the standard tool of the geomorphologist in the near future!

FURTHER READING

The following books supplement the subject matter of this chapter and provide valuable additional reference material. ZENKOVICH, V. P., 1967, *Processes of Coastal Development* (Oliver and Boyd, Edinburgh), is extremely useful because it provides many Russian examples rarely mentioned in other standard works. A mathematical approach to the action of waves on beaches is: MEYER, R. E., 1972, *Waves on Beaches and Resulting Sediment Transport* (Academic Press, New York), but this text should be consulted by readers with a limited mathematical background only after they are fully conversant with the basic equations cited in the other reference sources listed here. KING, C. A. M., 1972, *Beaches and Coasts* (Arnold, London), second edition, is a comprehensive reference on world-wide coastal studies and includes detailed discussions on beach processes.

For those interested in the hydrodynamics of the nearshore zone the following book evaluates problems pertaining to sediment and water mass movement, as well as problems allied to coastal engineering: HAILS, J. R. and CARR, A. P., (eds.), 1975, *Nearshore Sediment Dynamics and Sedimentation* (Wiley, New York). A complementary text dealing more specifically with North American examples is: DAVIS, R. A. and ETHINGTON, R. L., (eds.), 1976, *Beach and Nearshore Sedimentation*, Society of Economic Palaeontologists and Mineralogists, Spec. Pub. 24.

A detailed study of the physical processes of beaches and associated sedimentary deposits, integrating material from engineering, geology, geography and oceanography is: KOMAR, P. D., 1976, *Beach Processes and Sedimentation* (Prentice-Hall, Englewood Cliffs).

All students of coastal geomorphology should refer to the *Shore Protection Manual*, 1973 (CERC), which is a useful source of tables and graphs.

4
Aeolian Processes

Coastal dunes have been mentioned only briefly so far because the processes leading to dune formation, and destruction, are discussed in more detail in this chapter. In deserts, the products of exfoliation resulting from insolation, chemical weathering and the abrasive action of wind-blown sand are transported and redeposited by wind and, less frequently, by water. Particular attention will be focused here on the action of wind in moving sand grains. The entrainment and transportation of loose particles by wind also occurs on sandy coasts exposed to prevailing onshore winds and over outwash plains adjacent to glaciers and ice-caps in high latitudes where katabatic and cyclonic winds predominate. Only brief reference will be made to very fine-grained sediments, such as silt and clay particles. Thus, for the purpose of this chapter, the term *aeolian* pertains to wind-transported deposits and to the erosive action of wind.

Aeolian bedforms, which are similar to those formed under water, display a regular repeated pattern of size, shape and spacing, and have a hierarchical arrangement. For example, ripples invariably occur on dunes, and dunes, in turn, are commonly found on larger features known as *draas* in north Africa. The processes leading to the formation of bedforms will be discussed by selecting examples from different arid regions. Reference will also be made to the gaps in our current knowledge of aeolian processes.

Because this chapter deals only with basic principles it should be read in conjunction with standard texts and the references cited in the bibliography so that the reader can gain an overview of general trends in arid zone research and can appraise objectively the disagreements about effective processes reported in the literature.

4.1 ENTRAINMENT AND TRANSPORT: THRESHOLD SPEED AND GRAIN SIZE

The wind moves sand by a combination of creep and saltation. The bouncing motion of saltating grains is initiated by granular impact and wind tunnel experiments have shown that the grains travel at a velocity parallel to the ground which is comparable to that of the wind speed. Individual grains have a low parabolic trajectory because their density is about two thousand times greater than that of air in which they are travelling (Figure 4.1). There is also evidence to suggest that grain or particle shape has a marked effect on the bouncing properties of grains and hence the relative flatness of their trajectory. On impact with the ground, saltating sand grains can move grains six times their size because of their kinetic

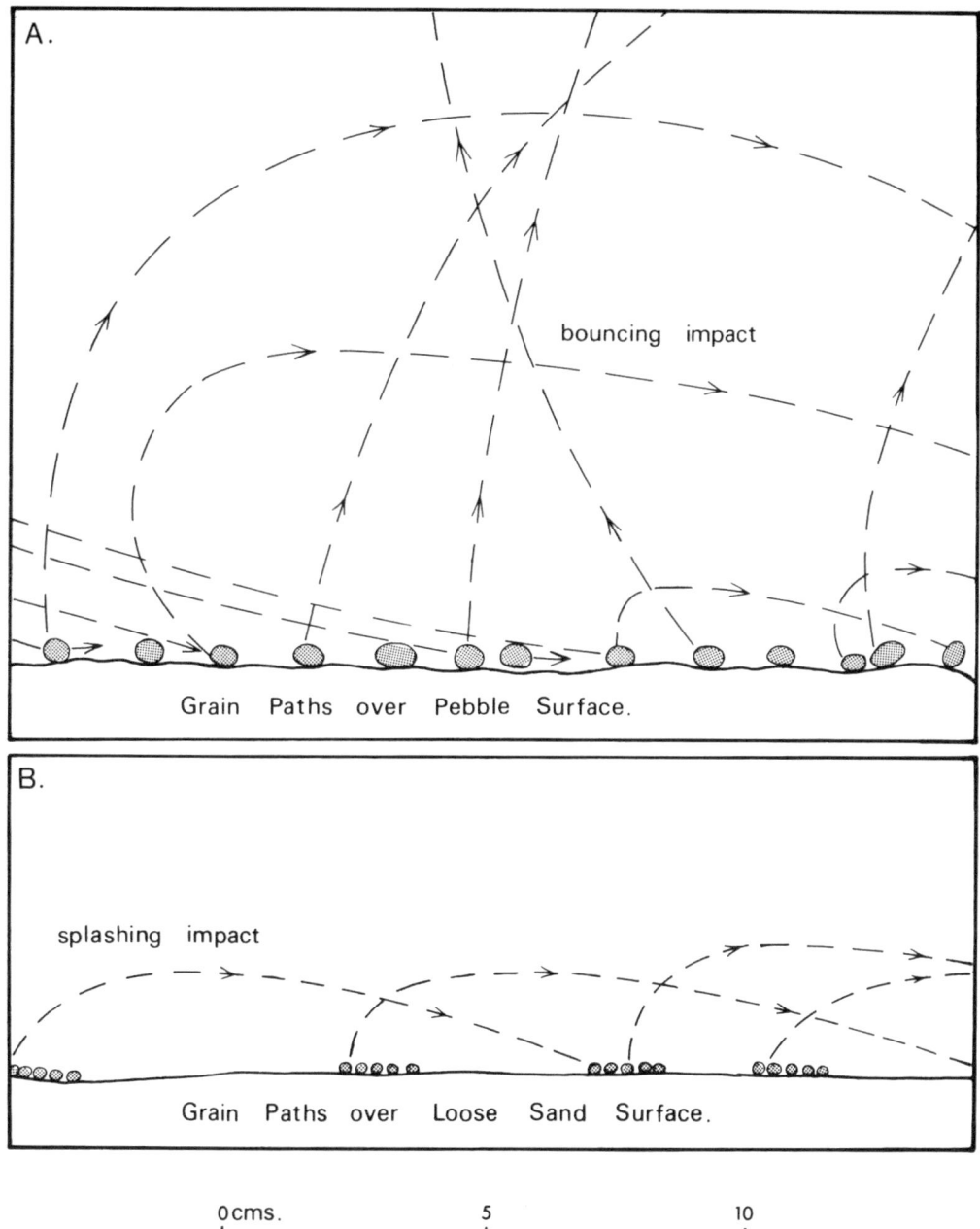

FIG. 4.1. DIFFERENCE BETWEEN SALTATION OVER SAND AND OVER PEBBLES
(Reproduced from Bagnold 1954, with permission of Chapman and Hall Ltd).

energy. Thus, grains 1 mm in diameter can induce surface creep of particles some 6 mm in diameter.

Figure 4.2 shows the variation of the threshold velocity (V^*_t) with grain size for air. The grain size is drawn on a square-root scale in order to show the relation $v \propto \sqrt{d}$ for the larger

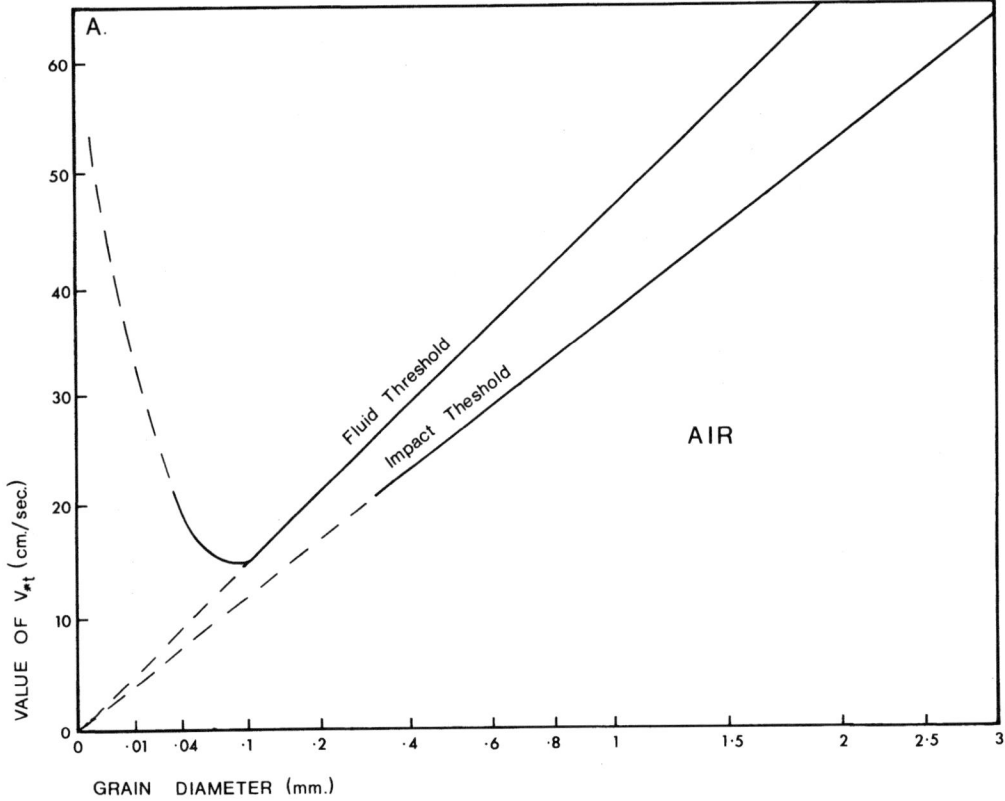

FIG. 4.2. VARIATION OF THRESHOLD VELOCITY WITH GRAIN SIZE FOR AIR
(Reproduced from Bagnold 1954, with permission of Chapman and Hall Ltd).

grains. The graph shows that the wind velocity gradient is zero near the ground. This means that after suspended grains smaller than about 0.03 mm or 30 μm (microns) settle on the ground they cannot be entrained as individual particles because, to cite Bagnold (1941), 'they sink into a viscid surface layer of air and are out of reach of the disturbing influence of the eddies of turbulence. The surface of the ground acts as a sort of dust trap.' It is shown in Figure 4.2 that grains smaller than about 80–100 μm have a higher threshold velocity than larger particles. Therefore, a greater threshold velocity is necessary for fine-grained material such as loess on account of increased interparticle cohesion, greater moisture retention and lower values of surface roughness (see also Horikawa and Shen, 1960; Smalley, 1964).

Once they are mobilized at the surface, however, silt-sized particles are transported in suspension because of their small size and low density. Under convective conditions, silts may be lifted hundreds of metres into the air and sometimes thousands of kilometres from their source. Known under a variety of local names, such as brickearth in south-east

England and *limon* in France, the generic term *loess* is now widely used to describe accumulations of such wind-transported silt.

In the dry valleys of eastern Antarctica, very high katabatically-enhanced wind velocities, added to the relatively high density of the air at very low winter temperatures, result in higher than normal thresholds for both direct dislodgement and for impact-induced movement. It appears that a grain 2 mm in diameter can be lifted to a height of 2 m at a wind velocity of 36.05 m/s at an air temperature of −70 °C (the velocity required at 0 °C being 45.42 m/s), so that the critical wind velocity in the Antarctic winter is about 10 m/s less than in the subtropical deserts (Figure 4.3). The same factor also raises the

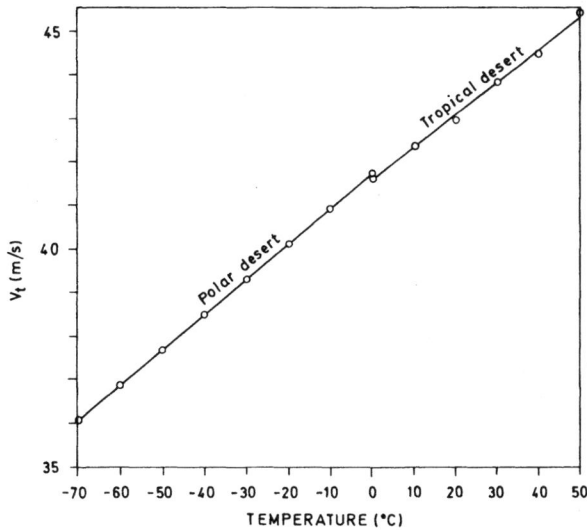

FIG. 4.3. THRESHOLD WIND VELOCITY REQUIRED TO LIFT A 2 MM GRANULE TO HEIGHT OF 2 M IN TEMPERATURE RANGE 50°C TO −70°C.
(After Selby, Rains and Palmer 1974).

upper grain size which can be moved by creep and thrown into a saltation mode. Given that saltating grains move by impact clasts up to six times the diameter of those making up the saltating load, the upper limit of wind-traction load in Antarctica appears to be about 18 or 19 mm. Pebble ripples containing this size of clast have been described from this region by Selby, Rains and Palmer (1974).

According to Bagnold, the threshold value of $V^{\ast}$ for a fine dune sand of diameter 0.025 cm is approximately 20 cm/sec and the corresponding value of $V^{\ast}d/v$ is near the limiting value of 3.5. Entrainment by wind, or the critical velocity at which particle movement begins, may be expressed in the form:

$$V^{\ast}_t = k_1\sqrt{\frac{\tau_o}{\rho}} = k_1\sqrt{\frac{\sigma-\rho}{\rho}} \; gD \tag{4.1}$$

in which τ_o is the shear stress, σ and D are the density and diameter of the grains respectively, and ρ is the fluid density. The threshold value of V^* should vary as the square root of the grain diameter provided the general coefficient k_1 is a constant approximately equal to 0.1 provided the grain Reynolds' Number

$$\frac{V^*d}{v} = 3.5$$

Although wind erosion is related to the textural properties of surface materials and to the properties of the wind, near the surface, there is still debate as to the actual cause of sediment entrainment (see, for example, Bagnold 1941, 1953; Chepil and Woodruff, 1963). Bagnold, and Chepil and Woodruff, maintain that sand particles are entrained because of the drag velocity of the wind which, in turn, is related to the drag exerted on the wind by friction at the surface, τ. Thus:

$$V^* = \sqrt{\frac{\tau}{\rho}} \tag{4.2}$$

where ρ is the density of air (1.22×10^{-3} gm/cm^3).

Both Bagnold (1941, 1953) and Kawamura (1951, 1953) have provided theories on the relationship between sand transport rate and the force exerted by the wind. These are now fairly well substantiated by field and wind tunnel measurements. Kawamura postulates that the shear stress at the sand surface (τ_o) consists of two components, τ_s due to the impact of saltating sand grains and τ_w due directly to the wind. Thus

$$\tau_o = \tau_s + \tau_w \tag{4.3}$$

However, under conditions of steady sandflow, or in the equilibrium state, τ_w is found to equal $\tau_{o(\text{crit})}$ which is the critical shear stress, so

$$\tau_s = \tau_o - \tau_{o(\text{crit})} \tag{4.4}$$

Also, because τ_s is equal to the loss of momentum from the wind

$$\tau_s = W(\overline{u_2 - u_1}) \tag{4.5}$$

where W is the amount of sand falling in unit time over unit area of sand surface, and $(u_2 - u_1)$ is the mean value of the difference between the final and initial forward velocities of the sand particles. As τ is related to the drag velocity (V^*) and to the density of air (ρ):

$$\tau = \rho V^{*2} \tag{4.6}$$

More information is still required about the surface shear stresses on migrating dunes, threshold values of the surface shear stress which suspends dune sand grains, the trajectories of suspended sand particles and the application of such data to dune stabilization programmes. The shear stresses τ_o can be determined from eddy correlation measurements when

$$\tau_o = \overline{\rho u' w'} \tag{4.7}$$

Here u is the component of motion parallel to the dune face, w is the component perpendicular to the dune face and ρ is the density of air. The over-bar indicates a time average and the primed quantities are instantaneous departures from their time averages. The values of τ_0 can be related to the initiation, rate and direction of sand transport at monitoring stations.

4.1a Grain size

Aeolian sands are commonly bimodal, with a mean size which is rarely <0.20 mm and seldom >0.45 mm, and are better sorted than fluviatile and beach sands. Most desert and coastal dune sands are positively skewed (i.e., they have a fine tail) but desert dunes derived from poorly-sorted alluvial sediments are not so well sorted as coastal dunes composed of sand previously sorted by wave action.

It was established in the preceding section that sand is selectively removed according to its grain size and by varying wind velocities from dune fields and deflation plains to leave behind a concentrated lag deposit of coarse grains. In some areas of the Oman, pebbles, originally left behind as a lag deposit, form ridges transverse to the prevailing wind direction on inter-dune deflation plains (Glennie, 1970). As the wind velocity increases, the coarser grains are also removed until the coarsest material is moved either by saltation or by surface creep. According to Bagnold (1956), the largest grains, of the order of three to seven times the diameter of those in saltation, are moved by surface creep up the windward slope of dunes so that they occupy exposed crestal positions. Thus, by a combination of deflation and surface creep, ridges are formed, the wavelength of which increases in time as the surface grading grows coarser with the removal of finer material (see also Table 4.1). Sands comprising dune crests are usually better sorted than the sand on the flanks, regardless of grain size. This fact may be attributed, in fact, to the reworking of crest sands by winds of differing strength and direction. Also, dune height and spacing are controlled, in part, by grain size.

Even if the grain size properties of dune sands do not differ markedly there can be consistent differences in the properties of the dune sub-environments. Folk (1971a) investigated the longitudinal dunes of the northwestern edge of the Simpson Desert, Australia, and concluded that the dune crests are coarsest, best sorted and most positively skewed; lee flanks are finest, most poorly sorted and less positively skewed; and windward flanks are intermediate in these respects. Very similar relative differences between the crest, windward, and leeward flanks of transverse dunes in Texas have been recorded by M. B. C. Waitt (unpublished).

Warren (1971, 1972), based on his work in the Ténéré Desert, Niger, relates bimodal sands in different aeolian environments to what he terms the *protectionist theory*. It is suggested that sands in saltation include some grains that are sufficiently fine to occupy the inter-particle voids between the coarse particles of the creep load. When they strike the bed the fine grains penetrate the voids and are thus protected from further bombardment. The coarser fraction of the saltation load, which is too large to fit into the voids, remains on or near the surface where it is bombarded by other particles and moved onward. This sorting process is associated with distinct dune patterns in the Ténéré Desert where flat areas with bimodal sands are located downwind of most seif dunes. According to Warren, the high threshold velocities that are required to move the sands of the fine mode do not produce bimodal sands. The protectionist theory has been advanced by C. Verlaque to account for the bimodal sands on the horns of barchans at In Salah (Sahara Desert) and by Sharp (1963) to explain the presence of cores of fine sand with ripples.

Amstutz and Chico (1958) have demonstrated the sorting action of wind in a grain size study of barchan sands from the coastal plain of South Peru, about 130 km southwest of Arequipa. These sands, collected from the tails, crests and horns of four barchans in the vicinity of the Carretera Pan, are mainly composed of quartz, with some feldspar, biotite, augite and hornblende. However, the percentage concentration of these heavy minerals changes not only from one barchan to another but also from one part of a barchan to another part, thus illustrating the selective sorting or sizing action of the wind at different points on the barchans.

Thick lenses or seams of wind-deposited heavy minerals occur in association with wave-concentrated deposits in open cuts and along foredune sections on the coast of New South Wales (Hails, 1964). At Point Plomer (Figure 4.4) the thickness of aeolian sand is considerable and here a long period of wind action, together with selective sorting has resulted in the burial of a former beach ridge.

FIG. 4.4. OPEN CUT AT POINT PLOMER NEAR CRESCENT HEAD, NEW SOUTH WALES
Wind-deposited heavy mineral seams overlying wave-concentrated deposits. (Photograph: J. R. Hails).

4.2 ABRASION AND DEFLATION

Wind erosion is controlled to some degree by smoothness of the ground surface which, in turn, is determined mainly by particle size of surface debris and the type and percentage cover of vegetation and the size and frequency of obstacles.

4.2a Surface roughness

Rough surfaces composed of pebbles and cobbles influence wind velocity near ground level and create a layer of almost motionless air immediately above the surface. Large areas of cobbles are termed *stony deserts* and are invariably known as *serir* in Libya, *reg* in Algeria and *gibber plains* in Australia. Surface roughness also causes air turbulence. In many deserts two layers of particles are present in the air. A lower one consists of sand grains and extends from a few centimetres to a metre or so above ground level. This grades upward into a second layer which contains silt and clay particles extending to considerable heights. These respective layers can be compared with the bed load and suspended load of a stream. Figure 4.5 shows the effects of obstacles of various heights on wind-velocity profiles, which appear as straight lines because height is plotted on a logarithmic scale. The profiles would be curved on an arithmetic scale and therefore similar to the velocity profiles of a stream. In Figure 4.5, V_1 is slower than V_2 and the two profiles are shown for winds blowing across obstacles of average heights, a and b. A single point above ground level represents the top of the dead-air layer where all velocity profiles for winds blowing across obstacles of the same average height intersect. Generally, the thickness of the dead-air layer is one-thirtieth the average height of the obstacles. The obstacle-height bars in Figure 4.5 are unrelated to the air-velocity scale along the abscissa.

4.2b Ventifact formation and deflation

Sand grains, moving predominately by saltation, may effect a notable degree of erosion of rock surfaces. This is perhaps best seen in polar deserts where vegetation is absent and the abrasive action of the blowing sand may be enhanced in winter by blowing snow which, at very low temperatures, attains a hardness similar to rock particles (Figure 4.6). For example, in the 'dry valley' region of southern Victoria Land, Antarctica, rocks of the Beacon Sandstone Series have surfaces smoothed and grooved by wind blown debris abrasion. In addition, the finer-grained rocks, especially the Ferrar dolerite, show a high degree of smoothing and faceting by sand-blast, the smoothed faces indicating the direction of the formative wind.

Removal of fines produces a residual boulder pavement in which stones known as ventifacts are common. *Ventifact* is a general term for pebbles and cobbles of a deflation lag that have facets cut on their windward surface by the abrasive action of wind-blown sand. The German term *dreikanter* is frequently used for three-facetted pebbles shaped by sandblasting and a pebble with one sharp edge forming the intersection between two facets is known as an *einkanter*. Whether or not all ventifacts have been shaped entirely by wind abrasion is still debatable because of the effects of mechanical weathering or insolation (rock splitting); thus, their nature and origin is in dispute. For example, Sugden (1964) has suggested, as a result of his study of facetted pebbles in some recent desert sediments of southern Iraq, that the facets of ventifacts originate by fracturing. Glennie (1970) found no preferred orientation of facets on angular pebbles in the Oman and 'none that could be referred to known prevailing winds either existing at present or inferred for earlier times'.

The formation of desert varnish (a surface coating of manganese and iron oxide) following rock splitting seems to be a more likely mechanism for the formation of many ventifacts. Although desert varnish is removed by sandblasting, the possibility of earlier shaping by this mechanism cannot be entirely discounted. Many ventifacts, though, have retained essentially the same position throughout the time they have undergone wind erosion so their features may be correlated with wind directions. Interestingly, field and laboratory data (wind-tunnel experiments on rocks, minerals and artificially prepared mineral aggregates) tend to support the hypothesis that wind erosion involving suspended dust-size particles as both abrading and burnishing tools is an effective process like sand abrasion (Whitney and Dietrich, 1973).

Ventifacts occur at most levels on the floor and sides of ice-free valleys west of McMurdo Sound in southern Victoria Land, Antarctica; in fact, they occur on surfaces of a variety of

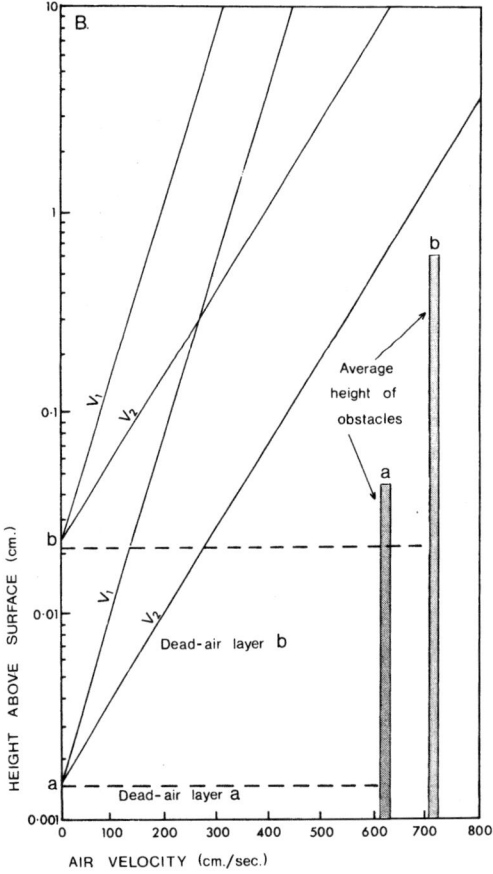

FIG. 4.5. GRAPH OF WIND-VELOCITY PROFILES, SHOWING INFLUENCE OF OBSTACLES OF VARIOUS HEIGHTS (Reproduced from Longwell, Flint and Sanders 1969, with permission of John Wiley and Sons, Inc.)

ages. According to Lindsay (1973a), who has worked in Wright Valley, the morphology of lag gravel ventifacts indicates that the distribution of wind-polished faces is determined mainly by the shape of the original unpolished rock fragments. During the initial, more active stages of development, the ventifacts tend to be orientated either transverse or parallel to the wind direction. Salt weathering is also a major factor in shaping ventifacts in Antarctica and results in pitting and flaking on the underside of most rocks. By this mechanism, a ventifact may be reduced to a hollow shell which ultimately collapses under its own weight.

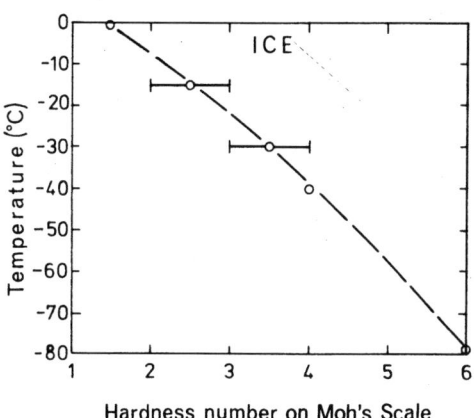

FIG. 4.6. HARDNESS OF SNOW GRAINS AS FUNCTION OF TEMPERATURE (after Selby, 1977). Values reach the equivalent of orthoclase feldspar (hardness 6) in the Antarctic winter.

More questions than answers have emerged from recent studies of ventifact formation and therefore more attention should be given to aerodynamic processes, particularly negative and positive turbulent flow, and vorticity. Other important factors to be considered in ventifact sculpture and evolution include rock hardness, mineralogy, angle of grain impact, the nature and density of saltating particles, the basal area to height ratio of particles and, of course, surface roughness. For a more detailed account of earlier work on ventifacts the reader is directed to the references in Cooke and Warren (1973).

Deflation is the removal of material from a beach or other land surface by wind action. In deserts, the process of deflation removes the finer material and leaves a lag of coarser pebbles and boulders. Deflation basins or hollows are topographic features produced by the removal of sand and finer material from a disintegrating surface in such a way as to leave a complete rock rim surrounding the area of removal. *Desert pavements* or *desert armours*, which consist of a surface lag of coarse particles, are produced by the deflation of sediments composed of silt, sand and pebbles. The sand and silt are removed by either wind or sheet erosion until the original surface has been lowered sufficiently to expose a continuous cover of pebbles which act as a pavement or protective cover.

Differential rates of deflation in horizontal strata are related to variations in lithology and degree of cementation, so as to produce marked topographic expression in some deserts. Glennie (1970) refers to caliche cemented protective layers which result from the evaporation of carbonate-rich flood or groundwater. If these layers are eroded or broken, the underlying softer and more poorly cemented sediments are removed at a faster rate so that 'mushroom-shaped' features are produced.

4.3 BEDFORMS

Invariably, surfaces of aeolian sand deposits are formed, on a varying scale, into regularly repeated patterns or *bedforms* which are similar to those formed under water (see Allen, 1968a). However, when grouped together, bedforms are a large and multifarious group of natural landforms which include wind-formed draas, dunes, gravel ridges, ripples, sastrugi and yardangs, as well as water-formed antidunes, bars, beach cusps (see Chapter 3), braids, flute scours, megaripples, ripples, river meanders, sand ribbons and tidal current ridges. According to Wilson (1972), the elements of bedforms can be divided into four groups that can be identified by their respective wavelengths – draas (0.5–5 km), dunes (10–500 m), and aerodynamic and impact ripples (0.01–10 m). These, in turn, can be subdivided into transverse and longitudinal elements.

It is thought that aerodynamic ripples, dunes and draas are formed by the action of regular secondary flows in the wind which occur because of aerodynamic instability. But, as Wilson points out, it is not known whether these secondary flows exist before the development of the bedform or whether some arise from concurrent interaction with it from small beginnings. Wilson's work on the ways in which dune patterns, resulting from combined transverse and longitudinal secondary flow, interact with the bedform itself have been summarized by Cooke and Warren (1973). They give details of flow patterns over aklé ridges and describe dune patterns (aklé) produced by combined 'wave' (two-dimensional) and vortex (three-dimensional) flow patterns with both in-phase and laterally displaced vortices (Figure 4.7g). According to Cooke and Warren, such patterns can account for the odd elongation of barchan horns, the crossing patterns in sand seas, and the displacement of barchanoid and linguoid elements in aklé dunes (see p. 179; Figure 4.7c, d and Table 4.4).

Sand movement and deposition in deserts is not a random phenomenon but occurs in patterns, analogous to drainage systems that are closely related to wind systems and not to topography (Wilson, 1971). Thus, it is possible to predict how and where sand deposition will develop, given a particular sandflow system with a reasonable number of sandflow resultants in a region.

FIG. 4.7. FORMATION OF VARIOUS TYPES OF DUNE

A – (1) Pur-Pur dune or draa-sized barchan in Peru (after Simons, 1956). (2) Oblique elements in barchans (after Clos-Arceduc, 1967). B – Schematic diagram of barchan dimensions (after Long and Sharp, 1964). C – Dune network pattern: sinuous ridge transverse to wind showing *linguoid* (facing wind) and *barchanoid* crescentic sections. D – Dune types (after Cooke and Warren, 1973). E – Dune trends typical of central Simpson Desert, Australia, with systematic opening of tuning-fork junctures to the SSE (after Folk, 1971). F – Formation of seif dunes, following helicoidal wind-flow theory proposed by Bagnold. Paired helical vortices from parallel to the prevailing wind (after Folk, 1971). (L) original level of poorly sorted fluvial plains; (F) poorly sorted fluvial debris; (D) windrift dunes, often cored by alluvium; (R) reg, deflation flats on which form bimodal lag deposits. Arrows show direction of wind flow; dashed arrows the direction of sand movement on ground. G – Three-dimensional pattern of 'vortex (Taylor-Görtler) flow'. Note relationship of flow to ground flow lines and to longitudinal dunes (after Cooke and Warren, 1973).

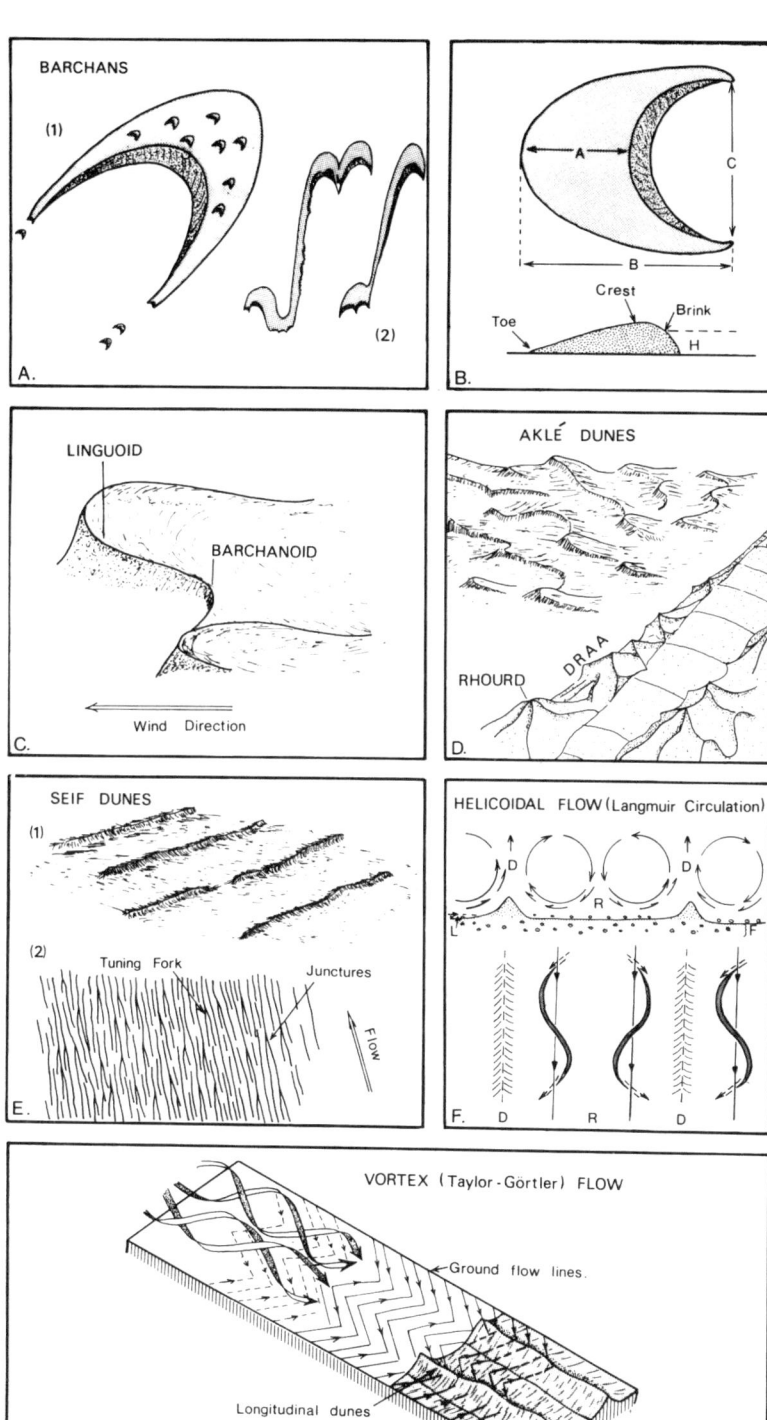

A. BARCHANS (1) (2)

B. Crest, Brink, Toe, H, A, B, C

C. LINGUOID, BARCHANOID, Wind Direction

D. AKLÉ DUNES, RHOURD, DRAA

E. SEIF DUNES (1) (2) Tuning Fork, Junctures, Flow

F. HELICOIDAL FLOW (Langmuir Circulation), D, R, L, F, D, R, D

G. VORTEX (Taylor-Görtler) FLOW, Ground flow lines, Longitudinal dunes

According to Bagnold's (1941) formula, the potential sandflow rate, or the amount of sand that a wind can transport over a bed of fairly uniform grain size, may be expressed as:

$$q = 1.4 - 10^{-5}V'^{*3}\sqrt{d} \tag{4.8}$$

where d is the sand grain diameter (cm) and V'^{*} the wind shear velocity (cm/sec). This agrees with sandstorm records and dune migration rates (Finkel, 1959) and has been verified experimentally by Williams (1964).

Sandflow maps of the Sahara, interpolated from meteorological and bedform data, have been studied by Wilson. He maintains that sandflow resultants can be measured from wind velocity data by using Bagnold's formula (Equation 4.8), from sandstorm duration and direction records, and from size and orientation of bedforms such as dunes and ripples. Therefore, at any point a mean potential sandflow resultant can be calculated for each specified combination of grain size (see p. 167), time interval and sampling area, and of course all these may be different.

4.3a Sand ripples and ridges

The fundamental distinction between small-scale sand features and dunes has been summarized in Table 4.1. Sand ripples, which develop with the axes of their crests transverse to

Table 4.1
SMALL-SCALE SAND FORMS: TRANSVERSE RIPPLES AND RIDGES
Conditions Under Which They Occur (After Bagnold, 1941)

Condition	Sand form
1 Deposition	
(a) Rapid	Relatively flat surface with no small-scale surface relief
(b) Slow, or equilibrium sand flow	*Ripples* – wavelength depends on wind strength, and height/wavelength ratio depends on *width* of surface grading. For almost uniform sand this ratio is extremely small, increasing with greater variation in grain size
2 Removal	
(a) Wind above ultimate threshold strength for largest grains	*Ripples* – as for 1 (b)
(b) Wind below ultimate threshold	*Ridges* – size and wavelength increase indefinitely with time, rate of growth depending upon quantity of coarse material available, and upon intensity of oncoming saltation

Note: In all cases, finest material collects in troughs and coarsest material on crests, a characteristic which distinguishes small-scale features from dunes where reverse process is invariably the case.

the wind, become fully developed while still small and therefore cannot be considered as small dunes. The five mutually interactive factors that are responsible for controlling the height shape and wavelength of transverse ripples and ridges are listed in Table 4.2.

Experiments by Bagnold (1941) with varying wind strength and sand of almost uniform

Table 4.2
FACTORS CONTROLLING FORM OF SMALL-SCALE SAND FORMS

Factors (mutually interactive)	Process
1 Wind	Imparts primary motive power to sand grains in *saltation*
2 Saltation	Causes *surface creep* (movement amongst surface grains by impact); controlled wind, according to its intensity
3 Surface grains	Move differentially, according to size, to form *ripples* and *ridges*
4 Surface relief	Causes variations from place to place in rate of ejection of new grains into saltation
5 State of sand movement	Either deposition, removal or steady sand flow. Also controls size and grading of surface grains

grain size have shown that ripple wavelengths increase with the wind gradient V'^{*}, but that the ripples flatten out and disappear when the wind exceeds a certain strength. It can be seen from the figures in Table 4.3 that the ripple wavelength is a physical manifestation of the length of hop made by the average sand grain in its journey downwind. As Bagnold has

Table 4.3
COMPARATIVE FIGURES FOR WIND GRADIENTS, CALCULATED RANGES FOR A GRAIN PATH, AND FOR MEASURED DISTANCES BETWEEN ACTUAL RIPPLES PRODUCED ON SAND SURFACE (after Bagnold, 1941)

V'^{*} (cm/sec)	19.2	25.0	40.4	50.5	62.5	88
Calculated range (cm)	2.5	3.0	5.4	8.0	11.6	27
Measured ripple length (cm)	2.4	3.0	5.3	9.15	11.3	—

Note: Range calculated theoretically from evidence of wind velocity measurements.

stated, transverse ripples in sand arise because of instability in the grain flow. They form with a wavelength similar to the mean path length of the saltating sand grains and are produced by the impact of the grains on the bed. It is for this reason that they have been termed *impact ripples*.

The *ripple index* is the ratio of wavelength to height which is commonly between 15 and 20 for aeolian sand, although this index can be as high as 50 or 60 when ripples flatten out during high winds. Sharp (1963) has recorded indices between 12 and 20, with a mean index of 15, for granule ripples in the Kelso Dunes of the Mojave Desert, and believes that the ripple index varies inversely with the grain size of the sand and directly with the wind velocity.

Bagnold (1956) has demonstrated that a tangential stress is applied to the surface of a bed of sand grains during the formation of ripples, and that some grains are eroded. These grains are redeposited because they cannot be suspended without an increase in the applied stress. As with sand ridges, the coarsest grains collect at the ripple crest so that it rises into the region of stronger wind. It follows, then, that the height of a ripple depends upon the ability of the coarsest grains to remain in their crestal position for a given wind strength as well as upon the range in grain size of the sand forming the ripple. Large ripples are characteristically asymmetrical and associated with varying grain sizes. According to

Sharp, the degree of asymmetry varies directly with grain size but, as already stated, with a uniform grain size only small ripples are formed; for a uniform sand of grain diameter 0.25 mm ripples disappear when V'^* exceeds 65, or about three times its threshold value. Thus the ripple height and shape depend upon the surface sand grading. Figure 4.8 shows ripples on the windward slopes and flanks of active transgressive dunes in the Coorong, South Australia, although ripples may be absent from some dunes because of strong winds or a sand supply of uniform grain size.

4.3b Seif or longitudinal dunes

Seif (the Saharan native term for 'Sword-' or 'Knife-edge') dunes are mainly erosional forms which dominate the inland deserts of the world. They occur, for example, in Arabia, Arizona (USA), Australia, parts of Europe, Egypt, Libya and Southwest Africa (Northern Namib desert). Following the first description of longitudinal dunes in the Simpson Desert (Australia) by Captain Charles Sturt in 1849, Blanford (1876), working in the Indian (Thar) desert, was the first person to establish that such dunes parallel the dominant wind direction. Several workers, though, have proved subsequently that secondary side winds modify the dune crests and produce asymmetrical longitudinal dunes (see, for example, Aufrère, 1930; Enquist, 1932; Madigan, 1930; Wingate, 1934). No true barchan dunes are known in Australia and longitudinal forms are almost the only type present throughout that continent. Madigan (1936) and other workers (D. King, R. C. Sprigg, and Veevers and Wells (1961), for example) have demonstrated how these dunes parallel existing dominant south-southeasterly winds over Australia.

Longitudinal dunes commonly occur on relatively level plains of alluvial origin, which are underlain by deep water tables and where there are extensive deposits of unvegetated sandy material that can be scoured by winds moving in one dominant direction. In 1953 Bagnold advanced the theory that these dunes may be formed by helicoidal air flow (Langmuir circulation). This type of flow can be observed on the surface of windswept water and during the hydraulic erosion of mud bottoms both in nature and in the laboratory (see McLeish, 1968; Allen, 1969). Strong support for Bagnold's hypothesis has been provided subsequently by Houbolt's (1968) study of linear sand ridges, or longitudinal sand ribbons, in the North Sea. According to Houbolt, these trend parallel to the currents that produce them, and are formed by a spiral motion superimposed upon a tidal current that moves in an overall linear direction because of the absence of flotsam over the ridge crests and the presence of lines of flotsam over the axes of the inter-ridge areas. It must be borne in mind, however, that the direction of the tidal currents is reversed four times each day, unlike the winds in seif dune areas.

Provided, then, that hot desert floors exist over which there is strong tropical air movement, paired helical (roller) vortices will form with horizontal axes trending parallel to the prevailing wind. It can be seen in Figure 4.7f which shows the direction of sand movement on the desert floor (dashed arrows) that the regs are eroded by deflation which is caused by descending or centrifugal winds. Bimodal sands (see p. 167) are concentrated in such inter-dune troughs by the selective sorting of ground-surface winds which sweep sand from the regs and heap it on the flanks and the crests of the dunes. The secondary side winds, just mentioned, maintain sand movement but are not significant in orientating the

FIG. 4.8. A – RIPPLES ON WINDWARD FLANKS OF ACTIVE TRANSGRESSIVE DUNES, COORONG, SOUTH AUSTRALIA (Photograph: J. R. Hails). B – RIPPLES AND GRAINFLOW TONGUES ON COASTAL DUNE, NORMANVILLE, SOUTH AUSTRALIA (Photograph: V. A. Gostin).

dune chains. The occurrence of tuning-fork junctures in the dunes, which invariably open upwind, is associated with the helicoidal air flow (Folk 1971a, b).

It is pertinent to mention here that Bagnold (1941) originally proposed a two-wind-resultant theory for the formation of longitudinal dunes. He postulated that side winds convert a barchan form into a longitudinal dune by the elongation of one horn which then fuses with the next dune *en echelon* with it, the eventual dune form representing the resultant of the bi-directional wind – for example, prevailing versus storm winds, seasonally-variable winds, or morning versus evening winds. Corroborative evidence to support this theory has been reported by McKee and Tibbitts (1964), who consider that seif-dune structure near Sebhah Oasis in south-western Libya is controlled largely by winds from two directions (the southeast in the mornings and the northeast in the evenings) about 90° apart. On the other hand, Glennie (1970) is not convinced by their arguments because there is little available information on the strength and direction of wind at ground level and no continuous, 24-hour-day records, although the general pattern of wind directions over most deserts is now known.

The fact that local surface winds move diagonally up the flanks of longitudinal dunes in a converging feather-barb orientation appears to resolve the difference of opinion between the one-wind and bi-directional wind theories of longitudinal dune formation (Folk 1971a, b). So far, though, no satisfactory explanation has been given for the uniform lateral spacing of the dune ridges and for the presence and uniform orientation of the tuning fork junctures systematically opening upwind (Figure 4.7e). The 30° to 50° juncture angle in desert dunes is postulated by Folk (1971b) to match the flow lines of sand particles in the flats and dune flanks. Seif dunes are narrow and the intervening deflation flats (regs) broad because the boundary between two adjacent sets of helical vortices is extremely narrow compared with the width of the rollers.

Clos-Arceduc (1966) relates the rhythmic spacing of seif and oghurd (star) dunes to the stability of oscillatory seiche-type air currents. Although Folk (1971b) agrees with this concept in principle, he favours the fixed position of the helical vortices rather than an oscillatory seiche type of motion. He argues that after a dune is formed it produces its own rising air current by a self-accelerating mechanism which builds a larger dune, *ad infinitum*. If Folk is correct, it can be implied that the dunes in most deserts have developed under present wind regimes and therefore are not relics of any wind-belts that have shifted since Pleistocene times (see also Brookfield, 1970). According to Folk, good evidence of fixation of the helical vortices is the fact that the dunes in the Simpson Desert have cores of old alluvial sediment that have not shifted laterally in position since dune formation began.

Such an insinuation, of course, conflicts with the view that many major dune systems were formed in the Pleistocene epoch when very high winds prevailed, especially during glacial periods. Fairbridge (1964) equates active dune formation in desert areas with polar glaciations and refers to vegetated seif dunes disappearing below sea level in northern Australia, Arabia and West Africa. Both Flint (1959) and Fairbridge (1964) draw attention to remnants of linear dunes that have been traced in the Congo and the southern Sudan. These are probably Pleistocene in age, and now occur of course in the tropical low pressure zone of the world between 10° N. and 10° S. of the equator. It is now generally accepted that the shorter distances between areas of high and low pressure during a Pleistocene glaciation gave rise to a much stronger global air circulation.

Long, wide, but very low, parallel sand ridges in the Wankie National Park of Western Rhodesia, which are thought to be the eroded stumps of former longitudinal dunes, have been described by Flint and Bond (1968). These writers are of the opinion that the former dunes were degraded, in the main, by sheet erosion which was effected by a high-intensity rainfall, a partial vegetation cover and an unconsolidated, permeable sandy terrain. Deflation basins between the ridges are underlain by a mixture of sand and clay that originated from the ridges. Flint and Bond attribute the origin of the ridges to Pleistocene climatic change when a very wet period succeeded a very dry period which, in turn, has been followed by a short, recent dry period.

Three different types of stabilized dunes, which cover about 56 980 km^2 of central and western Nebraska, are believed to have formed during early Wisconsin time when wind action was particularly strong and widespread. According to Smith (1965), the oldest dunes are broad massive forms, generally elongate, with many variations in ground plan, that may have originated as compound transverse dunes and associated forms under desert conditions. Those of intermediate age are narrow, approximately symmetrical subparallel linear ridges of intermediate length, superimposed on the older dune topography. They are interpreted to have developed as non-migratory upsiloidal forms transitional into *longitudinal* types, in the presence of some vegetative cover, under semiarid conditions. The youngest dunes are blowouts of various shapes and sizes, omnipresent on all older dunes, which are believed to have been formed during the Hypsithermal interval (Postglacial Climatic Optimum).

4.3c Barchan dunes

Barchans, formed by unidirectional winds on hard desert pavements with sparse supplies of sand, are crescentic dunes characterized by a slip face, or avalanche slope, with an angle of about 34° (the angle of repose for dry sand) and horns that point downwind (see also Table 4.4). Often, both horns are distinctly elongated, but in many areas they are elongated more in one direction than another, a phenomenon which has been attributed to an asymmetrical wind pattern, to an asymmetry in the sand supply or to a dipping desert pavement (see, for example, Rim, 1958; Holm, 1960; McKee, 1966; Norris, 1966; Lettau and Lettau, 1969). Barchan dunes advance by either the addition of sand at the crest or by the removal of material from the toe of the slip face which causes grains to roll down the steep lee side until the slope of the slip face has readjusted to the angle of repose. In contrast, the gentler windward convex slope of barchans has a variable inclination. Individual sand grains on the slip face or avalanche slope have a preferred orientation with their largest and shortest axes approximately parallel to the azimuth of maximum dip, but they are imbricated so that their long axes lie between the horizontal and the maximum dip of the bedding (Rees, 1968; Glennie, 1970).

Active barchans travel at different speeds according to their size, shape and location. The height of the slip face is now generally considered as the most important factor affecting the movement of barchans, although variations in local wind regimes, sand supply, topography and vegetation cannot be ignored. Both two-dimensional and three-dimensional flow patterns have been observed in barchans.

Several interesting and informative studies have been made of barchan-dune movement

Table 4.4
COMPARISON BETWEEN BARCHAN, TRANSVERSE, SEIF AND OTHER DUNES

Characteristics	Barchan Dunes	Transverse dunes other than barchans	Seif dunes	Other dunes	Coastal dunes
Morphology	Crescentic form with slip face (avalanche slope) and horns. Asymmetric profile with gentler slope on convex side and steeper slope on concave or leeward side. Example: *Pur-Pur* dune in Peru – *draa*-sized barchan with slip face, and covered with dune-sized barchans	Sinuous ridges consisting of alternating *barchanoid* (facing away from wind) and *linguoid* (facing wind) crescentic sections. Linear ridges; symmetrical upper part in cross section with slopes of 20° to 25°. Smooth windward slope, steeper lee slope. Broad rounded crests	Long, narrow ridges with axes parallel to dominant wind direction. Short, isolated ridges in plan. Some symmetrical in cross section. *Hook-shaped* variety, the tail of one ridge fitting into the head of another	**Oghurd** Star-shaped (stellate) or pyramidal in form **Aklé** Network of sinuous ridges comprising alternate *linguoid* and *barchanoid* elements	Parabolic dunes, having approximately form of parabola in ground plan with concave side towards wind; loosely termed U-shaped dunes. Long-crested transverse dunes in arid/semi-arid regions. Barchan forms where water table is exposed by deflation. Aklé forms in coastal sand dunes of Oregon, Washington, California and Mexico
Dimensions	5 to 400 m wide, horn to horn; 2.5 to 250 m long, parallel to wind; 0.5 to 152 m high	Several kilometres long (8 to 50 km); 165 to 270 m high	200 m to 10 km apart in transverse direction. Up to 50 km in length beyond Y-shaped junctures. 5 to 200 m high	1000 to 2000 m across at base	Large, fixed transgressive dunes on Moreton Is, Queensland, Australia, attain heights > 375 m above mean sea level. Isolated sandhills > 150 m high and 5 km across base

Features	Crest, brink, slip face (avalanche slope), and lower lee slope	As for barchan dunes; crests irregular in plan and profile; crestal hollows formed by crestline bifurcation and joining. Slip face height 2.5 m. Lee slope 30° to 33° (metastable); slumps are common	Tuning-fork or Y-shaped junctures. Long, sharp and sinuous crestlines.	Steep, radial, branching arêtes of sand with pointed summits; smaller dunes have one peak	*Barchanoid* and *linguoid* elements enclose hollow which may be bounded by small longitudinal ridge. Barchanoid element of one ridge replaced downwind by linguiform element in next ridge	See Morphology
Orientation	Gentler convex side facing wind; horns directed downwind; foresets dip downwind at angles of up to 34°	*Not* longitudinal to well-defined prevailing winds	Generally parallel to prevailing/dominant winds; fairly constant transverse spacing. Connected with three-dimensional vortex flow patterns. Low, rounded upwind end of dune contrasts with sharp downwind end. Some small seifs aligned obliquely across sides of *draa* size longitudinal dunes, e.g. Saudi Arabia.	Probably built by alternating winds from several directions. Also probably form when smaller barchans overtake larger ones during migration	Aklé patterns are associated with relatively unidirectional winds and result from combined two-dimensional and vortex three-dimensional) flow patterns	Parallel and transverse to predominant/prevailing winds
Factors Affecting Rates of Movement						
(a) Main Factor	Shape and size, particularly height of slip face.					
(b) Other Factors	Local wind regime; sediment supply; topography; location	Local wind regime; sediment supply; vegetation, location	Local wind regime; sediment supply; vegetation, location			Local wind regime; sediment supply; vegetation, location

in different parts of the world. Notable changes in the form and position of barchan dunes on the west side of the Salton Sea, Imperial Valley, California, have been recorded over the 22-year interval 1941–1963 by Long and Sharp (1964) who paid considerable attention to the dimensions of the dunes as shown in Figure 4.7b. Most of the dunes in this area of the Imperial Valley are individual barchans, the smallest of which are 9 m horn to horn and about 2 m high. In contrast, the largest ones are several hundred metres across and 8–12 m high. Some of the barchans have a low a/c ratio, or in other words, are relatively thin with a transverse dimension that is large relative to the distance from brink to toe (see Figure 4.7b). Morphologically complex and areally larger dune masses, in which barchan features are often obscured, occur where individual dunes have coalesced and fused as they have migrated.

In an attempt to evaluate the rate and direction of barchan movement, Long and Sharp considered the influence of dune area by devising an index which was obtained by assuming that a barchan covers half the area lying between an ellipse and a circle of radius equal to the minor semi-axis of the ellipse. They calculated the area *(A)* of the barchan by the formula:

$$A = \frac{\pi c(b - \tfrac{1}{2}c)}{4}$$

in which b and c are the dimensions shown in Figure 4.7b. Dune shape was also considered in terms of dimensions parallel and transverse to the axis of symmetry, the a/c ratio. According to Long and Sharp, it is important to know whether the barchan is growing or is in a steady-state condition when evaluating the influence of dune shape on movement. They also point out that the difference between the crest of a barchan dune and its brink merits attention in future research programmes, because dunes with a separate crest and brink may behave differently from dunes in which the two features are one and the same. Their study has shown that an increased sand supply, resulting from an unusually dry period over the 22-year interval, has been primarily responsible for the movement of the barchans and that the shape of growing barchans may also influence the rate of movement.

Hastenrath (1967), following the earlier work of Finkel (1959), made a study of the aerodynamics and mechanics of barchan dune movement in the Pampa de La Joya near Arequipa, southern Peru; the micrometeorological and climatological part of this study is the subject of a separate report which is scheduled to be published shortly. The barchan field extends for 100 km along a broad, almost level plateau, about 1200 m above sea level, which consists of a wind-erosion pavement of fine, closely packed gravel resting on sand. Hastenrath measured the wind length (L_w), slip face length (L_s), horn width (W) and height *(H)* in metres; the angle formed by the desert surface and centreline on the windward slope of the barchans (sin α_o); the total dune displacement over a given number of years in metres (D_1); rate of displacement in m/year (D_3). He related the variety in dune shape to the size of the dune sand. By plotting the bulk specific gravity (in g/cm^3) of dune samples as a function of the distance upwind, Hastenrath found that there was a tendency for the bulk s.g. to decrease downwind. The size distribution of barchans in the dune field indicates that dunes are formed, grow to a maximum size and then shrink again and eventually disappear in the downwind part of the dune field. The sand transport by dunes is a sorting process according to grain size distribution and mineralogical composition.

The main conclusions of Hastenrath's study may be summarized as follows. First, under relatively uniform wind conditions with only a moderate supply of sand and with a flat underlying surface, barchans may move as an aerodynamic equilibrium (steady-state) form without disturbing their size and shape. Second, bulky objects or an array of pebbles on the windward face of a barchan disturb the air flow over the dune so that it may be easily destroyed. Third, dune movement is related to the duration of the wind speed above a certain threshold value rather than to variations in the mean wind speed. It follows, then, that marked variations in the displacement rate may occur with only small changes in the average wind speed. Finally, the diurnal katabatic wind system in the Pampa de La Joya contributes both to the westward turning and the shrinking of barchans at the downwind end of the dune field.

Barchans are of limited extent in the Thal desert, Pakistan, compared with longitudinal and transverse sand ridges. They occur mainly in severely devegetated abandoned channels in close proximity to levee banks with an abundant supply of sand and in flat areas of abandoned cultivation. Thus the barchan dunes are believed to be recent, as a result of man's activities.

Although aeolian deposits are relatively rare in Antarctica, reversing barchan dunes have been recorded by Lindsay (1973b) over an area of about 2 km² along the northern side of lower Victoria Valley near the terminus of Packard Glacier. Lindsay's investigations have provided some insight into aeolian processes in a cold desert. Statistically, he has not identified any significant difference between sand samples from either flank of the dunes or between their windward flank and crest. Furthermore, grain-size analyses have shown that the dune sands in the area are similar in texture to some barchans of more temperate climates. During summer when the temperature rises above freezing the surface sands are saturated because of snow-melt, and erosion therefore occurs slowly as the wind dries the sand. Consequently, dune morphology is controlled by two major factors. Firstly, the net dune migration is small as the winter stratigraphy of interbedded snow and sand units is partly preserved. Secondly, during summer, dry sand moves from an interdunal area on to dunes with an immobile, oversteepened windward slope. Therefore, under such conditions, sand is deposited near the crest of the dune. This mechanism produces an oversteepened lee slope which, in turn, develops into an avalanche slope as described by Bagnold (1954). Approximately 30 crescentic dunes, up to 10 m high at the crest, occur in Victoria Valley and these have their axes of symmetry orientated east-west with their horns to the west in response to predominantly easterly summer winds.

Air and sand movements to the lee of dunes. The size and strength of eddy currents developed to the lee of dunes, whether they are transverse, barchan or coastal forms, are significant in moving sand, particularly with winds gusting in the 60–80 km/hr range. It has been mentioned already that deposition on the lee slope of dunes occurs as sand is blown over the crest, and when material deposited rapidly near the crest slumps and slides down the lee slope. The action of eddy currents also appears to be important despite the fact that the movement of sand and air currents to the lee of dunes is a controversial subject since the pioneer work of Cornish (1897, 1900) and Bagnold's (1941) *wind shadow* concept. In fact, later research by Cooper (1958) on the coastal dunes of Oregon and Washington, USA, and by Sharp (1963) in the Mojave Desert, California, USA does not support the eddy concept.

On the other hand, Hoyt's (1966) observations on air movement in the dune fields of South West Africa partly substantiate the importance of eddy currents to the lee of transverse and barchan dunes. He found that a current of air moving up the windward slope of a dune continues over the lee slope in various directions, depending upon the speed and direction in reference to the slope, the grain size of the material and the height of the dune. However, for most of the time, particularly under strong wind conditions, the air current moves out over the lee slope, then down toward the toe of the slope and returns up the lee slope in the form of a large eddy. In some cases, according to Hoyt, the lee eddy appears to remove an increment of sand from the surface in front of an advancing dune which keeps this particular area relatively free of sand until it is eventually buried by the dune. This process of deflation influences indirectly the bedding characteristics of the dune deposition at the toe of the lee slope.

The effectiveness of the lee eddy seems to depend mainly upon wind velocity, sand supply, grain size, sorting, moisture and dune height. These factors are interrelated and any given combination of them will determine the quantity of sand the lee eddy can actually move.

4.3d Ergs

An *erg* may be arbitrarily defined as 'an area of aeolian sand deposits which covers at least 20 per cent of the ground surface and which usually has bedforms with a 3-order hierarchy of ripples, dunes and draas (large sandy bedforms)'. The sand cover in ergs is not completely continuous and parts of the underlying substrate are exposed between the draa and dune ridges. Grove (1958) noted that active ergs larger than 12 000 km^2 are located within the 15 cm annual rainfall isohyet. According to Cooke and Warren (1973) it has been calculated that 99.8 per cent of aeolian sand is confined to ergs covering an area larger than 125 km^2 and 85 per cent is in ergs greater than 32 000 km^2. The Rub al Khali erg in Arabia covers an area larger than 560 000 km^2. Ergs are usually confined to basins and their absence from highland areas is largely attributed by Wilson (1973) to acceleration and divergence of the winds which makes the sandflow undersaturated. Ergs, then, are seldom in equilibrium with their current mean sandflow system.

In most deserts, ergs generally form downwind of source areas composed of alluvial deposits which are undergoing rapid deflation. Ergs with thick aeolian sand deposits and draa development occur in the Sahara, Arabia and Asia in marked contrast to the thinner dune ergs of Australia which lack draa development. Because they determine the threshold velocity of sand movement, regional variations in grain size also probably control the effective wind regime in an area. Wilson's (1973) investigations in northern Algeria show that patterns of sandflow, bedform height, as well as spacing and orientation, vary significantly with position within the ergs on account of grain size.

The development of an erg with bedforms differs from one without such features in four main ways (Wilson, 1971): deposition takes place in metasaturated zones; erg growth is initially by lateral extension rather than thickening; erg shape and the actual sandflow rate and deposition rate pattern are time dependent until bedform growth is complete; bedform migration influences the erg's development.

4.4 LOESS

The aeolian differentiation of loess, marked by a decrease with distance in grain size, heavy mineral content and proportion of minerals of low stability, has been demonstrated by Fedorovich (1972) while clay mineralogy has been used as a source indicator of loess in the Middle West of the United States. Glass, Frye and Willman (1968) demonstrate a northwesterly source (dominated by vermiculite and dolomite) and a northeasterly source (dominated by illite), the whole varying vertically as well as horizontally because of drainage diversions in the Pleistocene. Stratification is not usually evident in primary loess because of its grain-by-grain accretion but it develops a vertical columnar structure with weathering. The characteristics of 'primary' loess (i.e. loess which has not been re-deposited by other agencies) are well documented and are consistent with an aeolian origin. Two thirds to three quarters by weight of some periglacial loess may lie in the silt range (0.002–0.06 mm) with varying amounts of fine sand and clay (Figure 5.10). As the clay size material in loess is transported in silt-sized clusters, loess is moderately well sorted in dynamic terms, but when it is disaggregated in the laboratory it appears poorly sorted. Deposition by differential settlement through air produces a microfabric characterized by abundant voids between the angular to subangular silt grains, yielding low dry densities (1200–1600 kg/m^3) and high porosities (45–55 per cent).

In the dry state, loess has considerable bearing strength and maintains near-vertical terrace and river bluffs. The strength of loess is controlled by the particle interaction which has three components. A calcite cement may be present, especially around root hairs and fine partings between aggregates. Clay bonds may also be present at points of contact between silt grains. Finally, at the low moisture contents (2–15 per cent) prevalent over much of the year in continental climates, soil water is held under high tension, water suction levels reaching several thousand bars. The strength of loess declines spectacularly as moisture content rises, however. When flooded under a load, for example, a loss of volume of up to 20 per cent may result, producing collapse (hydroconsolidation). In prolonged, monsoonal-type rains, whole hillsides of loess may collapse and flow away as a slurry. This is a major process in the erosion of the thick Pleistocene loesses of China.

4.5 CONCLUSIONS AND PROSPECTS

In the field, there is considerable difficulty in measuring adequately, particularly over long periods, turbulence, strength and direction of the wind at or near ground level. This means that wind and sandflow patterns are generalized because of a lack of representative and reliable data. As wind turbulence is difficult to measure and model mathematically its possible effects on bedforms and its significance in sediment entrainment can be evaluated only superficially at present. In this context, the fundamental principles of sediment entrainment have yet to be established by empirical work.

Apart from the first metre or two below the surface of dunes, little is known about the internal structure of these bedforms owing to the instability of dry sand. Despite the common occurrence of dune blowouts on coasts, few projects have assessed how variations in vegetation cover influence the nature of wind erosion. The relative importance of eddy currents in moving sand behind frontal dunes merits more attention than it has received so far, and the relationship between air-flow patterns and the size and shape of deflation hollows in coastal dune systems needs examination before dune stabilization programmes are implemented. There is also ample scope for investigating the threshold values of the surface shear stress which suspends dune sand grains.

Information on the rate of dune migration in general and the displacement of individual dunes in particular can be easily obtained from aerial photographs by using the latest photogrammetric and remote sensing methods but, of course, these are no substitute for process studied in the field. The need to depart from *ad hoc* studies and to undertake co-ordinated inter-disciplinary research projects is now widely acknowledged because of the complexity of monitoring aeolian processes and the need to acquire instrumented data. Some of the gaps in our knowledge have been mentioned briefly in this chapter. In many cases the combined expertise of the geomorphologist will be required to verify theoretical studies and to provide critical data for the development of models. Undoubtedly, a major problem will be the cost of conducting programmes in remote arid areas. This, however, can be overcome if there is a balance between basic and applied research.

FURTHER READING

Although published more than 30 years ago, the standard text in this field remains: BAGNOLD, R. A., 1941, *The Physics of Blown Sand and Desert Dunes* (Methuen). A welcome recent addition to the literature on deserts, including a useful summary of the theories of sediment transport is: COOKE, R. U. and WARREN, A., 1973, *Geomorphology in Deserts* (Batsford). A useful treatise on sand movement by wind is: HORIKAWA, K. and SHEN, H. W., 1960, 'Sand movement by wind'. *US Army Corps of Engineers, B.E.B., Tech. Memo.*, 119. Students wishing to widen their understanding of sediment movement in general should consult: ALLEN, J. R. L., 1968, *Current Ripples: Their Relation to Patterns of Water and Sediment Motion* (North Holland).

5
Cryonival and Glacial Processes

Realization that an enormous area of the land surface of the earth has been overrun by glacier ice at least once and probably many times in very recent geological time, spread relatively quickly after the publication of *Études sur les glaciers* by Louis Agassiz in 1840. Despite many diversions created by poorly founded and sometimes fantastic arguments, the glacial theory was well established by the beginning of the present century, and some of the most perceptive observations had been made and the essential relationships between process and landform by way of sedimentological properties established by the 1930s. Further progress was inhibited by the need for a stronger observational basis, especially in subglacial locations, and by the rudimentary state of glaciological theory. With the rapid rise of the science of glaciology in the 1950s and 1960s, glaciological theory overtook that of glacial geomorphology and sedimentology and it is only in the last two decades that work in the field has been tailored to formulation of general theories of glacial erosion and deposition.

These theories have greatly stimulated field work and they themselves are currently being tested and refined in the light of this field research. These changes, amounting almost to a revolution in thinking in glacial geomorphology and sedimentology, have been brought about by many professional geologists and geomorphologists, but the work of W. V. Lewis and his associates in the 1950s (Lewis 1960) and of G. S. Boulton since 1968 have been fundamental. General application of these ideas and of recent glaciological theory to perennial problems of glacial landform genesis has been attempted in the book by D. E. Sugden and B. S. John (1976).

Awareness of the nature and extent of the cryergic (frost) and nival (snow) processes dawned early but application of the knowledge to geomorphological problems was slow to develop. Certainly, knowledge of the extent and thickness of perenially frozen ground existed in the seventeenth century. Establishment of the study of cryonival processes on a scientific basis was delayed until after the first World War, fundamental contributions being by Taber (1929, 1930) and Beskow (1935).

Application of these principles to the understanding of landform and sediment genesis was relatively slow, however, and had to await the growth in awareness of the strategic and economic potential of the polar lands during and following the second World War. Steady but notable progress has been made in the past thirty years, as may be judged from the major contribution by Washburn (1973).

5.1 FORMS OF ICE IN THE LANDSCAPE

At the present point in geological time, atmospheric temperatures regularly fall below the freezing point of water at some time in the year over more than one quarter of the land area of the globe. Sedimentological, organic and geomorphological evidence suggests that this was true of between one third and one half of the land surface during the cold phases of the Pleistocene epoch (2.5 million–10 000 years ago).

Water in the solid phase is an important geomorphological agent. It exists in the landscape in two major forms: as finely segregated aggregations in soil or rock and massive, essentially pure bodies of ice, the generic term for which is *ground ice*; and as large tabular or elongate masses of polycrystalline ice variously mixed with rock debris resting on the land surface and extending locally into standing water bodies and known as *glaciers*.

Ground ice includes a wide range of forms (Table 5.1). The geomorphological importance of ground ice lies in its disruption of the ground surface by volume increase with the phase change water-ice-water and the associated modification of structural, strength and pore water properties of rock and soil. These freeze-thaw, or *cryergic* processes are associated with those arising from the action of snow, the *nival* processes: the generic term *cryonival* processes is applied to the whole suite.

Glaciers exhibit a very wide range in their shape or morphology and in their volume (Figure 5.1). Small glaciers occupying a distinctively equidimensional valley head or valley side alcove are known as *cirque glaciers*. When snow supply exceeds loss within the cirque, the glacier will overspill its threshold and move downslope, typically following valley lines as a *valley glacier*. Control of glacier morphology by valley size and shape produces *reticular glacier systems*, small glaciers joining trunk glaciers as tributaries, high level overspill of ice producing *diffluent glaciers* and, where overspill occurs across a major drainage divide, *transfluent glaciers*. Where valley glaciers emerge from a narrow valley on to an open plain, an *expanded foot* may develop. *Piedmont glaciers* develop where ice supply is more abundant, the lobe taking on an equidimensional form and flow pattern quite distinct from the glaciers which supply it.

Large accumulations of low-gradient ice masses in the divide regions within dissected mountain massifs are known as *icefields*. Icefields have a gently concave or flat surface and contrast with the convex-upward or dome form of the ice cap, which may be of considerable size (up to 50 000 km^2). Masses of ice cap type which exceed 50 000 km^2 in area are usually referred to as *ice sheets*. Ice sheets may reach such great thicknesses that they overwhelm the land surface over wide areas, their surface morphology bearing no relationship to detailed bedrock form, although particularly large mountain masses may pierce the ice sheet and stand out as *nunataks*. Icefields, ice caps and ice sheets evacuate ice from their basins as *outlet glaciers*. When outlet glaciers are not constrained by rock outcrops marginal to the ice sheet, they assume the broad, equidimensional dome-like form of piedmont glaciers (Figure 5.1).

Where ice sheets or large outlet glaciers reach tidewater a floating *ice shelf* may form. The contribution of land ice to ice shelves may be considerable where large, rapidly moving ice streams are involved, but in other cases local snowfall may be the dominant

FIG. 5.1. DIFFERENT TYPES AND SIZES OF ICE BODY IN PART OF EASTERN ANTARCTICA
On right, seasonally open water (light hatching) gives way to fast or bay ice (medium hatching) and to shelf ice (dense hatching). Terrestrial ice masses include margin of Antarctic ice sheet on left, large outlet glaciers such as Taylor Glacier and large piedmont glaciers. In partly ice-free mountain ranges are many small valley glaciers and equidimensional cirque glaciers (as in Asgard Range and Kukri Hills). Some transfluent glaciers lie south of the Ferrar Glacier.

supply source together with some accumulation from below by freezing of water on the underside of the shelf. Wastage is by the breaking away of large slabs or icebergs (calving) and by bottom melting. As floating ice masses, ice shelves exert only local and very limited stress on rock surfaces, although they may be significant transporters of debris where they are supplied with substantial ice and debris from ice streams.

5.1a Ground and soil ice

When atmospheric temperatures fall below 0 °C any water at atmospheric pressure on or near the surface will freeze. Water may remain in the solid phase for only brief periods in some climates while in others prolonged freezing of both soil water and ground water may occur. The length of the period of freezing, i.e. between freezing and thawing events, is known as a freeze-thaw cycle. Climatic environments characterized by freezing and thawing may be classified according to the number of freeze-thaw cycles, the severity of the mean winter temperatures and the proportion of the annual precipitation occurring in the solid form (snow, hail, rime). Ground and soil ice includes a wide range of types summarized in Table 5.1. Very low year-round air temperatures and the lack or thinness of an insulating snow cover result in ground temperatures remaining below freezing down to depths which may exceed 100 m. Such perennially frozen ground is often referred to as *permafrost* (Figure 5.2). The upper surface of permafrost is called the permafrost table, the zone of seasonal thaw being called the *active layer*. Unfrozen zones *(talik)* may occur within permafrost.

Table 5.1
GROUND ICE: TYPE AND PROCESS

Type	Process
Needle ice (pipkrake)	Strong radiative cooling of bare earth surfaces
Pore ice	Rapid freezing of saturated soils
Segregated ice	Slow freezing of partly saturated soils, ice clusters forming in some voids but not in others
Intrusive ice	Growth of sheets and lenses of ice by ice adhesion as water is injected under pressure from adjacent rock and soil
Vein ice	Freezing of water and water vapour in subvertical cracks produced by thermal contraction or desiccation
Extrusive ice	Overflow water from springs forced on to surface before freezing
Buried ice	Burial by debris of snowbanks or glacier, sea, lake or river ice

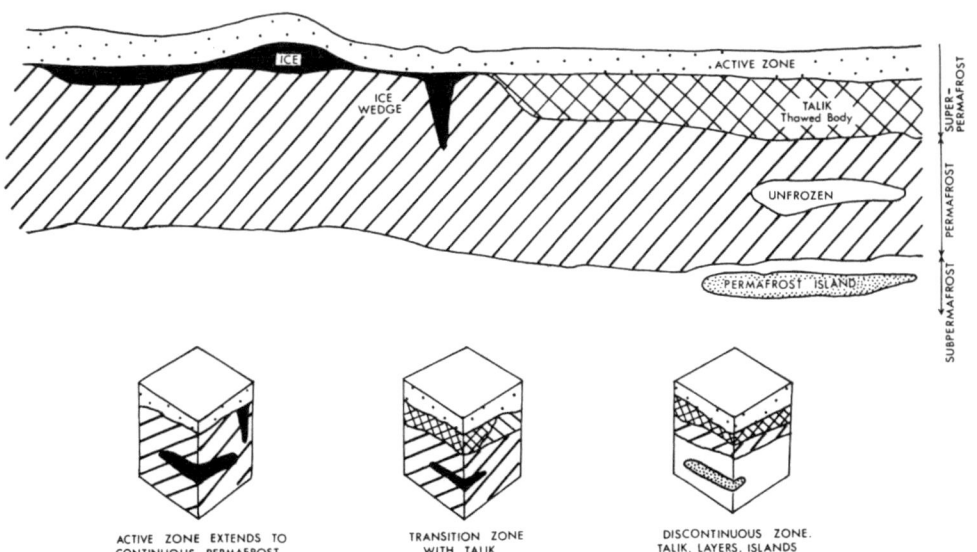

FIG. 5.2. SCHEMATIC DIAGRAMS OF PERMAFROST BODIES
(Re-drawn from Stearns, 1966 with permission of C.R.R.E.L.).

5.1b Snow and firn

In a climate in which annual accumulation of snow on the ground is greater than the loss resulting from ablation (a general term for losses due to melting, evaporation and sublimation), snowbanks accumulate and grow in size and thickness. In other words the hydrological budget (input minus output) is positive. Thus the growth of a snow bank represents the development of a *storage* within the hydrological cycle, similar storages with varying time lags between freezing and release by thawing occurring in the case of freezing groundwater and soil water. Snowpatches may be perennial, semi-perennial or merely annual features of the landscape.

Freshly fallen snow has a very low density (0.1 Mg/m^3 and less), the abundant voids being air-filled (high porosity). As snowbanks thicken, simple compaction in the lower layers tends to produce more uniform, equi-dimensional grains as the surface area/volume ratio is reduced, porosity declining to a minimum of about 40 per cent. At points of contact between grains, pressure melting occurs so that the grains fuse. This *sintering* process localizes the air spaces into pockets, so increasing the density, the crystals growing by molecular diffusion. Growth in grain size is more rapidly achieved in conditions of temperature oscillation about 0 °C. Rapid fusion of crystals raises the density to 0.4 Mg/m^3 and higher when the material is referred to as *firn*. In maritime areas of high snowfall and mild average temperatures, such as coastal southern Chile, coastal Alaska, the western Alps, western Pyrenees and the Cantabrian Mountains and New Zealand, this *firnification* may take place in a single winter. In continental areas with very low temperatures and low snowfall amounts, such as the north Greenland and Antarctic plateaux, the process may take decades.

5.1c Glacier ice

As compaction continues in a bank of firn, a large proportion of the air is expelled, the remaining air-filled voids becoming restricted to sealed bubbles within an interlocking crystalline mass of glacier ice. Crystals may exceed 1 cm in diameter and densities exceed 0.85 Mg/m^3. Compression of this low density ice (white or cloudy ice) results in further plastic deformation of crystals. Re-orientation of crystals and associated local liquefaction occurs, the interstices becoming sealed on re-freezing. The resulting blue glacier ice may reach a density of 0.91 Mg/m^3. Densities never approach very closely to 1.0 because of included air, dust and salts. The greatest crystal sizes appear to result from release of pressure and are found in stagnant glacier margins. The smallest crystals are found in zones of high stress, notably in the lowermost 30 cm of a glacier just above its bed. While normal stress is dominant during firnification, at glacier ice densities the stress field approximates a hydrostatic pattern. As a result crystal growth occurs in all directions and a random crystal orientation is characteristic. In zones of high shear stress, however, as on the margins and bed of valley glacier, crystal orientations may become strongly preferred in a direction normal to that of the loading stress, i.e. close to horizontal in the lower part of a glacier on a gently-sloping bed.

Parts of a glacier in which mean annual accumulation exceeds mean annual loss due to ablation are said to lie in the accumulation zone. This is a zone of stratified firn and snow, together with superimposed ice from re-frozen meltwater. Its lower boundary is termed the *equilibrium line*. Below the equilibrium line, ablation exceeds accumulation. No firn is formed in this *ablation zone*. The relative size on any single glacier of the accumulation and ablation zones is indicative of the state of a glacier's hydrological budget or *mass balance*. When a negative mass balance prevails, supply is less than ablation and, conversely, the mass balance is positive when accumulation exceeds ablation. Thus changes in the mass balance are expressed in variations in the altitude of the equilibrium line. The distribution of glaciers reflects this variable balance. Snow accumulates by direct fall, by snow drifting and by avalanching from adjacent slopes. When supercooled water vapour comes into contact with a cold surface it freezes in the form of rime. Ice may also accumulate in the form of superimposed ice (see above). Heat is provided for the ablation processes by direct solar radiation, molecular conduction from air or water surfaces, turbulent eddy transfer (convection) and that released in condensation. The latter two are by far the most important. Solar radiation tends to be the dominant heat source in continental climates and on high tropical mountains, and turbulent eddy transfer predominates in maritime, cloudy regions.

The process by which snow turns to firn and firn to glacier ice is essentially *diagenetic* and is akin to metamorphism in the rock cycle. Above the scale of the individual crystal, glaciers may be studied, like metamorphic rocks, in terms of their *foliation*. In glaciers, foliation may consist of alternating bands of clear and bubbly (blue and white) ice or of fine-grained and coarse-grained ice. The processes by which glacier foliation is generated are obscure. There is a strong body of opinion which favours the view that this longitudinal, concave lenticular structure (Figure 5.3) is the result of stress deformation arising from the flow, especially as it appears to be best developed in glaciers under demonstrably high

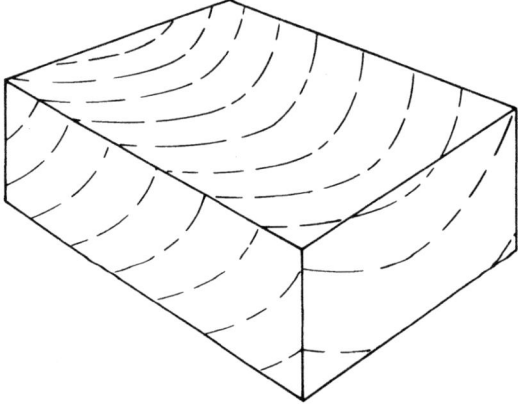

FIG. 5.3. BLOCK DIAGRAM OF GENERALIZED PATTERN OF FOLIATION IN VALLEY GLACIER FLOWING RIGHT TO LEFT

stress. High stress zones such as ice falls may impose the foliation which becomes revealed by surface ablation in the lower part of the glacier. However, examination of some glaciers in Norway has led to the suggestion that foliation is frequently derived from deformation of the sedimentary stratification of the accumulation zone, and that foliation which is not related to such pre-existing layering is uncommon. Understanding of the origin of foliation is important if patterns of concentration and redistribution of the suspended load of the glacier are to be understood properly.

5.2 SOME PROPERTIES OF MASSIVE ICE

A body of massive ice on or within a surface will deform with time under its own weight. The relationship between the rate at which deformation occurs (the strain rate) and the

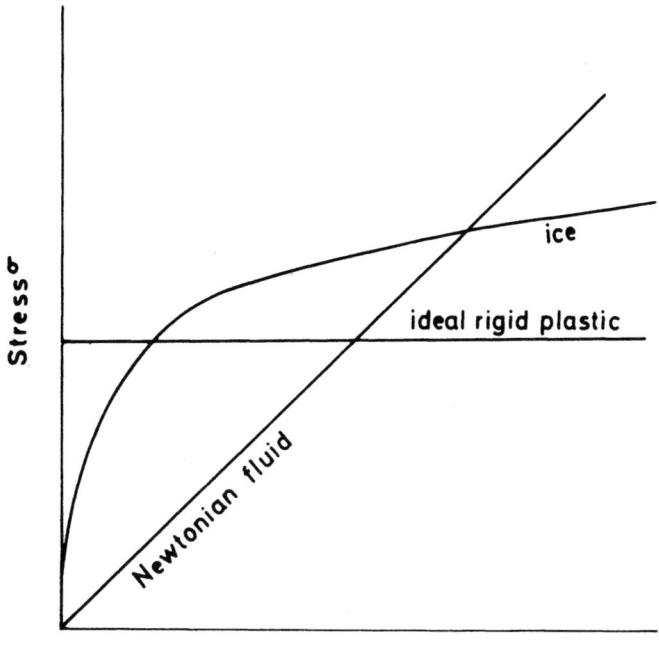

FIG. 5.4. REPRESENTATIVE STRAIN RATE/STRESS CURVES FOR NEWTONIAN FLUID, IDEAL RIGID PLASTIC, AND ICE. CF. FIG. 2.14

level of stress which produces it is shown in Figure 5.4. Ice is a polycrystalline solid whose behaviour when close to 0 °C is visco-plastic, deformation being predominantly by intra-granular creep parallel to the basal plane of the crystals. Creep curves may be divided into two types depending on the magnitude of the applied stress (Figure 5.5). If the creep rate

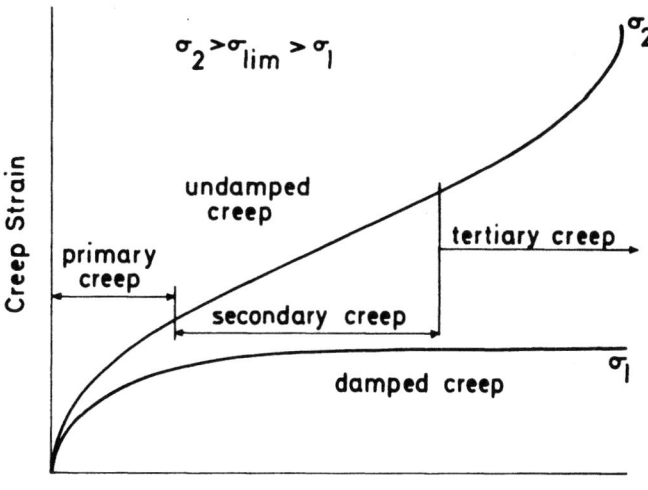

FIG. 5.5. REPRESENTATIVE ISOTHERMAL CREEP CURVES FOR FROZEN SOILS
(After Akili, 1970).

does not exceed a certain value, creep is said to be 'damped', and if this value is exceeded creep is 'undamped'. At low stress levels, the strain rate will decrease with time (primary creep) but if the strain rate becomes stabilized at a steady state value above zero, deformation will be constant with time (secondary creep mode). At higher stress values, this may give way to an accelerating creep rate (tertiary creep) which ends in rupture.

J. W. Glen showed experimentally that the final strain rate ($\dot{\varepsilon}$) across the basal plane of ice crystals is proportional to a power function of the shear stress (τ) across the basal plane in the form

$$\dot{\varepsilon} = A\tau^n \tag{5.1}$$

where A is a temperature-dependent constant and n is a constant (approximating 3–3.5) which is not temperature dependent. This applies to the steady state (secondary creep) mode and excludes the primary creep and accelerating creep modes. It can be seen from Figure 5.4 that polycrystalline ice has no yield stress, i.e. it is not a perfect rigid plastic and its behaviour is best termed visco-plastic.

As strain rate is so sensitive to ice temperature (constant A), especially close to 0 °C, the thermal regime of a particular glacier or massive body of ground ice will have a direct bearing on both the mode and the rate of internal deformation it displays. Thus, the climatic factor directly affects the process. This may be illustrated for both massive ground ice and glaciers.

Given that high stresses leading to the tertiary creep mode and rupture are likely to be rare and exceptional in ice bodies within soil, and given that ice creeps under relatively small stresses (as low as 0.03 kg/cm²), it seems likely that deformation owing to primary creep readily reaches a steady state value so that secondary creep may persist for long periods. The creep rate is influenced by the amount of contained debris, by the thickness of the ice body and by the degree of surface slope as well as by the ice temperature. A soil content in ice of 35 per cent may reduce the creep rate by an order of magnitude. Thus, perennially frozen ground made up of soil and ice of segregated type will display creep rates which are lower than those found in relatively pure intrusive or buried ice on slopes of the same steepness. Assuming relatively clean ice and an infinitely wide slope (a simplifying assumption to eliminate the necessity to calculate edge effects such as those produced by valley sides), ice 0.6 m thick will have a creep velocity of 0.3 cm/yr on a slope of 5° and this will rise to 2.0 cm/yr with an ice thickness of 3 m within a 30° slope (McRoberts 1975).

The geomorphic effects of creep in massive ground ice have not been assessed, although it is believed that they are widespread and most marked following an amelioration of climate or in locations where erosional incision has created new slopes. In both cases, the thermal balance in the ground ice is modified and the creep rate increased.

Glaciers move by a combination of internal deformation (intergranular and intragranular creep) and mass sliding on their beds. Glaciers which approximate a temperature of 0 °C throughout (isothermal ice) are known as temperate glaciers. Being at the pressure melting point throughout (except for a thin upper layer of colder ice in winter), temperate glaciers show high absolute values of internal strain and basal sliding. In relative terms, however, the high rate of pressure melting and production of meltwater and, in particular, the thin film (as thin as ~ 1 μm) of meltwater maintained at the sole, results in a high proportion of the total movement of temperate glaciers being by basal sliding (commonly 90 per cent).

When clean, polycrystalline ice slides over a smooth rock surface, the frictional force varies with ice temperature and with the velocity of sliding, reaching a maximum value of about 10^{-5} m.s^{-1} (Figure 5.6). Resistance to sliding of ice over rock consists of two components: friction arising from adhesion of ice and rock and the (friction) force required to 'plough' the irregularities of the rocks through the overriding glacier sole. Under most glaciers, however, the relationship is complicated because the debris concentration in the basal ice alters the area of contact between ice and rock, so affecting the value of the adhesion component, as well as the size range and the internal frictional components of strength of obstacles at the glacier sole (see p. 237).

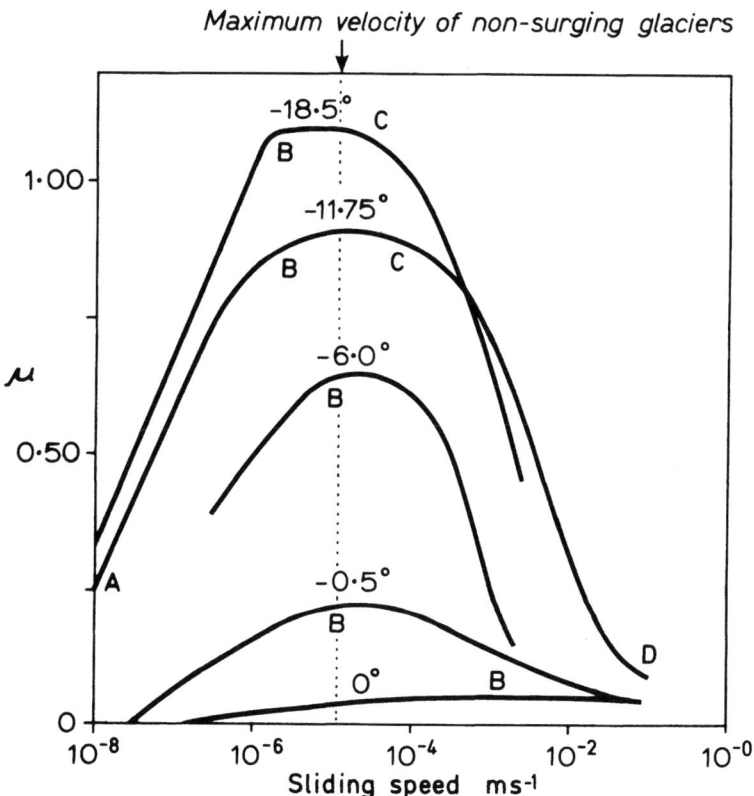

FIG. 5.6. FRICTION OF POLYCRYSTALLINE ICE SLIDING ON GRANITE OVER A RANGE OF TEMPERATURES
The velocity dependence of the friction is a result of three processes as follows: A–B, creep; B–C, plastic flow and fracture; C–D, frictional heating and melting of thin water films. From Robin, G. de Q. *J. Glaciol.*, 16 (1974), p. 185, following Barnes, P. Tabor, D. and Walker, J. C. F., *Proc. R. Soc. London*, Ser. A, 324 (1557), pp. 127–55. With permission of Dr Robin and Prof. Tabor.

A greater than average thickness of basal meltwater may develop as a result of a progressive thickening and steepening of the ice in the lower part of the glacier basin giving rise to very high basal shear stress values in the transition zone from the stagnant snout to the actively sliding area. Current theory predicts that such an increase in the basal shear stress producing sliding will result in a reduction in basal water pressure upstream of this zone, i.e. an up-glacier water pressure gradient (or at least a condition of zero gradient) will develop. In such a situation, down-glacier migration of water from upstream will be impeded and greater than average thicknesses of water will accumulate beneath the sole. As the water layer thickens, the effect is to 'lubricate' the ice-rock interface with the result that the glacier slides more easily and basal shear stress values decline. The friction may be reduced in this way over a sufficiently large proportion of the glacier sole to allow sliding velocities greater than 10^{-4} m.s^{-1}; they may sometimes exceed 10^{-1} m.s^{-1} (Figure 5.6). Glaciers sliding at such velocities are said to *surge*.

A glacier in which there exists an approximate balance between flux and storage of both mass and energy over a period of time is said to be in a steady state. Changes in the input will result in changes in storage which will eventually be balanced by changes in output so that the glacier tends to adopt a new steady state reflected in its three-dimensional geometry (length, breadth, thickness).

The nature and distribution of glaciers reflect variations in the factors controlling mass balance. Glaciers occur in the higher latitudes where even the low annual snowfall increments are preserved because of perennially low temperatures combined with low radiation inputs. At high altitudes, glaciers develop because of a combination of low temperatures and high snowfall amounts. At very high altitudes above mean upper cloud levels where humidities decline, temperatures are so low that the relatively high radiation inputs fail to result in complete ablation of thin firn and ice covers. Here, snow increments may be so slight and firnification so slow that the densities characteristic of glacier ice are never achieved. In maritime areas, on the other hand, such as those close to sea level in the subarctic or in mountain areas in middle latitudes, snowfall may be so heavy that even rapid summer ablation fails to remove all the winter's accumulation. Such glaciers exhibit high input and output values (high throughput) and are said to have a high activity index. They tend to have a high erosion potential. In contrast, dry polar glaciers tend to have a low or very low throughput (low activity index).

In continental polar situations, glacier temperatures may be below the pressure melting point by as much as 10–50 °C and, even at the sole, the negative temperatures persist so that the glacier is frozen to its bed and basal sliding is absent. In such polar glaciers all movement is by internal strain. These glaciers display greater regidity than those of temperate type, i.e. there is a more markedly elastic response to stress. Between the temperate and high polar glacier types there is a gradation of glaciological conditions in which only parts of a glacier persist at a temperature below the pressure melting point (Figure 5.7). Sub-temperate glaciers move mainly by basal sliding except for their outer, thinner margins in winter. Subpolar glaciers move predominantly by internal strain in their outer zones but by both flow modes in their thicker, central zones.

Generalization is clearly dangerous, however. Given a constant temperature at the ice-atmosphere interface, variations in the basal temperatures of a glacier are controlled by the ice thickness so that ice thickening in bedrock depressions may locally raise the basal ice temperature to the pressure melting point. This appears to be true of parts of the Antarctic ice sheet. Internal strain rate is also affected by the mass balance. In the case of two glaciers in equilibrium moving entirely by internal strain, for example, the larger glacier will have a greater input and greater loss than will be the case with the smaller

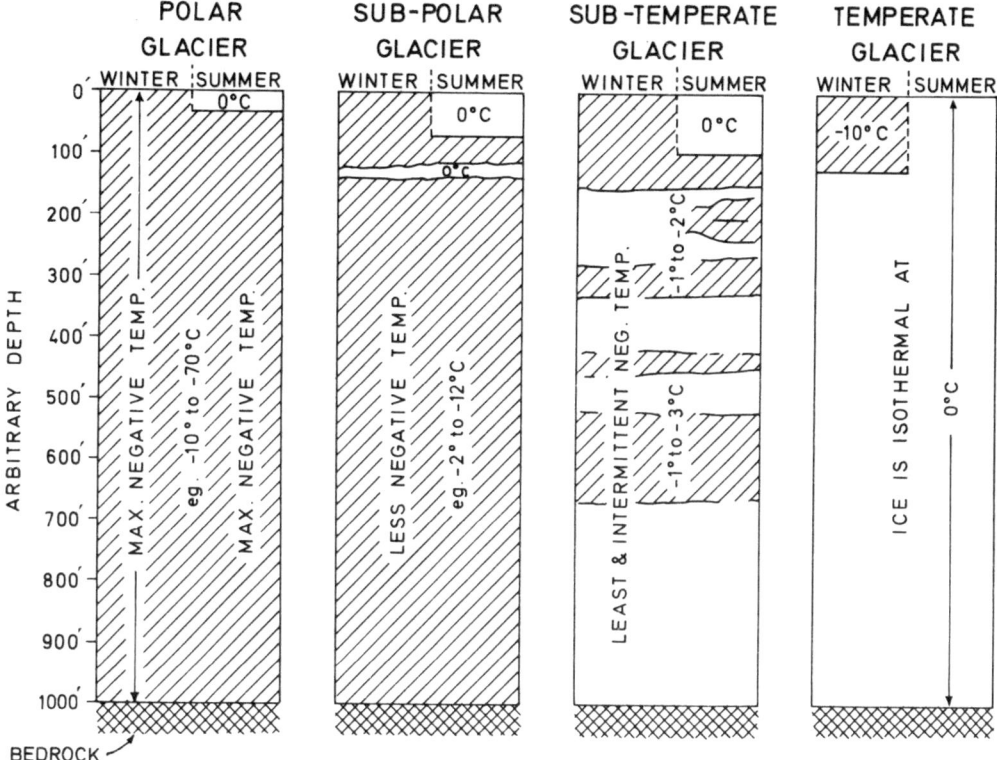

FIG. 5.7. THERMAL CLASSIFICATION OF GLACIERS ACCORDING TO MILLER (1973)

glacier, so that the throughput will be greater. Hence the absolute value of the strain rate will also be greater for the larger glacier.

In high stress zones as, for example, where convexities occur in the bed ('ice falls'), or where the glacier flows around a bend in a valley, an accelerating creep rate (tertiary creep mode) may prevail and rupture of the ice mass ensue. This occurs under both compressive and tensional stress. J. F. Nye (1952) has set out clearly a model of these two types of flow which he called compressing and extending flow, respectively (Figure 5.8). Compressing flow occurs where ice is being lost by ablation from the surface and where the bed is concave. Extending flow occurs where increments of ice are building up on the glacier and where the bed is convex. In general, then, extending flow is more characteristic of the accumulation zone and compressing flow of the ablation zone. However, shifts from extending to compressing flow may be seen in many glaciers especially in relatively thin valley glaciers flowing over an uneven bed.

The trajectories of maximum shear stress in a perfectly plastic glacier are known as 'slip lines': they are disposed tangentially to the bed, intersect the glacier's upper surface at 45° and converge downglacier in extending flow and upglacier in compressing flow (Figure 5.8). Movement occurs along the slip lines by plastic deformation or by brittle fracture. Brittle fracture in extending flow produces open, downglacier curving crevasses: in compressing flow thrust planes (analogous to reverse faults) dipping upglacier are the result.

A

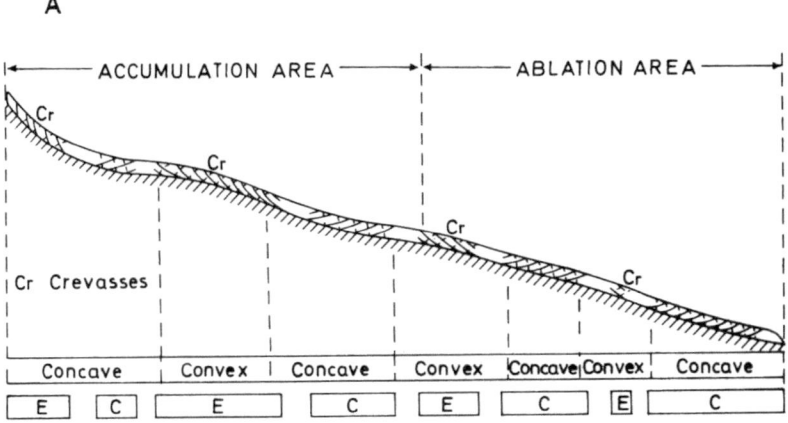

B

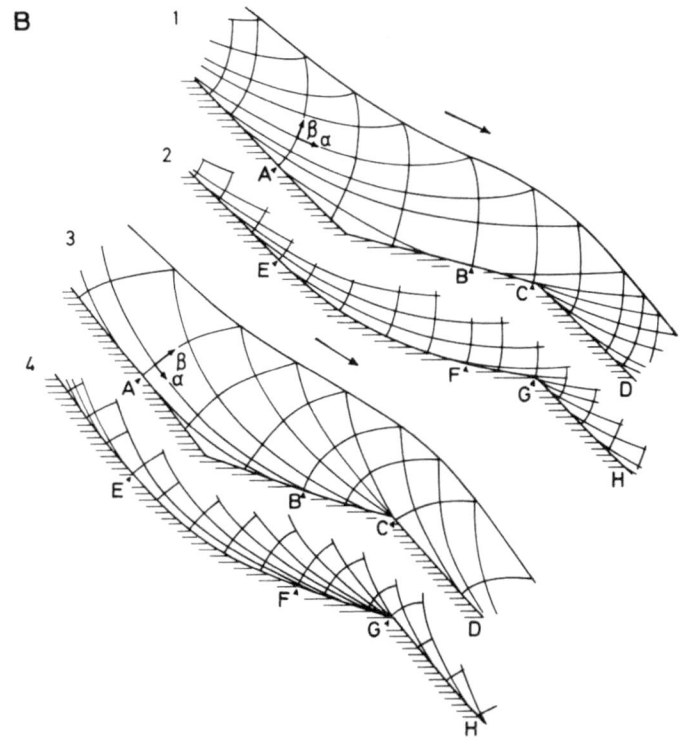

FIG. 5.8. A – LONGITUDINAL SECTION OF AN IDEAL GLACIAL VALLEY SHOWING ZONES OF EXTENDING (E) AND COMPRES-
SING (C) flow in a glacier moving over an uneven bed. Slip line field and crevasse zones also shown. After Nye
1952, by permission of the author. B – Slip-line fields are represented by two families of curves, the direction of
the curves at any point indicating the two directions (α and β) in which the tendency to shear is greatest. Slip-line
fields in compressing flow before (1) and after (2) erosion and in extending flow before (3) and after (4) erosion.
After Nye and Martin, 1968 by permission of J. F. Nye.

At any given value of ice thickness, bed gradient, bed roughness and valley sinuosity, the stress value at which ice enters the tertiary creep mode leading to fracture will vary with ice temperature and hence with climate.

The relative importance of each of the two modes of flow (internal strain and basal sliding) and the absolute value of the internal strain are subject to wide variation with climate, but they also vary within the same glacio-climate province because of the important effect of ice thickness on the pressure melting point of ice. It follows that knowledge of the thermal regime of glaciers is fundamental to an understanding of the processes of glacial erosion, debris entrainment by ice, and glacial deposition.

5.3 CRYONIVAL PROCESSES

The cardinal process in cryonival regions is the cyclic phase change to ice of water situated on or within rock and debris masses. It is useful to distinguish those processes that act without significant translocation of material (processes *in situ*) and those processes that result in debris transport. The latter processes have already been discussed: free-fall, sliding, granular creep, flow, solifluction and surface wash in Chapter 2, wind transport in Chapter 3 and plastic deformation in Section 5.2.

The *in situ* processes which affect both bedrock and soil are freezing and thawing, salt crystal growth and the cracking of the land surface as a result of contraction at very low temperatures.

5.3a Rock disruption

The action of gelivation (freeze-thaw) on rock surfaces or rock clasts within heterogeneous debris gives rise to frost-shattering or *gelifraction*. Marked oscillations of temperature subject rock to stresses which may be strongly directional under the influence of solar radiation and the surface form and structure of the rock mass. Most rocks in the dry state have considerable resistance to such thermal stress, and breakdown resulting from fatigue may take a very long time. The presence of water in the rock greatly accelerates fatigue and the effect of freezing is maximized by the 9 per cent expansion of water in pores, bedding planes and joints.

The soils, disturbed surficial debris (regolith) and many of the rock types found at the surface of the earth contain abundant void spaces. These voids include a variable proportion of passages called capillaries which are so fine that solutions may be held in them against the force of gravity. Such water may be held under varying degrees of tension, the energy required to remove it being termed the capillary potential. Total rock porosity and mean pore size are important variables in rock shattering regardless of rock hardness. Both the rate of water absorption and the percentage water content are influenced by the pore size because of the effects on water movement of capillary tension. As a result, only rarely are all voids in a rock mass filled with water and one result of this is that considerable volumes of air are trapped in the capillary pores. Water in the larger pores freezes first as its pressure is close to atmospheric and, as freezing progresses into the finer interstices, pore pressures rise while, in turn, the freezing point of the remaining water falls by 0.8 °C for each atmosphere of pressure increase.

As only the air in the air-water-ice system is compressible, considerable air pressures may be generated as the volume of water varies with the water/ice phase change. Slow freezing prolongs the duration of the phase change and extrusion of ice from the larger pores attenuates the pressure effects. With relatively rapid freezing, however, relief of stress by slow growth and extrusion of ice is minimal and a crack or series of cracks develops, stress being greatest at the advancing tips of the cracks. When cracks attain a length at which the rate of liberation of energy owing to stress relief in the rock becomes greater than the rate of energy consumption by work done at the advancing tips, the fissure will extend without any additional rise in the applied stress. The value of the applied stress at which small cracks reach this critical length is known as the *tensile strength*. As pore pressures rise in the closed or partially closed system which develops in the pores in conditions of rapid freezing, the tensile strength of the rock may be overcome and shattering of the rock mass results.

Because the porosities and pore diameters of different rock types vary, frost shattering susceptibility may be expected to vary with rock type. For example, all chalks are known to be frost susceptible. Experiments by Lautridou have demonstrated notable variations in the rate of fragmentation under laboratory conditions of several types of limestone (Figure 5.9). Atmospheric weathering preparatory to freeze-thaw action greatly increases the yield of frost-shattered debris.

The concentration of salts (particularly common salt, gypsum and magnesium sulphate) in solutions within rock pores plays a part in rock disintegration which may locally exceed gelifraction in importance, particularly in polar deserts where it has been fairly well documented. In ice-free valleys in southern Victoria Land, Antarctica, for example, boulders of granite, granite gneiss and dolerite develop deep concavities (alveoles) in their surfaces (cavernous weathering) and a variety of elongate forms ('taffoni'). Some of the largest and best developed alveoles occur under the margins of the boulders where shade is greatest and where salt is concentrated in the surrounding mixed regolith by local melt of thin snow mantles and upward migration of salt-rich soil solutions in the conditions of strong evaporation which characterize polar rock desert. Crystallization of salts is a major cause of alveolation. Salt is known to be locally very highly concentrated in the very thin surface of seasonal melting or 'active' layer (commonly only 8–10 cm thick) in the Antarctic dry valleys.

The following conditions appear to be necessary for salt weathering to be effective: a supply of salts; availability of sites protected from severe washing or blowing out so that the salts may accumulate; and cyclic changes in humidity, temperature or both which include the crystallization point of at least one of the salts present.

At least three processes arising from the action of salt coatings on rock surfaces have been proposed: differential coefficients of thermal expansion, hydration, and the force of crystal growth (Evans 1970). It has been suggested that, as most salts have a much higher coefficient of thermal expansion than the common rocks, heating of a rock containing concentrations of salts within cracks may propagate the crack owing to differential crystal expansion at a time (of drying) when the force of crystal growth will have reached a peak. The importance of this process requires further testing. Expansion in crystal size by cyclic hydration of anhydrous salts is well documented, especially in the case of gypsum/anhydrite. It has been used in explanation of granular disaggregation for many years, especially

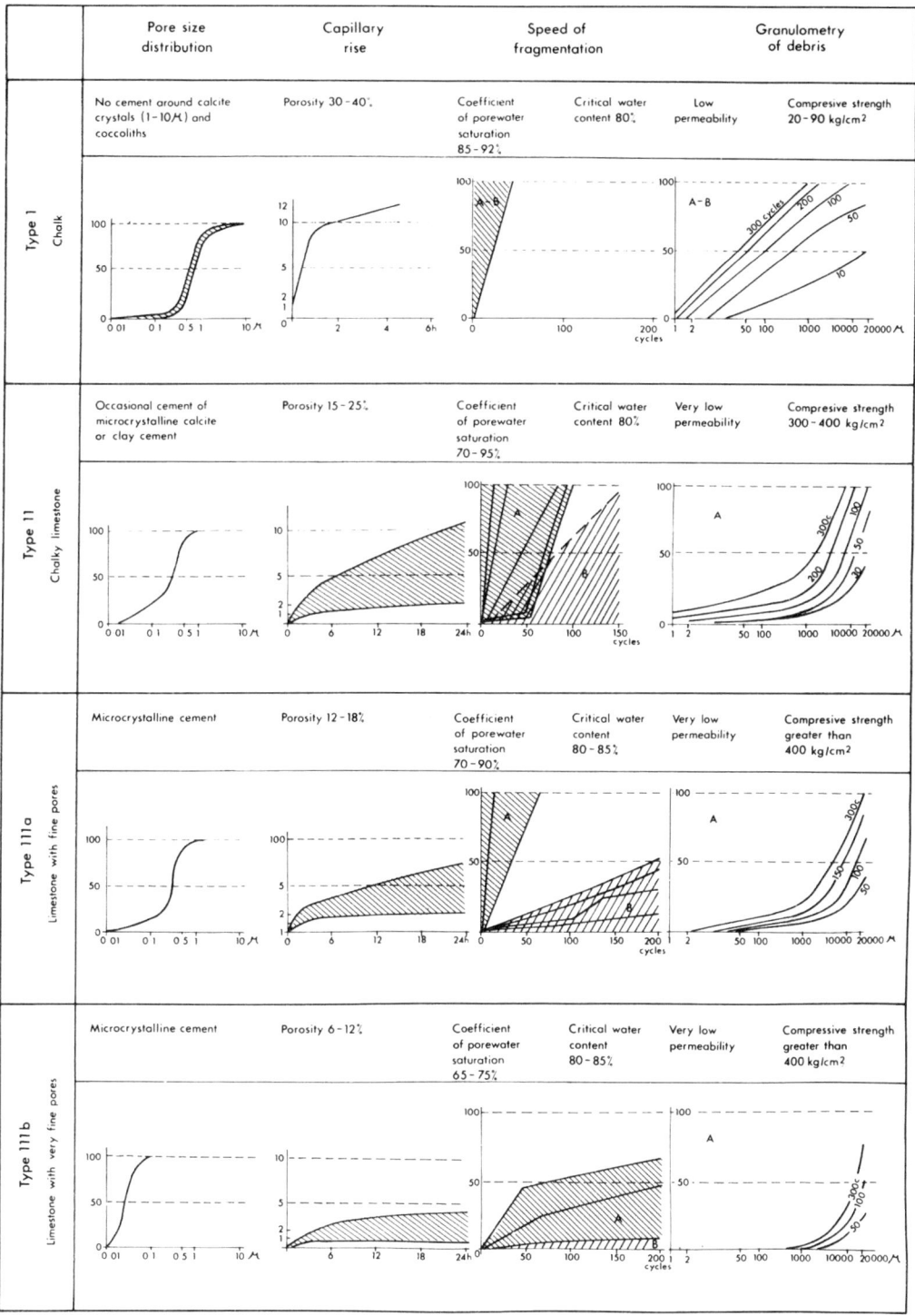

FIG. 5.9. RESPONSE OF FOUR TYPES OF LIMESTONE BLOCKS (1–2 KG) TO FREEZING AT −5°C
A – abundant water supply, water depth 4 cm. B – water depth 1 cm (after Lautridou 1972).

in desert regions, but there are obstacles to its acceptance as a general explanation. Evans notes that while hydration is probably very important in the case of carbonates and sulphates, it does not explain the very numerous examples where sodium chloride is known to be the principal salt present. Evans sets out a strong case for considering the force of crystal growth to be the dominant process. It is known, for example, that with a supersaturation of 1 per cent, calcite may crystallize against a pressure of 10 bars, which approximates the tensile strength of many rocks. As with frost shattering, disruption arising from crystal growth will be most evident in porous and poorly cohesive rocks. In addition, the process will be most widespread in strong evaporation conditions conducive to high levels of supersaturation at times of crystallization.

The widespread, if often very slow chemical weathering in periglacial environments has been somewhat neglected because of the predominance of mechanical weathering products in such regions. Under a wide range of vegetation type and cover, biological and chemical processes have contributed to the development of soils in northern Eurasia and North America, including podsols, bog and tundra varieties. The interrelationships of soil-forming processes and cryogenic processes in the forest-steppe of Trans-Baikalia has been demonstrated by Makeev and Kerzhentsev (1974). In winter, the organic litter making up the snow-free surface is broken up by frost. In spring, biological activity is inhibited by low soil temperatures but by the end of May, the soils thaw to a depth of 80–100 cm and the upper 20 cm reaches temperatures of 8–10 °C. During summer rainfall events, water and solutes move down the profile. With freezing in autumn, the translocated salts and organic material are re-distributed. Cracking in winter can be followed in summer by inwashing of surface humus to some depth, frost action in the succeeding winter resulting in contortion and even inversion of the soil horizons (producing an ACB profile locally). This interdigitation by frost action of inwashed humus and ferruginised and bleached podsolic horizons is known to produce extremely complex soil profiles.

In the extreme polar desert environment of the dry valleys of eastern Antarctica, there is accumulating evidence of chemical weathering in soils and rocks. Slow hydration of mica is producing clay minerals in the soil under the present climate. The presence and mobility of liquid water films of high ionic strength on mineral grain surfaces in these soils is strong circumstantial evidence of chemical alteration, the mobility of the ions and solutes occurring during infrequent and brief periods of replenishment of soil water by snowmelt and localized melting of ground ice. It seems clear that the weathering sequence of the principal rock-forming minerals is the same in these conditions as it is in more temperate regions: the *rate* of chemical weathering, however, is very much slower.

Long continued gelifraction on surfaces of low slope breaks up bedrock into angular masses known as blockfields *(blockmeer, felsenmeer)*. Disruption along planes of weakness in bedrock by frost wedging is followed by gelifraction of the disturbed blocks so that a progressive comminution occurs. Owing to differences in rock structure, susceptibility to frost-shattering and length of period of exposure, the grain-size distribution of material making up such a locally-derived blockfield may vary. This, together with the action of ancillary processes such as removal of fines by suffosion, produces a range in fabric from openwork to that typical of a diamicton (see 5.3b). This modification of the fabric further affects the susceptibility of the blockfield material to frost heaving. The best developed blockfields are found in enclaves which remained unglaciated in the late Pleistocene, such as in parts of Labrador and Norway, and so have experienced a very prolonged period of frost climate.

The chaotic appearance of many blockfields conceals some important systematic relationships. For example, the mean size of the frost-riven blocks frequently increases with proximity to the source outcrop. These residual outcrops or tors can be seen in all stages of disaggregation in the Antarctic dry valleys where salt action plays an important role in the case of granite tors. Here, both hillslope and summit tors show clear evidence of blockfield development by frost riving, frost shattering and exfoliation, variations in tor morphology from angular to rounded and variations in the incidence of chemical weathering being a product of microclimate as it affects the distribution of snowdrifts (Derbyshire, 1972). Retreat of the tor masses under frost attack with movement of the frost-riven blocks and associated fines by solifluction, nivation, suffosion and surface wash produces a cryoplanation or *goletz* terrace cut into bedrock which becomes progressively thickly veneered with debris as distance from the bedrock outcrop increases.

Blockfields themselves frequently show a distinctive micromorphology related to bedrock contours, presence or absence of a matrix of fines in the blockstream, and variation in mean diameter of the blocks. On slopes greater than about 5°, according to Washburn (1973), or 15° (Caine, 1968), the fabric of blockfields may become anisotropic owing to downslope movement such that the longest axes of the blocks are aligned and imbricated downslope as gradient increases. Local variations in fabric include the alignment of blocks transverse to flow on the risers of stepped features such as stone-banked lobes and terraces (see below). Such debris is best termed a blockslope and, where clearly elongate, a blockstream.

Talus development may be strongly seasonal; for example, Rapp (1960) found that talus development ceased entirely during the depths of winter in Spitsbergen, rockfall activity reaching a maximum in spring and early summer. Seasonal variation in the supply rate and in the incident process on talus slopes may explain the distinctively stratified slope deposits of Pleistocene age known as *grèzes litées* or *éboulis ordonnées*. Stratified scree is rhythmically bedded especially, though not exclusively, in calcareous terrains. It consists of layers of coarse angular to subangular talus with an openwork fabric, platy and elongated clasts lying parallel to the surface slope and alternating with layers of fines (sand to silt range). While there is general agreement that the angular scree is the product of gelifraction, the nature and origin of the fine layers has not been determined. Experimental work by Guillien and Lautridou (1970) suggests that both coarse and fine fractions may be produced by frost-shattering, but several workers have described the presence of far-travelled material especially in the fine fraction. Seasonal sorting of scree by meteoric waters or snowmelt in association with solifluction, summer mudflow deposition, differential creep, sorting by meltwater and pipkrake, or slope-wash sorting of debris in thin sheets over fresh ground layers and accretion of fine, wind-blown dust (loess) have been invoked in explanation of stratified slope deposits. Despite a large body of opinion to the contrary, long-lying snowdrifts and stratified scree slopes appear to be mutually exclusive. Accordingly, interpretation of stratified screes as a result of nivation processes (see p. 225) may be an oversimplification.

5.3b Soil disruption

The grain size of the products of gelifraction and mechanical salt weathering may show wide variation owing to differences in rock type (as affecting grain size, mineralogy and structure), location and disposition of the surface (micromorphology), salt and water supply and period of exposure to these processes. Frost-shattered rocks often show the influence of joint and bedding characteristics in the marked differences in clast size and shape of the debris produced. Sometimes, the frost-shattered mass of debris (gelifractate) may be dominated by angular boulders while, at others, large clasts are set in a finer matrix produced by granular disaggregation, free-fall impact, hydration, chemical weathering and inwash by surface waters. Such mixed deposits, possessing a fine as well as a coarse mode (*diamictons:* Figure 5.10), are the most widespread surface materials in cryonival regions. For the sake of brevity, they are referred to hereafter as soils.

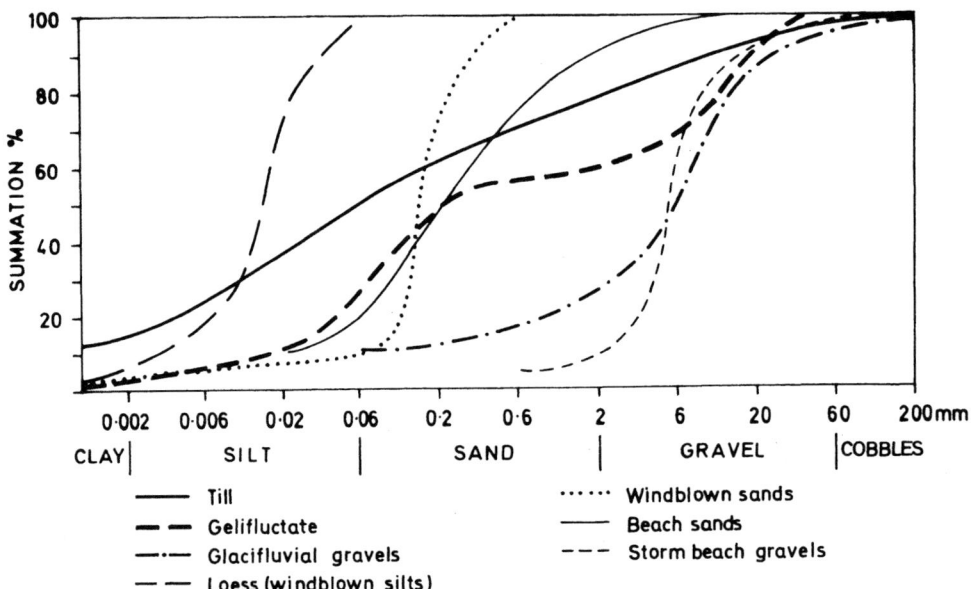

FIG. 5.10. PARTICLE SIZE DISTRIBUTION CURVES FOR SELECTION OF NATURAL SEDIMENTS
Better sorted sediments have steeper curves. Till and gelifluctate, extremely poorly sorted, are diamictons (with two maxima, one in gravel range and another in sand/silt range).

Water dispersed in the finer pores of rock and mixed debris cannot be considered in the same terms as free water: it is influenced by the surface electromagnetic energy of the soil particles and is referred to as adsorbed or 'bound' water. Adsorbed water freezes gradually from the outside of the layer towards the surface of the soil particle. Such thin films separating ice and mineral grains are essential elements of the theory of ice segregation and rock and soil disruption.

Freezing of the soil system involves a complex interaction between and within many sub-systems which is not well understood. However, it has been known for many years that moisture and heat produce a segregation of pure ice within some types of soil body (Taber, 1930), and considerable attention has since been given to heat and mass flow in freezing soils. Heat flow within soils is a major influence on water crystallization. The flows of heat and water in soil are intimately related. The heat of fusion of water is 80 cal/g which means that for each 80 calories conducted to the soil surface, 1 ml of water must be supplied to maintain the rate of segregated soil ice growth. Heat is transferred in soils by conduction and by the movement of solids, liquids and vapour (mass transfer mechanisms). Movement of water in a freezing soil occurs in the liquid phase as thin films and capillary flow, and by vapour diffusion.

The grain size distribution of the soil, in directly affecting properties such as void ratio and capillary potential, is of fundamental importance in its effect on the rate, incidence and type of ice crystallization. For example, in coarse granular sediments, bound water constitutes a very small proportion of the total, the free water migrating rapidly and freezing within the large voids. This results in little or no physical disruption of the particle arrangement (fabric) and the ice so formed acts as a cement. In fine-grained soils, however, or in diamictons with a matrix rich in the very small, colloid-size grains ($0.1–0.001$ μm), the situation is quite different. Because of the much larger surface mineral area of clays, unfrozen water per unit volume is high so that thin water films tend to be continuous. Once ice crystals develop, colloidal mineral grains tend to dehydrate and to become clustered, the density of the packing depending on the surface electrical energy of the particles and the type of ions involved. Hydrophobic colloids aggregate first; those which are hydrophilic aggregate later. Colloids with high surface energy will show relatively dense clustering, the reduction in bulk volume producing fissures and other voids in the soil which may become filled with ice. This is the process of ice segregation in which ice *schlieren* punctuate the mineral mass to form one very common cryogenic texture (Figure 5.11).

When soil temperatures fall below 0 °C, only the free water in the larger pores at or close to atmospheric pressure freezes, followed by freezing of capillary water and then by water adsorbed on the mineral particle surfaces. This phased freezing of the water component in soils (Figure 5.12) gives rise to supercooling in advance of the freezing front. (While water in bulk normally freezes at 0 °C, tiny droplets may remain in the liquid state at temperatures as low as -40 °C. Such water is said to be supercooled). As water freezes, the water content in the adjacent soil mass is reduced, the pressure difference between the ice and the water phases being partly expressed as a negative soil water pressure (i.e. tension). This soil water tension induces movements of water to the freezing front, the freezing temperature of the soil water being depressed at the same time. The hydraulic gradient produced by this reduction of pore pressures at the frost line maintains a supply of film water to the freezing front so that segregated ice may continue to grow.

A frost-susceptible soil is defined as one which is characterized by the formation of segregations (single grains, veins, wedges, thin horizontal films, thick sheets and a variety of irregular masses) of essentially pure ice within pores, fissures and other voids in the soil body. Volumetric increase in the soil is expressed by movement in the direction of heat flow (toward the ground surface) and is referred to as *frost heaving*. While freezing of the interstitial water may account for a maximum of only 9 per cent increase in the volume of the mass, development of segregated ice supplied with soil moisture in an open system commonly results in soil volume increases of 50 per cent, the amount being equivalent to the volume of segregated ice generated.

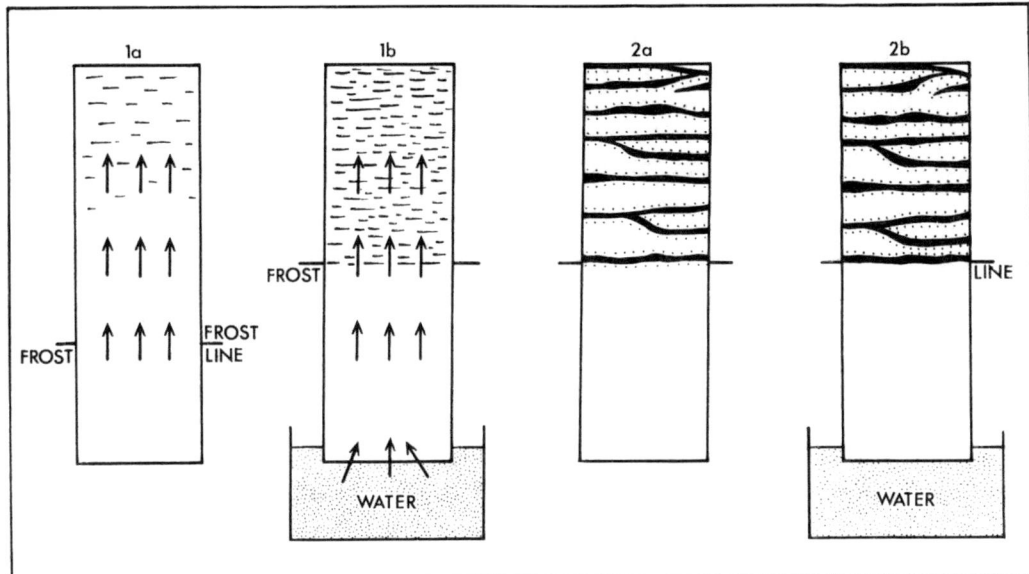

FIG. 5.11. DIAGRAM (REDRAWN FROM BESKOW, 1947) SHOWING DIFFERENCE BETWEEN FREEZING OF (1) SILT AND (2) A 'FAT' CLAY, (A) WITHOUT, AND (B) WITH, FREELY AVAILABLE GROUND WATER
In soil 1a, the small amount of available water flows quickly to freezing line and forms ice layers in top part of soil. The water is soon exhausted so that ice bands are thin and sparse and formation ceases quickly. Total amount of heaving is small. In case of soil 1b, continuous supply of water ensures abundant segregated stratified ice and considerable heaving. In case of soils 2a and 2b, little difference evident because fine grain size and higher capillary adhesion of clays inhibits water flow. Conditions depicted assume that the soil cylinders are capillary saturated to the same extent from the beginning of experiment.

Frost heaving pressures are related to capillary potential by way of grain radius, pore size and the grain geometry, and Penner (1966) generalized the relationship in the form

$$\Delta P = [2\,\sigma_{in}\,\cos\theta\,(1+B')]/r \qquad (5.2)$$

(where σ_{in} is the ice-water interfacial energy, θ is the contact angle between ice-water and water-solid surfaces, r is the particle radius and B' is the ratio of r/r_2 to the cosine of the contact angle, r_2 being the pore radius). Penner concluded that it is the smaller particles in a soil that determine the magnitude of the heaving pressure, so that, in a mixed deposit, the surfaces of the ice lenses undulate. This has important implications in patterned ground development (see p. 215). Although there is no *abrupt* dividing line separating frost-heaving and non-frost heaving soils, nevertheless, soils rich in silt grade particles (0.06–0.002 mm) are more susceptible to the growth of segregated ice than grades coarser and finer than this. The general relationship between heaving and soil particle size is shown in Figure 5.13. The effect of this on a mixed deposit is that when the freezing front moves from sand to silt, the rate of pressure increase rises sharply but when the freezing front moves from silt into sand the pressure levels off (Hoekstra, 1969; Figure 5.14). The relationship breaks down for clay-rich soils because of the very high particle surface force effects, most of the water being in the adsorbed condition. It has been shown that, although the general trend is for surface heaving to increase as frost depth increases through the

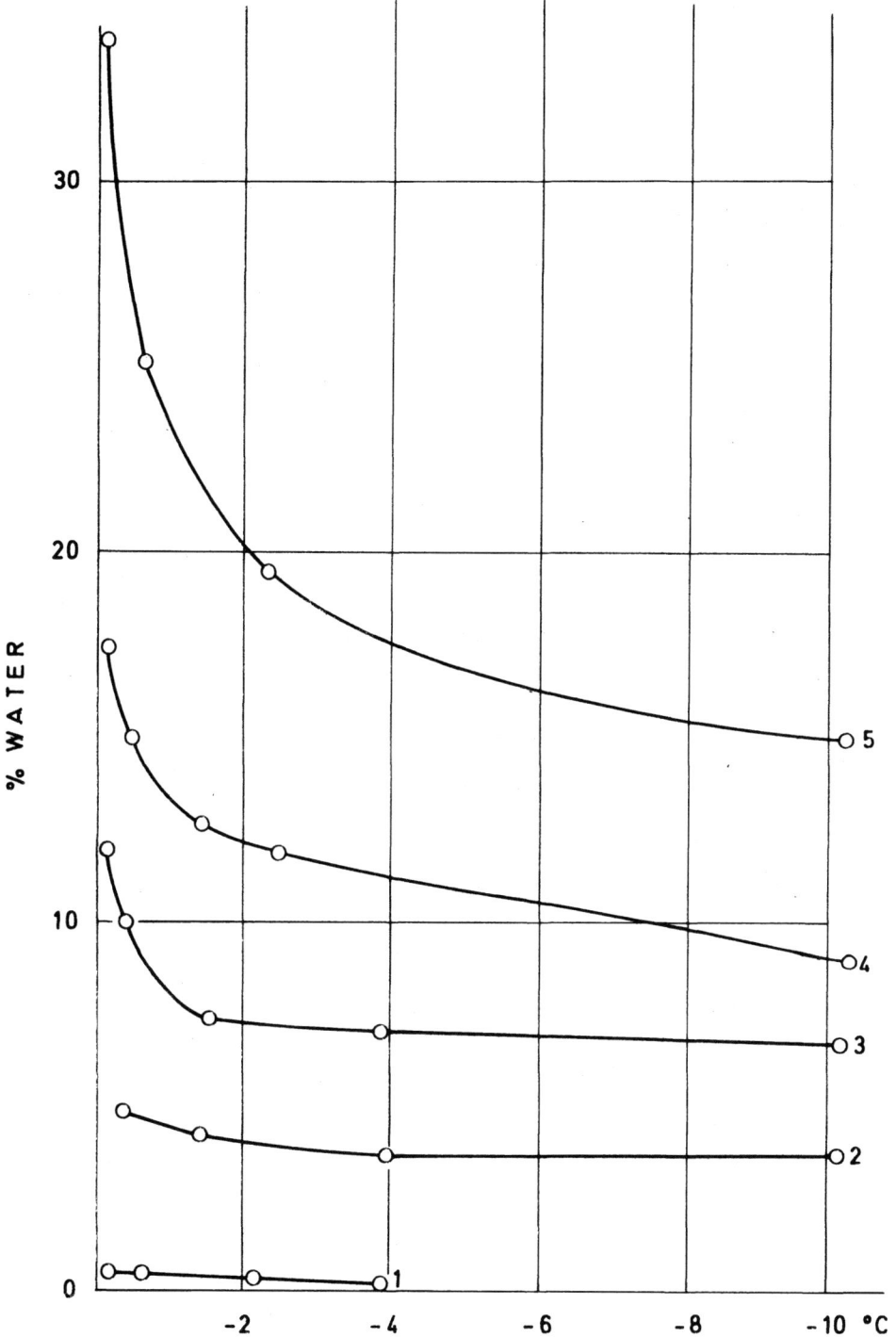

FIG. 5.12. UNFROZEN WATER CONTENTS IN TYPICAL NON-SALINE SOILS
1 – quartz sand; 2 – sandy loam; 3 – loam; 4 – clay; 5 – clay containing montmorillonite (after Nerscova, Z. A. and Tsytovich, N. A., 1966).

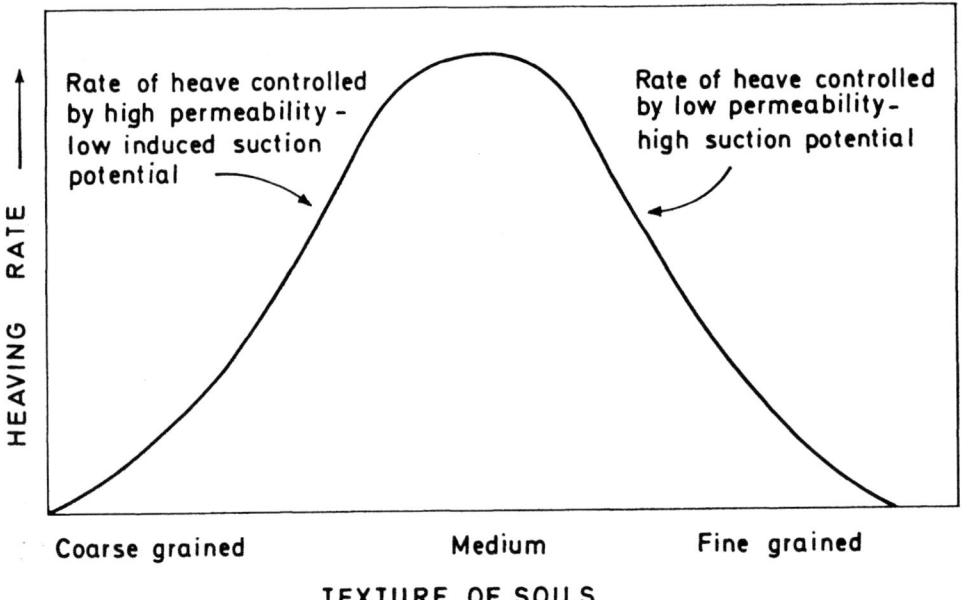

HEAVING RATE

Rate of heave controlled
by high permeability -
low induced suction
potential

Rate of heave controlled
by low permeability-
high suction potential

Coarse grained   Medium   Fine grained

TEXTURE OF SOILS

FIG. 5.13. SCHEMATIC REPRESENTATION OF RELATIONSHIP BETWEEN HEAVING RATE AND PARTICLE SIZE RESULTING FROM
ICE LENS GROWTH
(From Penner, E., 1968, *Soils and Foundations*, 8, pp. 21–29, with Author's permission.)

winter, there is a close and quite rapid response to weather variation, warm spells in winter reducing the ground temperature gradient and hence the heaving rate.

Frost penetration is also influenced by the thermal characteristics of the materials making up the soil. As ground freezing occurs over a relatively wide range of temperatures for reasons already discussed, latent heat of fusion is a factor in temperature changes below 0 °C within the soil. Accordingly, the apparent specific heats of freezing soils (specific heat plus latent heat) vary quite widely. Thus, the thermal properties of the soil materials constitute a major factor affecting the depth of frost penetration. It follows that frost susceptibility and frost penetration are inextricably related. Surface cover also affects frost penetration values. For example, a snow depth of 40 cm or more reduces the effective freeze-thaw cycles to one (the annual cycle) and the type, condition and density of vegetation, in affecting the radiant energy flux, influences frost penetration, although much research is required to quantify the relationship.

In addition to the flow of water and heat during soil freezing, ions, salts, colloids and coarser mineral particles also migrate. The repeated passage of a freezing front through a soil body is a major source of vertical sorting of soil particles. The most widespread and familiar example of this process is that associated with the growth of needle ice or pipkrake (Figure 5.15), needles of clear ice commonly exceeding 5 cm in length. Growth of pipkrake is favoured by supercooling of the soil in conditions of marked radiative heat loss from the surface, sufficient supply of soil moisture to initiate ice segregation, and continuity of heat flux from, and water flux to, the freezing plane. Given steady state conditions, there is only one surface temperature that will balance the heat flux equation (involving soil, sensible and latent heat flux components and the thermal radiation bal-

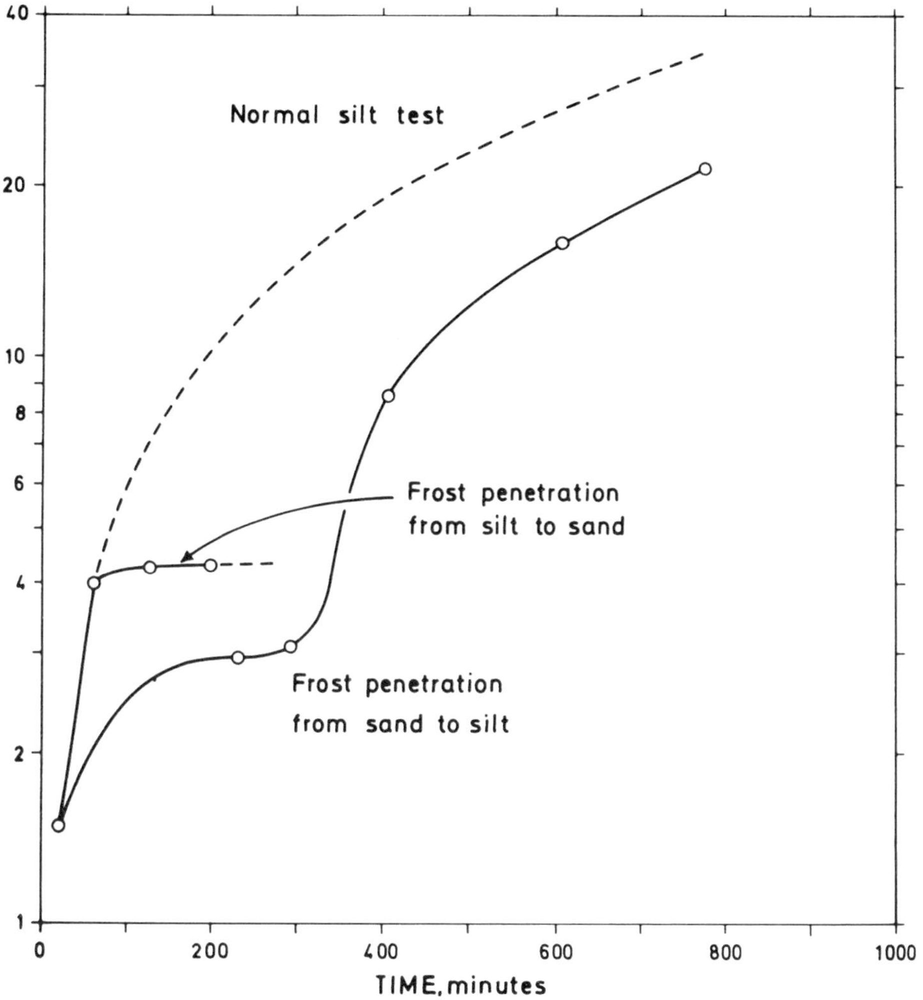

Normal silt test

Frost penetration from silt to sand

Frost penetration from sand to silt

40

20

10

8

6

4

2

1

0   200   400   600   800   1000

TIME, minutes

FIG. 5.14. FREEZING PRESSURES VERSUS TIME IN STRATIFIED SEDIMENTS
(After Hoekstra, 1969).

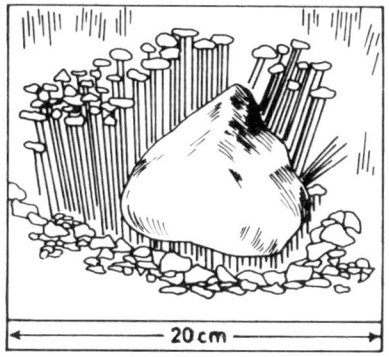

20cm

FIG. 5.15. UPLIFT OF SOIL PARTICLES BY NEEDLE ICE (PIPKRAKE) ON UNVEGETATED TILL IN CENTRAL LABRADOR-UNGAVA
(After photograph by E. Derbyshire).

ance) and known as the equilibrium surface temperature. This has been found to be at least as low as −2 °C for ice nucleation to occur, constituting the first necessary condition for needle ice growth.

As ice segregation is known to cease at a critical value of soil water tension, tensions below the critical value for the particular soil (a function of pore size) constitute the second necessary condition for pipkrake growth. The third condition requires that the heat flow rate from the freezing plane to the surface is balanced by the heat of fusion of water arriving at the freezing plane from below. Any increase in this heat flux will increase the soil water tension at the freezing plane. In this case the soil water will tend to freeze *in situ*, producing dirty needles or an undifferentiated frozen soil layer. Outcalt (1971) has developed an algorithm for needle ice growth based on these three limiting conditions (Figure 5.16). Wind may be a secondary factor in needle ice growth. The occurrence of

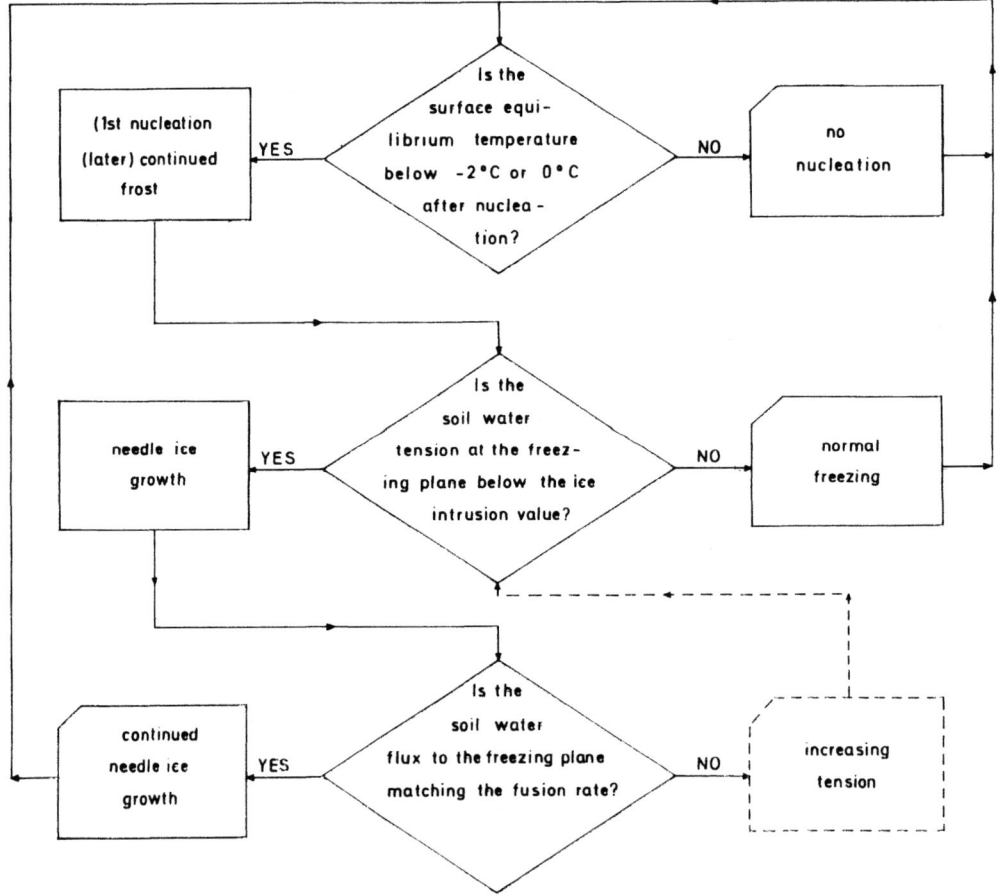

FIG. 5.16. ALGORITHM FOR NEEDLE ICE GROWTH (AFTER OUTCALT, 1971)
Soil capping ice needles will develop before step three in the sequence and polycyclic or 'tiered' needles are produced by periods of stress (critical conditions) at steps two and three. Rocks up to 15 kg in weight are known to have been lifted by needle ice and, in the frozen state, needle-ice segregated surfaces have bearing strengths considerably in excess of those applied by treading (~0.4 kg/cm²).

finely striped areas of needle ice has been explained by some as the product of the azimuth of sun and shadow in the melting phase. Beaty (1974) provides limited experimental data suggesting that temperature inversions with shallow cold air drainage produce striped needle ice parallel to the airflow direction. There is clearly a need for further experimental work to determine the role of wind in needle ice growth.

Heaving of stones to the surface from within the body of a soil has been widely observed in regions of frost climate and a general explanation for the phenomenon has been that ice lensing is concentrated beneath larger stones which have a higher thermal conductivity than the mixed debris containing them, the base of the stone reaching the freezing point before the containing soil at the same level, the ice lens so formed thrusting up the stone differentially. This has been called 'frost push'.

Upward migration of clasts within soil may also be effected by adhesion of parts of a stone to the heaving ice-soil mixture as the freezing front penetrates from above. This is called 'frost pull'. Both 'frost push' and 'frost pull' components are involved in vertical sorting of clasts in freezing soils. Particle shape will clearly influence processes such as the re-settling of clasts into voids on thawing and the tendency of tabular or rod-shaped clasts to become erected within frozen soil. This well-known phenomenon of vertical stones within frost-heaved soils has been explained by a variety of processes, the most promising being the sequence of force-couples generated during freezing. In a mixed deposit with randomly oriented clasts, 'frost pull' due to cyclic downward growth of segregated ground ice will favour an increase in the dip of elongate clasts during upfreezing, aided by heaving of voids *above* stones. As the heaving layer thickens to embrace the lateral margins of the stone so upward movement into such voids is enhanced, the void developing beneath the stone becoming filled with ice and suffering compression owing to heaving so that a 'thrust' component follows a 'pull' component of movement (Figure 5.17). Another factor causing variation in the proportion of these two processes may be depth below the ground surface and number of frost cycles experienced. On the other hand, it is difficult to envisage how the 'frost push' mechanism alone might result in erected stones. However, the differential nature of segregated ice growth as well as the downward growth of seasonal ice above a permafrost table result in cryostatic pressures (hydrostatic pressures within unfrozen lacunae) reaching 4 kg/cm^2, and these may act in a lateral as well as a vertical sense. A good deal of experimental work on stone disposition and sorting remains to be done.

It is known that fines migrate in front of an advancing freezing plane and that particle migration varies with grain shape and size and hence with mineralogy, as well as with freezing rates. The freezing rates required to maintain migration of selected minerals of various diameters ahead of a freezing front have been determined by Corte (1962; Figure 5.18). He demonstrated that fine particles move continuously in front of the freezing interface under a wider range of freezing rates than do coarser particles, the range of freezing rates being greatest for mica mobility and least for quartz mobility. It is inferred from this that the larger the range of freezing rates experienced by a particular soil, the more susceptible it will be to segregated ice formation.

Soils outside regions of perennially frozen ground freeze and thaw from the surface downwards. In permafrost conditions, on the other hand, the active layer freezes from both top and bottom but thaws only from the top. This is important in that the second case allows the development of a closed or partially closed system for part of the year in which

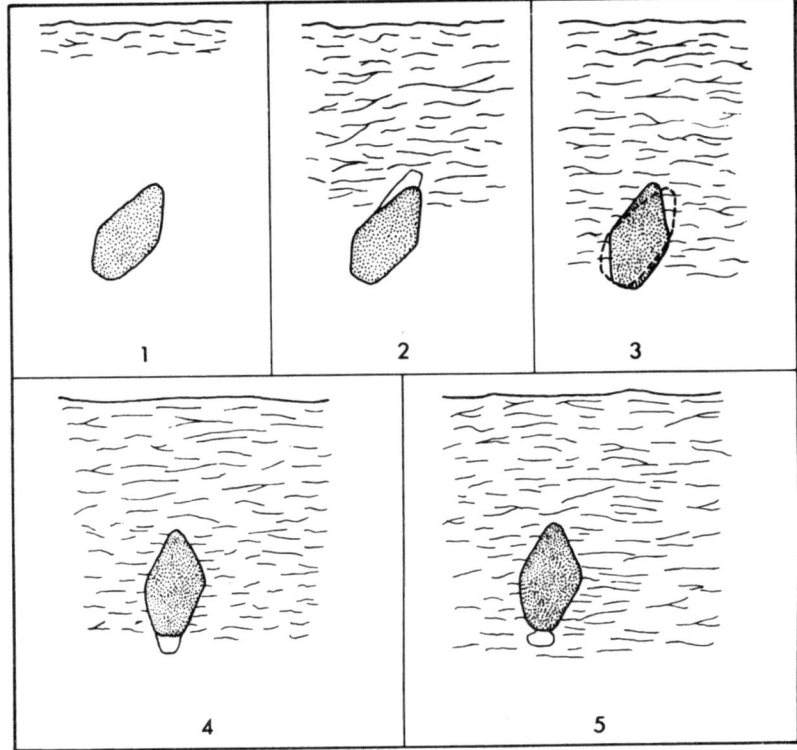

FIG. 5.17. ERECTION OF STONES BY DOWNWARD FREEZING OF SOIL, INVOLVING BOTH FROST 'PULL' AND FROST 'PUSH'. Growth of segregated ice *schlieren* produces void above stone (2). Stone moves into that space aided by frost 'pull' arising from ice adhesion (3). Void is left behind below stone (4). This becomes confined if deeper penetration of segregated ice occurs (5).

cryostatic stresses may reach higher values than in an open system. This may be reflected in distinctive soil structures (involutions) and surface morphology (hummocks: Figure 5.20). The same effect may be produced by volume changes in soils subjected to temperature alternations below 0 °C, the soil at no time being allowed to thaw.

Vertical sorting has been shown to occur in both permafrost and non-permafrost types of regime under cyclic freezing, the coarse particles moving against and the fines moving with the freezing direction. This proceeds until such time as a 'filter system' becomes established, i.e. until translocation of grains produces a situation where material in one layer can no longer pass through an intermediate layer into a third stratum. If the freezing front is inclined as, for example, on a river bank or terrace, migration of fines results in a lateral rather than a vertical sorting. In addition, the spatial variation in grain size within a natural sediment may result in differential volume increases such that a mounded microrelief is formed. This, in turn, may stimulate lateral sorting due to mechanical action from high to low points on the mound surface. The sorting mechanism is enhanced by slow freezing rates (Figure 5.19).

The tendency for mounds to form in frost-susceptible soils was investigated experimentally by Corte (1966a, 1972). A layer of crushed quartz (mainly of sand grade) was placed

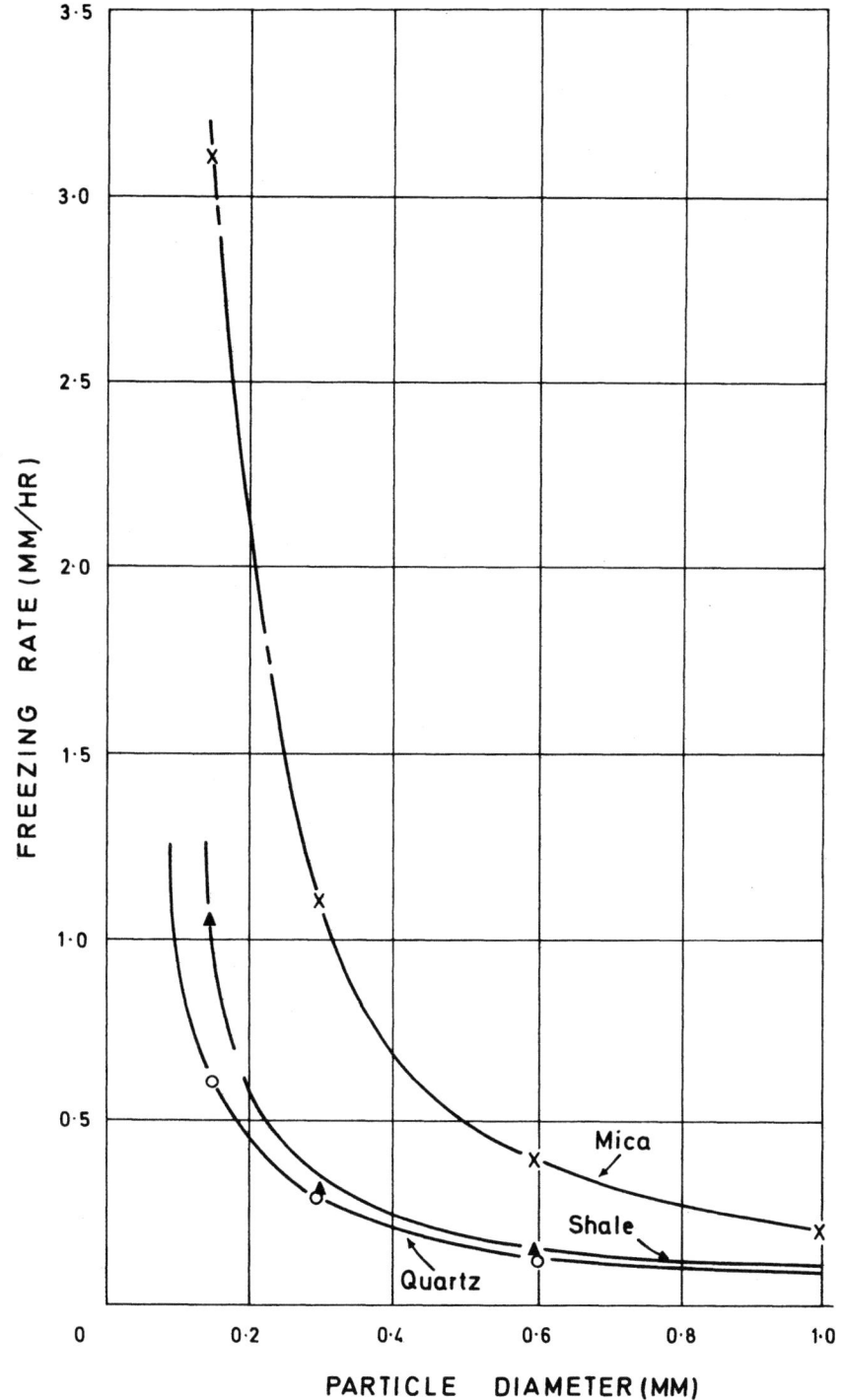

FIG. 5.18. FREEZING RATES REQUIRED FOR PARTICLES OF DIFFERENT DIAMETER TO MOVE CONTINUOUSLY IN FRONT OF FREEZING INTERFACE
(After Corte, 1962a).

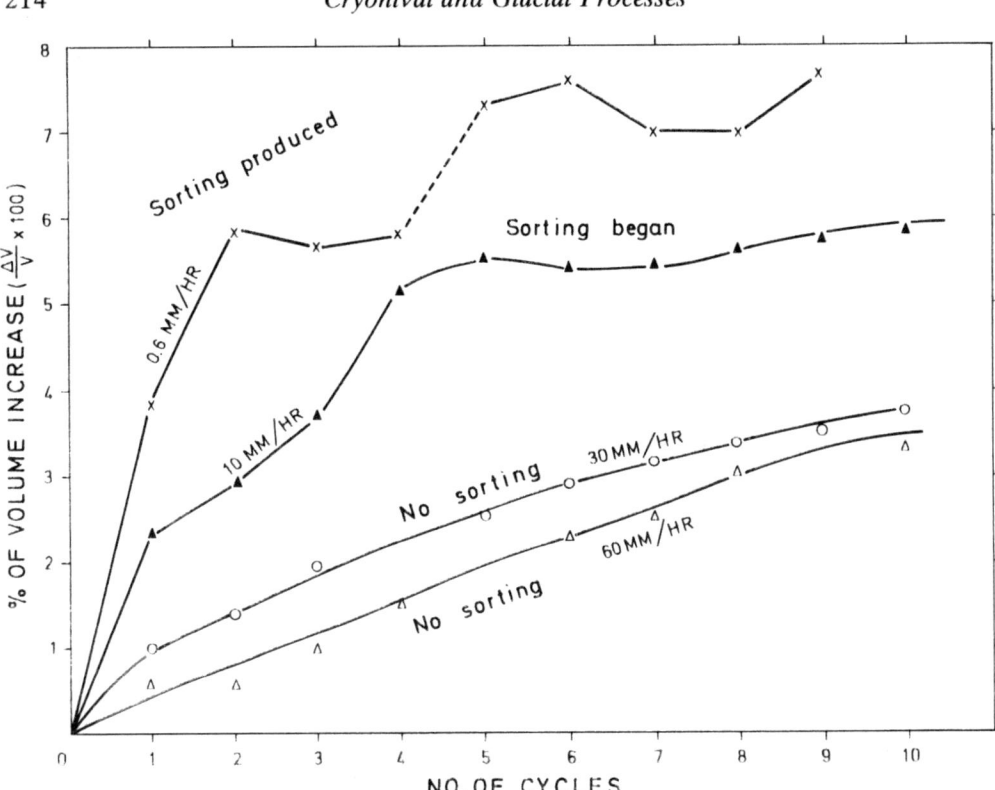

FIG. 5.19. VERTICAL SORTING AND RELATIONSHIP BETWEEN VOLUME INCREASE OF SOIL AND NUMBER OF FREEZE-THAW CYCLES, FOR FOUR RATES OF FREEZING
(After Corte, 1962b).

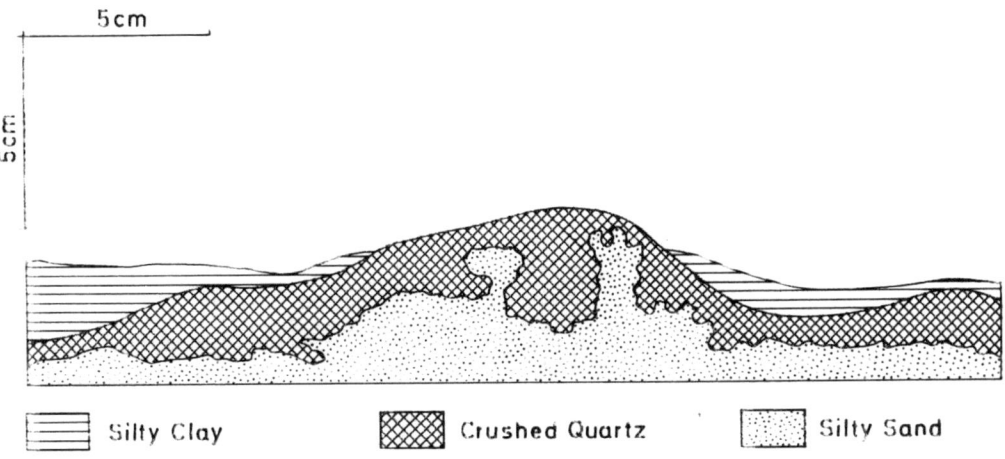

FIG. 5.20. CROSS-SECTION, SHOWING DISRUPTION IN ARTIFICIAL STRATIFIED DEPOSIT AT 57TH FREEZE-THAW CYCLE
(After Corte, 1972).

between layers of silty clay in a container insulated on its sides and base and in which water was uniformly available to the base of the sample. Freezing took place from above and proceeded to 5 °C at a rate of 1 mm/hr with thawing at a rate of 2 mm/hr. The soil heaved and sank under cyclic freezing. In the first 28 cycles, a single broad mound developed and volume increase reached 30 per cent. During freezing cycles 29 to 47, eighteen small mounds formed and maximum volume expansion in them was 24 per cent. Mound growth rate declined after cycle 37. Crushed quartz grains were extruded in the mound centres (Figure 5.20). The location of the mounds is explained by Corte in terms of the yielding of the upper silty clay as the crushed quartz layer began to freeze, the loci of failure being in areas with the highest content of unfrozen water.

The processes of small scale ice-segregation, heaving, frost creep, frost desiccation and dilation cracking and vertical and lateral debris sorting give rise to a complex micromorphology made up of domes, lobes and stripes, characteristic of periglacial terrains and to which the descriptive term *patterned ground* is applied. Round patches of unsorted, bare soil, most clearly seen where they punctuate a partly vegetated surface are known as nonsorted circles or, where centres of heaving are close together, nonsorted polygons or nets. The disequilibrium causing the raised circle or polygon may arise not only from differential frost heave but also from expansion caused by the presence of certain salts and swelling clay minerals, notably montmorillonite. Removal of a protective turf mat by wind scour may be an important initial process in the genesis of these forms. Both involutions and vegetated non-sorted hummocky ground are known from areas of swelling-clay soils in temperate and tropical regions, so that they cannot be used alone as indicators of former frost climates.

Small-scale non-sorted features known as hummocks *(thufurs, buttes gazonnées)* are particularly associated with moist vegetated and bog terrains in arctic and subarctic regions. Earth, turf or bog hummocks have a relief of up to 1 m. The uneven insulation offered by ground vegetation has been used to explain the build-up of hydrostatic pressures such that piercement of the organic layer by mineral soil occurs on freezing or liquefaction on thawing. Such inorganic debris mounds have been likened to non-sorted circles, but their essential formative processes are still to be determined.

Mounds of fines bordered by coarser material are known as sorted circles, and as sorted polygons when adjacent coarse margins merge. The size of the patterns is a function of the mean maximum clast size. All sorted patterned features appear to be derived from parent material of mixed grain size with at least 10 per cent in the silt/clay range, unsorted below the frost table, and bordered by 'gutter depressions' containing the largest stones along which summer drainage runs and in which tabular stones stand on edge as a result of lateral 'expansion-squeezing' (Goldthwait, 1976). Sorted circles formed in debris containing frequent coarse clasts attain diameters of 1–2 m, and may reach 10 m, those in finer debris generally being smaller. The amount of heaving becomes more unevenly distributed as the margins become increasingly deficient in fines of susceptible grade and the loading stress on the substrata increases at the sorted margins. Some cracking of the ground arising from thermal contraction is associated with all strictly polygonal forms of periglacial type. Chambers (1967) has suggested that movement in polygons may cease in an equilibrium stage when the coarse border material becomes 'anchored' at the base of the active layer.

Quantitative data on sorted circles and nets overlying permafrost near Thule, northwest

Greenland, have been presented by Schmertmann and Taylor (1965). The size and number of cobbles in the upper, coarse layer (0–15 cm) was found to increase systematically outward from the centre of the features. Orientation of the longest axes of clasts showed a clear tendency to point toward the centre of the feature so that a radial pattern was produced (Figure 5.21a). Dip of surface clasts showed a 0.5 probability of being 20° or

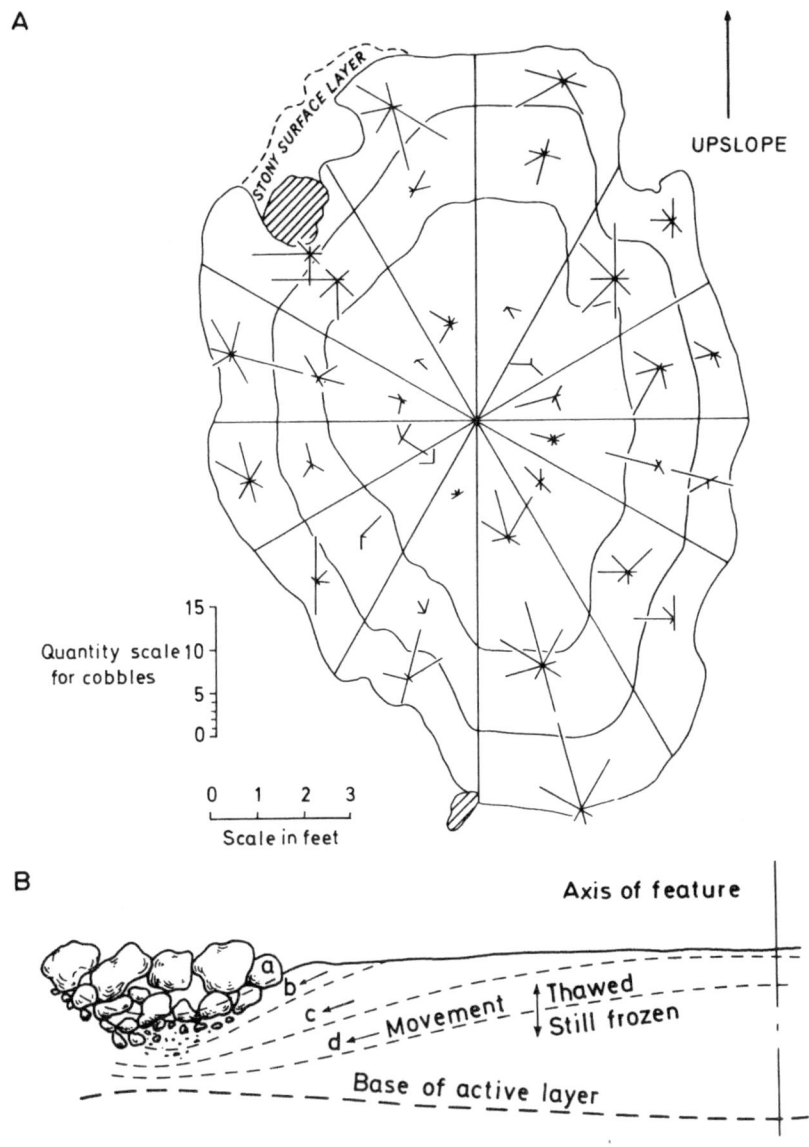

FIG. 5.21. A – ROSE DIAGRAM PLOT SHOWING AZIMUTH OF LONG AXES OF COBBLES IN SORTED CIRCLE IN GREENLAND
B – suggested solifluction movement of debris towards borders of sorted circle during initial penetration of frost table
(After Schmertmann and Taylor, 1965).

less. These characteristics are consistent with a period of gelifluction and creep moving debris outward from the features' centres during the spring thaw. Gelifluction is probably enhanced by the more rapid initial thawing of the coarse material of the circle borders so that an inclined seasonal frost/thaw plane is produced bearing little relationship to overall surface slope (Figure 5.21b). Differential heaving (1.5 cm/yr in the feature centres but only 0.9 cm/yr on the borders) is an important additional process, locally heaved slopes reaching gradients of 15°. Despite the clear lateral sorting in circles up to 10 m across, no significant textural variation of lateral type was found at depths greater than 60 cm below the surface of the features, although vertical sorting affected the full thickness of the active layer. The bulk of groundwater flow occurs in the border zones of the features.

Circles, polygons and nets are found on slopes up to 4° or 5° but the downslope component of strain due to gravity deforms the patterns on slopes steeper than about 4°, lobes or garlands developing on slopes between 3° and 6° and stripes between 4° and 11°, the range in critical values reflecting variations in percentage fines and moisture content according to Goldthwait (1976; Figure 5.22).

Patterned ground in the form of stripes is common on periglacial surfaces sloping at between 5° and 10° and may occur under some conditions on slopes up to 30°. A non-sorted series of stripes is produced where arctic alpine vegetation colonizes linear

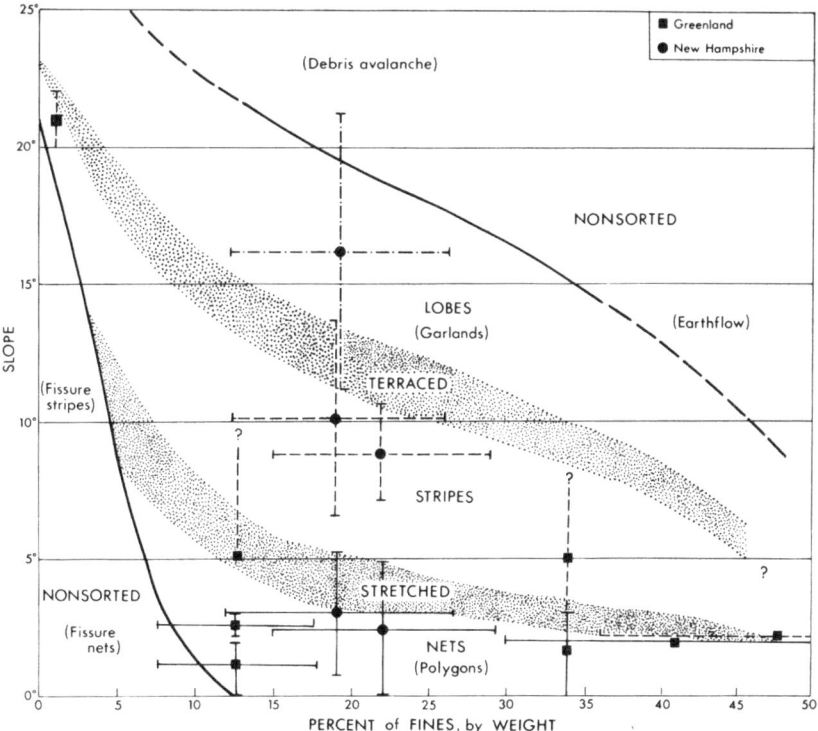

FIG. 5.22. SLOPE CHARACTERISTICS OF VARIOUS TYPES OF PATTERNED GROUND (POLYGONS, STRIPES, LOBES) Slope angles vary with percentage of fines (defined here as clay, silt and sand) and so with moisture content. Vertical broken lines show range of slopes observed. Horizontal broken lines show variations in content of fines (redrawn from Goldthwait, 1976).

zones running downslope across a surface of mixed debris in which size sorting is not systematically developed. The vegetation usually rests on a thin organic soil which rapidly wedges out at depth. A comprehensive hypothesis for the origin of non-sorted stripes has yet to be formulated.

Impeded deformation of circles and polygons by gelifluction and creep produces steps and lobes. In a detailed analysis of soil movement in terraces and lobes in the Colorado Front Range, Benedict (1970) found that turf-banked terrace forms are associated with areas of uniform snow accumulation which provides uniform soil moisture distribution. Lobate turf-banked features are common where spatial distribution of snow and soil moisture levels are more variable, the width of the individual lobes being proportional to the width of the soil drainage lines in which they form. Turf-banked lobes result from intense gelifluction beneath a vegetation cover which overturns and buries the turf. Morphological development of lobes may be cyclic, an optimum size being attained after which erosion modifies them. Movement of debris within lobes varies systematically, being greater along the lobe axis where ground water levels are highest.

Sorted stripes are made up of lines of alternating coarse and fine debris running downslope. The direction of stripe alignment faithfully reflects the direction of maximum surface gradient and the width of the stripes reflects the size range of the debris. Tabular stones tend to approximate the horizontal, their longest axes being aligned downslope. The coarse stripes, being drier, move by frost creep, while the fine stripes move much more rapidly (over twice as fast according to Benedict) by gelifluction. The fine stripes may be several times wider than the coarse stripes which rise above the level of the fine stripes to produce a ribbed microrelief which is present in the mixed debris below the surface sorted layer (Figure 5.23). A transverse component in clast ejection in the fine stripe appears essential to account for the concentration of clasts and the deficiency of fines in the coarse

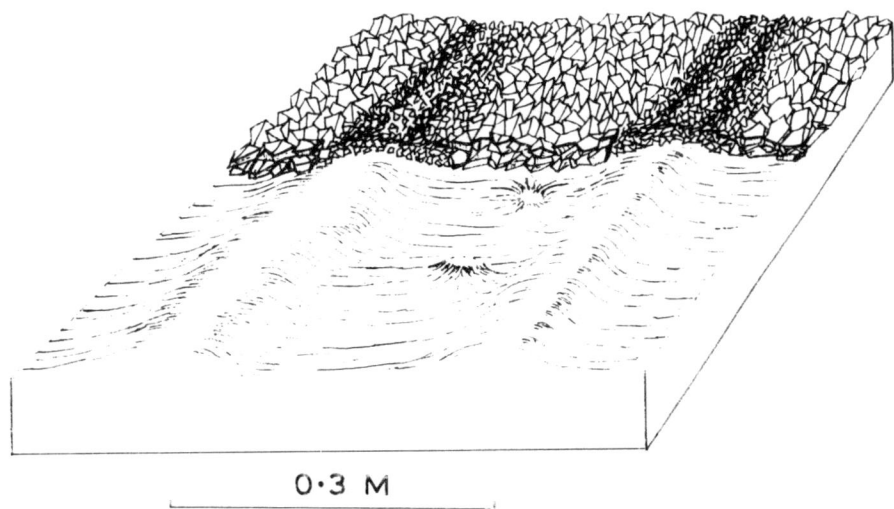

0·3 M

FIG. 5.23. BLOCK DIAGRAM SHOWING DETAILS OF FINELY STRIPED PATTERNED GROUND DEVELOPED IN VOLCANIC ASH AND SCORIA AT 1800 M ABOVE SEA LEVEL ON MOUNT RUAPEHU, NEW ZEALAND
The coarser debris making up the size-sorted stripes in the background was removed by hand to reveal a ribbed micro-relief in the silt-rich debris below. (From photograph: E. Derbyshire).

stripe, eluviation of fines during the spring melt probably being a secondary factor. Lateral sorting is also induced by the development of subvertical freezing planes adjacent to large clasts embedded in a fine matrix (Corte, 1966b). Quantitative studies by Mackay and Mathews (1967), suggest that, in the case of volcanic cinder soils, the critical surface sorting mechanism is associated with collapse of needle ice, frost heave, surface wash and drag from snow creep in some cases. The slower movement of the coarser component in stripes, also found by Benedict, is explained by Mackay and Mathews as the result of retardation of needle ice growth in coarse regolith deficient in silts. Thus, striped patterned ground may be produced by diverse processes, although the correlation between stripe width and debris size range appears to hold.

Generalized relationships between the cryergic processes and landforms are summarized in Figure 5.24.

In permafrost terrains, soil disruption may be highly localized and give rise to distinctive landforms. Lens-like bodies of intrusive ice may grow to such an extent that they lift the overlying sediments into a striking 40–60 m high conical or hemispherical mound known as a pingo. Muller (1959) distinguished two types of pingo which he called 'open' and 'closed' depending on whether sub-permafrost or intra-permafrost (talik) water has free access to the growing ice lens (Figure 5.25). It is possible to view the growth of ice lenses of the magnitude found in pingos as but a special case of the general theory of ice lens formation. Ryckborst (1975) has shown that thin ice lenses (up to 12 cm thick) can grow in sandy soils in a single winter, and he suggests that a lens up to 14 m thick may grow in silty soil in 75 years, and a thick ice lens giving rise to a 60 m high pingo may require about 980 years to accumulate. Pingo ice lenses grow at their bases in the active layer above the water table at a rate of between 50 and 300 mm/yr, summer melting removing 10–30 mm/yr from the top. A layer of low permeability (clay or clay silt) at the base of the pingo is essential to prevent downward growth of the ice lens while, at the same time, allowing upward translocation of 0.25 mm/day of water. As the pingo ice grows, the pressure of the

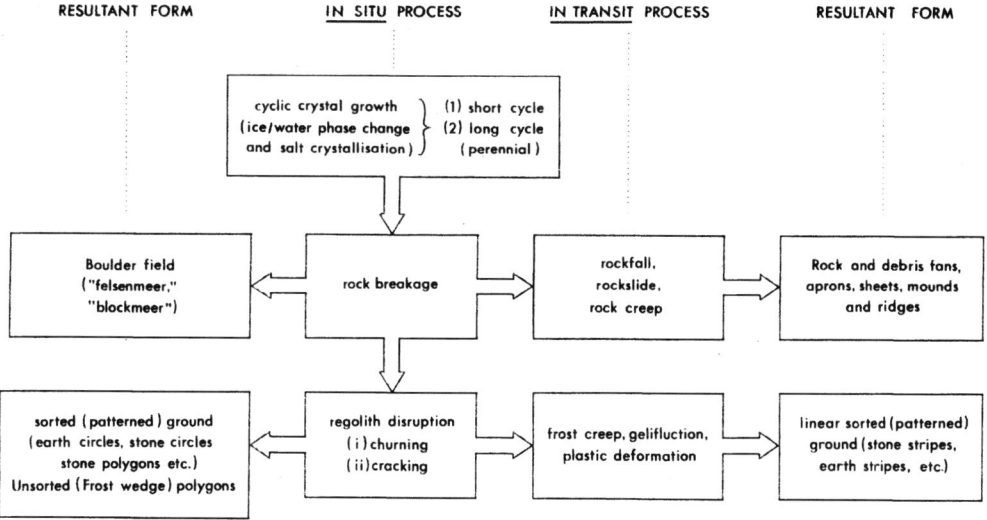

FIG. 5.24. GENERAL RELATIONSHIPS BETWEEN CRYERGIC PROCESSES, SEDIMENTS AND LANDFORMS

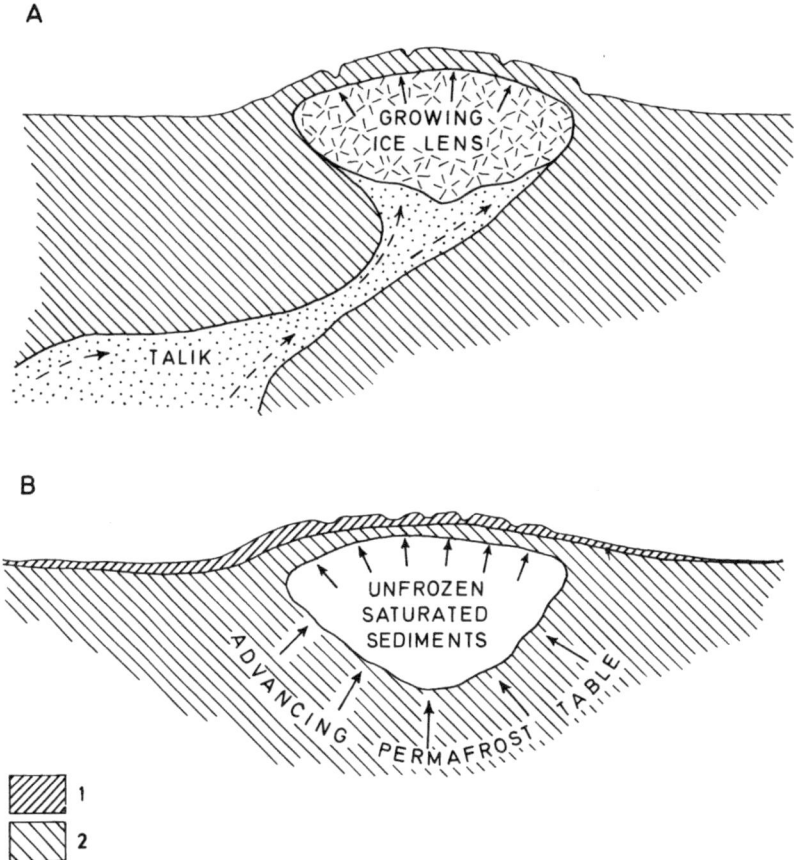

FIG. 5.25. A – OPEN SYSTEM PINGO WITH WATER SUPPLY FREELY ACCESSIBLE TO GROWING ICE LENS THROUGH TALIK.
B – CLOSED SYSTEM PINGO. 1 – sediments of former lake; 2 – perennially frozen ground.

ice lens and overburden will cause the freezing plane to advance, the pingo will freeze to the permafrost and begin to decay slowly as basal increments can no longer be added. The critical height appears to be about 60 m. As all pingos occur in groundwater discharge areas, Ryckborst sees no justification for a subdivision into several types of pingos and palsas.

Palsas are mounds of peat and segregated ice which may reach 7 m in height. Typically, they form within smooth fen surfaces in periglacial regions underlain by permafrost, their growth probably being enhanced by wind removal of an insulating snow cover from their crests. As the thermal conductivity of frozen peat is about four times that of unfrozen peat at the same water content, the domed dry surface of peat in summer insulates the segregated ice lenses but enhances heat loss in winter. The thickness of the unfrozen layers may be less under a dense tree cover than in forest clearings, because of shelter from both snowfall and solar radiation. Some palsas have a mineral core which conforms to and reaches to within half a metre of its surface. The clay bases of the wooded Manitoba palsa studied by Zoltai and Tarnocai (1971) are raised beneath the palsas (Figure 5.26).

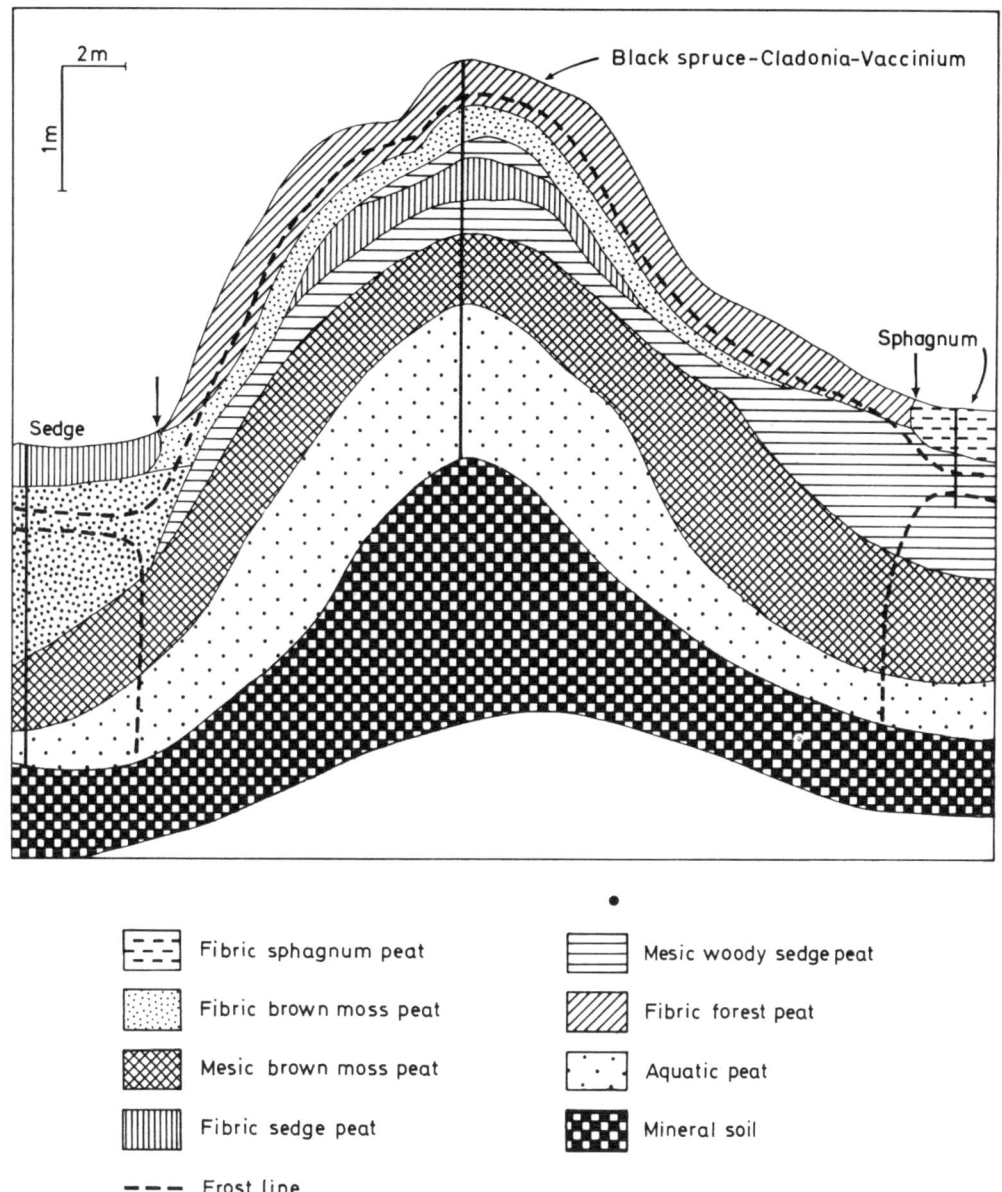

FIG. 5.26. CROSS-SECTION OF PALSA IN ARCTIC CANADA
(Reproduced by permission of National Research Council of Canada from Zoltai and Tarnocai, 1975).

5.3c Contraction cracking

In the frozen state, soil deforms plastically when loaded, but with marked temperature changes it displays elastic behaviour, contracting with temperature depression and relaxing with temperature rise. Horizontal failure planes in frozen soil have been explained by the rhythmical growth of ice lenses or *schlieren* with the advance of the freezing front (see p. 205). It has been noted, however, that the disposition of such ice lenses may be a composite pattern produced by the intersection of pre-existing fissures and the plane of advance of the freezing front (Beskow, 1947; Schumskii, 1964). Such discontinuities may themselves be formed by stress relief on freezing. According to H. Bertouille, lowering of temperature produces a tensile stress in the soil which may be relieved by the development of sub-horizontal fissures. If, on the other hand, temperatures rise within the frozen layers, the stress is compressive in the vertical plane and is relieved by failure along planes disposed at an angle to the vertical of $45° - \phi'/2$, where ϕ' is the angle of internal friction of the soil body. These so-called septiform cracks commonly act as a locus for ice vein growth.

The tension developed in the upper layers of a sediment upon severe drop in temperature may be sufficient to overcome the tensile strength of the soil, the stress being relieved by instantaneous vertical cracking. A type of vertical frost cracking in which the cracks slowly widen in winter and close again in summer has been described from altitudes between 3475 and 3530 m in the Front Range of Colorado. These cracks are attributed to variations in the effectiveness of frost heaving over short distances and are associated with an early summer phase of cracking produced by desiccation.

In perenially frozen terrains, boulders embedded in the permafrost may be pulled apart in the process of contraction cracking. The width of cracks of this origin depends on the thermal expansion coefficient of the soil, the degree of temperature decline and the distance to neighbouring cracks.

The hypothesis of contraction cracking in permafrost terrains originally set out by Leffingwell (1915) has been tested in mechanical terms and the result presented in a classic monograph by Lachenbruch (1962). Thermal contraction cracks form at the permafrost table and propagate upwards through the relatively weak active layer and downwards into the permafrost to depths of 6 m or more. With localized thawing in summer, the crack becomes filled with water and salts from above, to be re-frozen in the succeeding winter. Such ice veins are zones of weakness, their vertical folia becoming loci of further thermal cracking. Cracking appears to be initiated at the top of the permanently frozen layer and, once initiated, to persist in the growing ice-infill (termed an ice wedge) because it is of lower tensile strength (3 to 5 bars) than the surrounding frozen sediments (~ 10 bars).

Rapid cooling at low temperatures is required to generate the large thermal stresses necessary to crack frozen ground. The time lag in the penetration of the cold wave at depth, together with the weight of the overburden, results at any point at depth in horizontal stresses which are compressive. As the tension crack propagates downwards from near the surface, its depth of penetration is affected by this compression at depth as well as by the dissipation of the energy of strain by plastic deformation at the tip of the crack, an important influence here being the plasticity of the medium which is lower at the lower temperatures.

With cracking, the zone of stress relief so created has maximum values normal to the crack, the amount of stress relief being inversely proportional to the distance from the crack. Thermal tensional stress parallel to the crack is not significantly relieved, however. Thus, large differences in the horizontal tensional stress field (stress anistropy) are to be expected following cracking and any further cracking will tend to relieve this stress by aligning itself normal to the first crack to produce an orthogonal system of cracks known as ice wedge polygons (Figure 5.27). Such orthogonal systems may be apparently random, the absence of a preferred orientation resulting from complex stress anisotropy arising from factors such as small-scale variation in structural, thermal or moisture conditions. These patterns have been described as 'random orthogonal polygons' and probably represent the most extensive type in permafrost regions. Orthogonal crack systems may be clearly orientated as a result of a well-developed anisotropy in the soil fabric or where distinctive morphological or thermal conditions occur, e.g., receding margins of surface water bodies tend to produce a set of cracks parallel to the water limit and one set normal to it. J. R. MacKay found that about 40 per cent of all wedges in an ice wedge polygon system in Garry Island, Northwest Territories, cracked in any one winter, the succeeding increment of vein ice following thaw and re-freezing totalling up to 0.25 cm/yr.

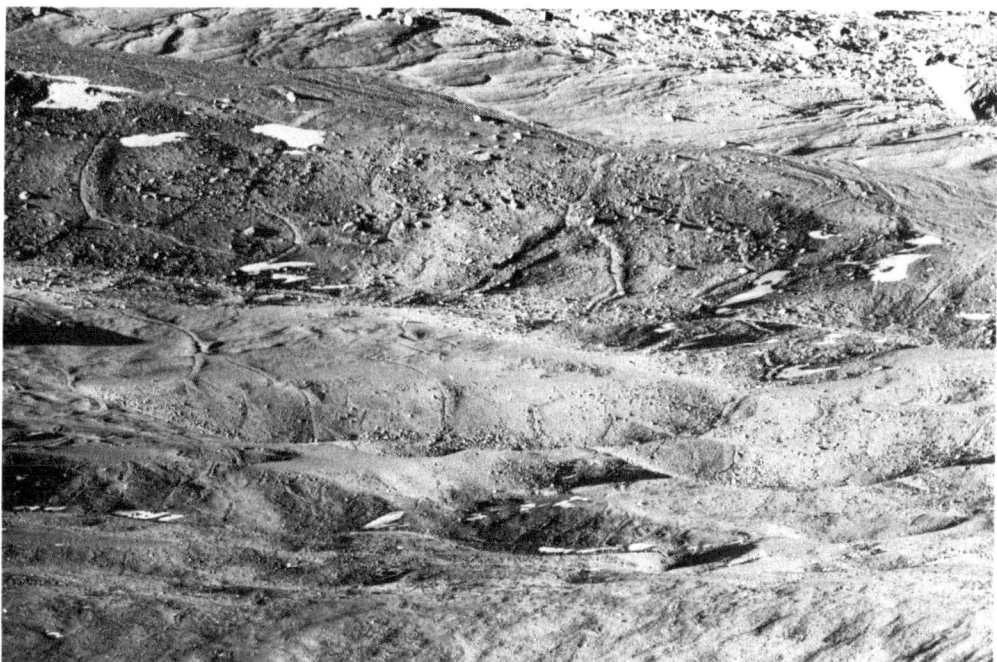

FIG. 5.27. ACTIVELY DEVELOPING THERMAL CONTRACTION POLYGONS ON UNVEGETATED SURFACE OF TAYLOR VALLEY, EASTERN ANTARCTICA
(Photograph: E. Derbyshire).

Ice wedges developed on a stable surface which is not undergoing active erosion or deposition will tend to increase in width by successive cracking and infilling until the compressive stresses of the surrounding sediments are relieved from adjacent or transverse cracks and growth in thickness effectively ceases. Such ice wedges are called *epigenetic*. Where cracking and ice wedge growth occurs at the same time as active accumulation of surface sediments, growth is said to be *syngenetic*.

Ice-wedge growth tends to produce raised polygon rims, separated by troughs above the ice-wedges axis. Upturning of the permafrost beds adjacent to the ice wedge is characteristic and is commonly explained by the expansion of melt-water in the seasonal crack within the wedge so that the ground is heaved apart in amounts which diminish with depth. The raised rims of ice-wedge polygons become accentuated with age, a fact which has been used to establish a chronology of geomorphic surfaces in ice-free areas of Antarctica (Black and Berg, 1963). According to these authors, raised centre (depressed rim) polygons are degenerate forms arising from decay of the ice wedges, where a significant melt phase has occurred, or from the differential development of a boulder lag deposit owing to prolonged removal of fines by deflation.

5.3d Degradation of permafrost

The degradation of perennially frozen ground by natural or artificially induced thawing gives rise to a suite of landforms including mounds, sinks, tunnels, caverns, ravines, lake basins, circular lowlands and beaded streams. By analogy with true (limestone) karst these landscapes are referred to as *thermokarst*. Disturbance of the thermal equilibrium at the ground surface as a result of changes in climatic parameters such as temperature and snowfall amounts, or because of modification of the vegetation cover by climatic change, wind action, natural fires or animal pressure, produces differential melting of ground ice. The permafrost table begins to fall and the active layer thickens and, as the segregated ice melts and infiltration from snowmelt rises accordingly, total moisture content rises to high levels. Seasonal frost heaving may temporarily increase as a result. The detailed morphological results of permafrost degradation may vary according to the rate at which mean annual temperatures rise. In the case of ice-wedge polygonal ground, for example, slow, shallow thawing causes ice wedges to melt before the segregated ice inside the polygon whereas rapid, deep thawing causes the segregated ice to thaw before the ice wedges because of the greater heat conductivity of the mineral element in the former. Thus, both high- and low-centred tundra polygons may occur in permafrost terrain.

When the ice wedges suffer accelerated melting owing to the action of both warm air and stream flow (thermal erosion), gullies or troughs develop over the wedges surrounding the polygons which become high-centred. Accumulations of surface water in polygon centres may enhance melting and result in a low-centred form. With further melting of the ice wedges, the ponds in low-centred polygons merge to produce larger lake depressions. Subsidence may result from the melting of massive beds of ground ice and also as a result of the eluviation of fines from a soil body as the segregated ice melts (thermal suffosion). Permafrost and wedge ice thaw more rapidly in coarse-grained sediments than in ice-rich fine-grained sediments and this is especially true where percolating ground water modifies the thermal regime. This is most likely in outwash plains, floodplains, and alluvial fans. Very severe disruption of the original sedimentary fabric may be effected in both freezing and thawing phases. While degradation of pingos may result in circular or ovoid lakes, perpetuated by the 'collar' of sediment rising several metres above them, the small size, lower ice content and high peat content of palsas makes them less likely to persist as landforms.

Degradation of permafrost may be rapid in situations where ground-ice is exposed in cliffs undercut by streams or by lake and coastal waves. Both heat transfer and mechanical action (thermo-abrasion) are involved. Detailed forms will vary according to the ratio between the thermal balance and the denudation balance. A shifting river channel causes the talik beneath it to migrate. The permafrost on the undercut slope degrades by thermal erosion and thermo-abrasion but decreasing soil temperatures on the slip-off slope result in the re-appearance of permafrost there. Rivers with basins entirely on permafrost and with headwaters in mountains with high snowfall are capable of considerable thermal erosion and thermo-abrasion to produce distinctive landforms. For example, stream undercutting of a permafrost bank results in debris fall and mudflow of the melting soil above. This lays bare part of the ground ice in the upper slope. There follows sliding of silt-blocks and further mudflows, producing large shallow hollows as the active layer is removed. Such features are common along streams in arctic Canada and in Siberia where they have been called 'thermocirques' (Czudek and Demek, 1970). The ice-rafting of lake-bottom sediments during floods may result in a change in albedo and thermal regime of the soil as the sediments are left in patches following thawing of the river ice. In the Noatak River delta in Alaska, this process has resulted in the development of thermokarst lakes with distinctively angular outlines (Ugolini, 1975).

Thermokarst forms may also develop as a result of the degradation of debris-covered glacier ice, although some workers believe that the term should be restricted to forms arising from decay of ground ice. Thermokarst development owing to the decay of buried glacier ice has been documented from several parts of eastern Antarctica including the eastern side of the Bunger Hills, Taylor valley in southern Victoria Land and the Mt Menzies nunatak in Australian Antarctic Territory (Derbyshire and Peterson, 1978).

5.3e Nivation

Residual snow and firn patches, both seasonal and perennial, concentrate a range of processes within a limited area and their sites display enhanced frost weathering, erosion and deposition. These processes, which include gelifraction, rockfall, frost creep, gel-ifluction, eluviation and abrasion of debris and rock are collectively termed 'nivation' when compounded in this way (Matthes, 1900). The constant wetting of the ground around and beneath the margins of melting snow patches enhances the efficacy of gelifrac-tion of bedrock and the heaving of soils. The erosional effect of nivation is to undercut the slopes of gully heads and other re-entrants, the nivation hollow or nivation cirque becom-ing enlarged, its form becoming progressively more circular in plan, owing to the tendency for ablation to round off the snow-patch. As the upper bounding slope or headwall becomes over-steepened in a broad arc, so the nivation cirque becomes more efficient as a locus of accumulation of snow and as a zone providing some protection from direct solar radiation.

Sediment transport mechanisms and resulting depositional landforms vary with the overall gradient of the nivation hollow. On gentle gradients, gelifraction debris falls to the base of the snow and is ultimately removed by subnival gelifluction to form snow-foot gelifluction terraces, eluviation of fines in the upper layers producing a stone pavement (Figure 5.28). Rockfall debris falling on to steeply sloping snowbanks tends to roll, glissade and creep over the snow surface, ultimately becoming concentrated at the foot. When the snowbank is stable in size and form year after year, such concentration gives rise to a moraine-like ridge called a protalus rampart. When considerable transport of fines occurs by supranival wash, the material in such ridges may be bimodal and so closely

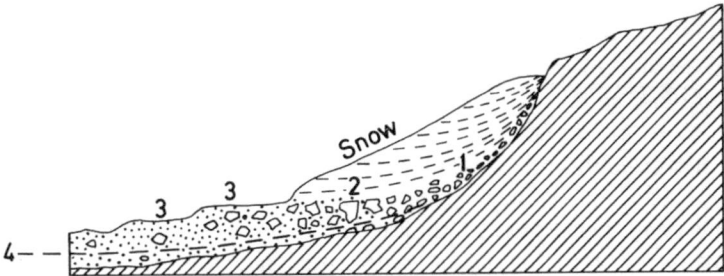

FIG. 5.28. CROSS-SECTION THROUGH SNOW BED AND NIVATION CIRQUE
1 – frost-shattered debris; 2 – boulder pavement; 3 – solifluction terraces; 4 – upper limit of frozen soil
(After Botch, 1946).

resemble till. Protalus ramparts which are rich in fines (and distinguished by the term nivation moraine) may readily be confused with moraine ridges of glacial origin, especially those of cirque type which may also be rich in angular, frost-derived boulders.

Some workers have suggested that gelifluction may act beneath snow patches although some snowbanks are so thick that they inhibit gelifraction and frost creep (but not eluviation and abrasion). The precise role of gelifluction within the suite of processes known as nivation remains to be determined. Williams (1957) described conditions in Norway in which unmelted snowbanks provide considerable percolating water in soils beneath them as well as acting as a load on the sloping soil surface. The abundant water weakens the loaded soil by reduction in effective stress (grain to grain forces), upward flow near the base of the snow patch reducing it further and washing out of fines occurring. Suffosion (Chapter 2) may enlarge pipes such that local soil collapse occurs. The convergence in both vertical and horizontal planes of subsurface water flow towards the lower margin of the snowbank, together with the increase in the stress exerted by the weight of the snow patch itself, is conducive to subnival soil failure by shearing in this zone.

A snowpack may move by creep and slow sliding motion as well as by avalanching. On Mt Seymour, British Columbia, Mackay and Mathews (1967), showed that the maximum downslope component of stress in the lowermost 0.45 m of an isothermal snowpack reaches a maximum (0.35–0.42 kg/cm^2) in the period February–March at the 0.4 m level but not until May just above the bedrock surface, producing differential creep. Steel poles and an adjacent tree showed deformation. There is accumulating evidence in support of the supposition of early writers (Chamberlin and Chamberlin, 1911; Lewis, 1925) that snowbanks are capable of limited abrasion and debris transport by mass sliding. It has been shown (Costin *et al*. 1973) that an annual snowpack which undergoes rapid firnification in spring on Mt Twynam (Kosciusko massif, southeastern Australia) develops shear stresses at its base of between 5.2 and 11.4 bars during slow mass sliding. Snow thicknesses range from 30 m to only 6 m, snow surface gradients being greater than 10° even on the lower slopes. Granodiorite bedrock is crushed and abraded and large stones leave behind trails of granular rock flour.

5.3f Rock glacier flow

Interplay of avalanche processes and those of nivation are common in periglacial mountains. Where avalanching of snow is associated with active talus development, snowdrifts may be buried and, with consolidation, turn to ice. Such interstitial ice within talus debris deforms by plastic flow, the associated movement of clasts being termed talus creep. Very slow downhill deformation of this ice-debris mixture (movements of about 1 m/yr), giving the talus a distinctively ridged, lobate surface reminiscent of true glaciers, has led to the term rock glaciers being applied to these deposits. Talus which has moved by 'rock glacier creep' is typically ridged along flow lines and also transversely to flow in its lower extremities. In its upper parts, the talus may be an openwork of very coarse joint-bound blocks unmodified since being wedged out by frost, but fines are commonly found at depth and towards the toe. Rock glaciers usually have sharp lower margins, frontal gradients being close to the angle of rest of the debris. The maintenance of an interstitial ice component (derived from water and snow of meteoric origin and from avalanche sources) as the critical component of flow within rock glaciers, requires mean annual temperatures below 0 °C on slopes where snow cannot accumulate to great thicknesses to produce glacier ice. Three morphological types of rock glacier are recognized, based on their shape in plan: lobate, tongue and spatulate. They are invariably associated with slopes beneath steep rock faces supplying the talus, and are commonly found within glacial cirques. Talus fans below straight escarpments may merge imperceptibly with a fourth type, the 'piedmont talus glaciers' of periglacial origin described by Smith (1973). Burial of avalanche snow by rock avalanching may be the critical process producing such forms.

Rock glaciers may also result from the slow decay of glaciers carrying a thick cover of supraglacial debris derived from rockfall. Continued thinning of the glacier in a cold but dry environment may result in this second type of rock glacier developing into the first. There is considerable evidence in favour of a glacial origin for rock glaciers. Many contain a thick core of glacier ice or pass headward into a true glacier, the zone between often developing into a distinctive 'spoon-shaped depression'. These characteristics together with the association of rock glaciers with cirques and their glacier-like morphology, have led some workers to consider all rock glaciers to be of residual glacier origin. Others consider that, while both types exist ('ice-cored' and 'ice-cemented' rock glaciers), they are transitional one to the other depending on the origin of the ice (glacier or meteoric-interstitial), the derivation of the rock debris (subaerial or englacial) and the climatic history. Some of the variables which need to be considered are shown in Figure 5.29.

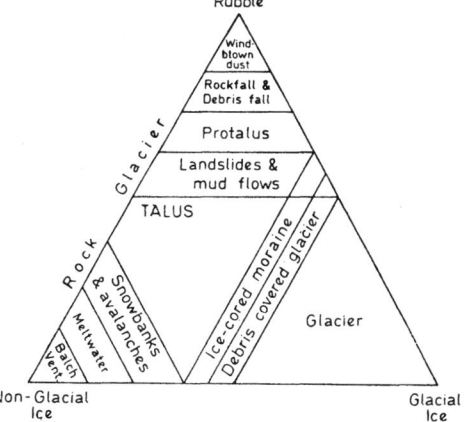

FIG. 5.29. CLASSIFICATION OF ROCK GLACIERS BASED ON ORIGINS OF MATERIALS (After Johnson, 1973).

The hypothesis that rock glaciers move solely by deformation of interstitial ice has been critically examined by Whalley (1974). Plastic deformation of a debris/ice mixture occurs readily at low debris concentrations under a shear stress which is a function of mixture thickness, mixture density and degree of slope (see Section 5.2). With high debris concentrations however, particle interaction greatly increases the yield strength. A second effect may be a component of cohesion derived from the adhesion of ice cement to rock grains, inhibiting the tendency of stressed sand bodies to dilate. The yield strength values of debris/ice mixtures may be up to ten times higher than that for ice ($\sim$ 1 bar) for a given slope. Whalley suggests that the component of flow in a rock glacier arising from yield of a debris/ice mixture is likely to be very small indeed, and is inadequate to explain known rates of flow, forward movements of one to two metres a year rather suggesting deformation of a glacier ice core, even though a thin ice core will not deform as a perfectly plastic body. Whalley concludes that the continuum is not from glaciers to interstitial ice-rock glaciers, but from clean glaciers, through rock glaciers with ice cores, to stagnating masses of ice, and ultimately to relict features devoid of any ice content.

5.4 GLACIAL EROSIONAL PROCESSES

Material eroded from the surface of the land and incorporated in glaciers and ultimately deposited by them is produced as a result of the action of two principal suites of processes. These are (i) subaerial (atmospheric) weathering, subaerial erosion especially by streams and rivers and tectonism (which may be grouped together as the preparatory processes), and (ii) erosion of rock surfaces by the glacier itself.

5.4a The preparatory processes

Subaerial weathering provides a source of relatively easily eroded materials for advancing glaciers and ice sheets. The deeply-weathered profiles characteristic of a wide range of climates from warm temperate to equatorial probably extended into present-day sub-polar latitudes prior to the Pleistocene glaciations as a result of prolonged warm-climate weathering throughout Mesozoic and much of Tertiary time. Occasionally, residual products of such weathering profiles (notably certain secondary clay minerals) are found in gelifluctates of Pleistocene age, and *in situ* profiles of deep weathering have been described in some areas overrun at least once by glacier ice, as in parts of eastern Scotland.

There is some evidence that unloading of bedrock by preglacial incision of rivers may be responsible for rock joints developing parallel to the land surface, so relieving the strain energy within the rock body. Such stress relief, or dilation, joints appear to follow the form of the preglacial relief and this process may be important in the loosening of coarse rock debris subsequently entrained by glaciers. Variable stress continues under an ice load, however, and the process may be perpetuated or even accentuated as the glacier grows (see Section 5.5b).

Tectonic activity is an important preparatory process in glacial erosion and the type of debris entrained by glaciers. Active uplift of mountain areas produces shatter zones, some of which develop preferentially and show a higher frequency of movement, a greater scale of movement, or both. These zones are made up of brecciated bedrock which is disturbed more readily by gelifraction and entrained more readily by the process of glacial plucking than is sound bedrock of the same lithology. This is an important source of variation in debris supply, glacial erosion rates and glacial valley morphology, and the effects can be seen in areas currently glaciated and those glaciated during the Pleistocene.

A period of rock disaggregation by gelifraction is particularly favoured by some continental European geomorphologists as an essential preparatory phase in the glacial deepening of valleys and fjords, for example (Tricart and Cailleux, 1962). While it is true that such a cryonival phase occurred in many areas, there are obvious difficulties in demonstrating beyond doubt that such conditions prevailed in specific cases. It is clearly demonstrable, on the other hand, that subaerial mechanical weathering in the rigorous microclimatic conditions adjacent to mountain glaciers (gelifraction, dilation, hydration and salt weathering) provides a great amount of rock debris which is carried on to glacier surfaces by the transportation processes (discussed in Chapter 2) of free fall, flow, slide and creep.

The products of erosion by rivers may survive one or a series of glaciations. In cutting their valleys, rivers provide loci of accumulation for firn and ice in their headwater reaches and avenues for downslope movement of the growing ice bodies. Glacial erosion modifies the detailed form of such preglacial river valleys but the regional valley network tends to preserve its integrity despite repeated inundation by thick glacier ice (Figure 5.30). The products of fluvial deposition are frequently incorporated in glaciers, transported and deposited again either directly by the ice or by the medium of glacial meltwater streams. Such deposits may undergo a series of cycles of re-incorporation in the glacier and re-deposition. Evidence of former fluvial processing of rock fragments in debris deposited by ice is provided by particle shape studies and by characteristic size-sorting curves (Section 5.7b).

5.4b Processes of glacial erosion

Glacial erosion involves detachment of fragments from rock surfaces, breaking down of those fragments into progressively smaller size (comminution) and the abrasion of the rock surface as the glacier sole, armoured with broken rock fragments, slides forward.

The role of the thin water film at the base of a temperate glacier has been shown to be important in the basal sliding process and in the entrainment of debris at the glacier sole. When a temperate glacier moves over an impermeable rock bed, the presence of a water layer may increase the sliding rate by between 40 and 100 per cent because the area of contact between glacier sole and bedrock surface is reduced by the buoyancy effect of the basal water film which is at pressures greater than atmospheric (because of the ice overburden pressure and the constantly changing, and variable scale of, cavities at the glacier bed). Thus, the *effective* normal stress on the rock head will be equal to the *normal stress* (σ_n: the product of the ice density (ρ), the ice thickness *(h)* and the acceleration due to gravity *(g)*) minus the water pressure (P_w) at the bedrock-ice interface:

$$\sigma_e = \rho gh - P_w \qquad (5.3)$$

The value of this effective normal stress may be very much less than that due to the ice pressure if the bed is impermeable so that a basal water layer is maintained. Water pressures (P_w), in addition to a tendency to decline in a downglacier direction, vary locally

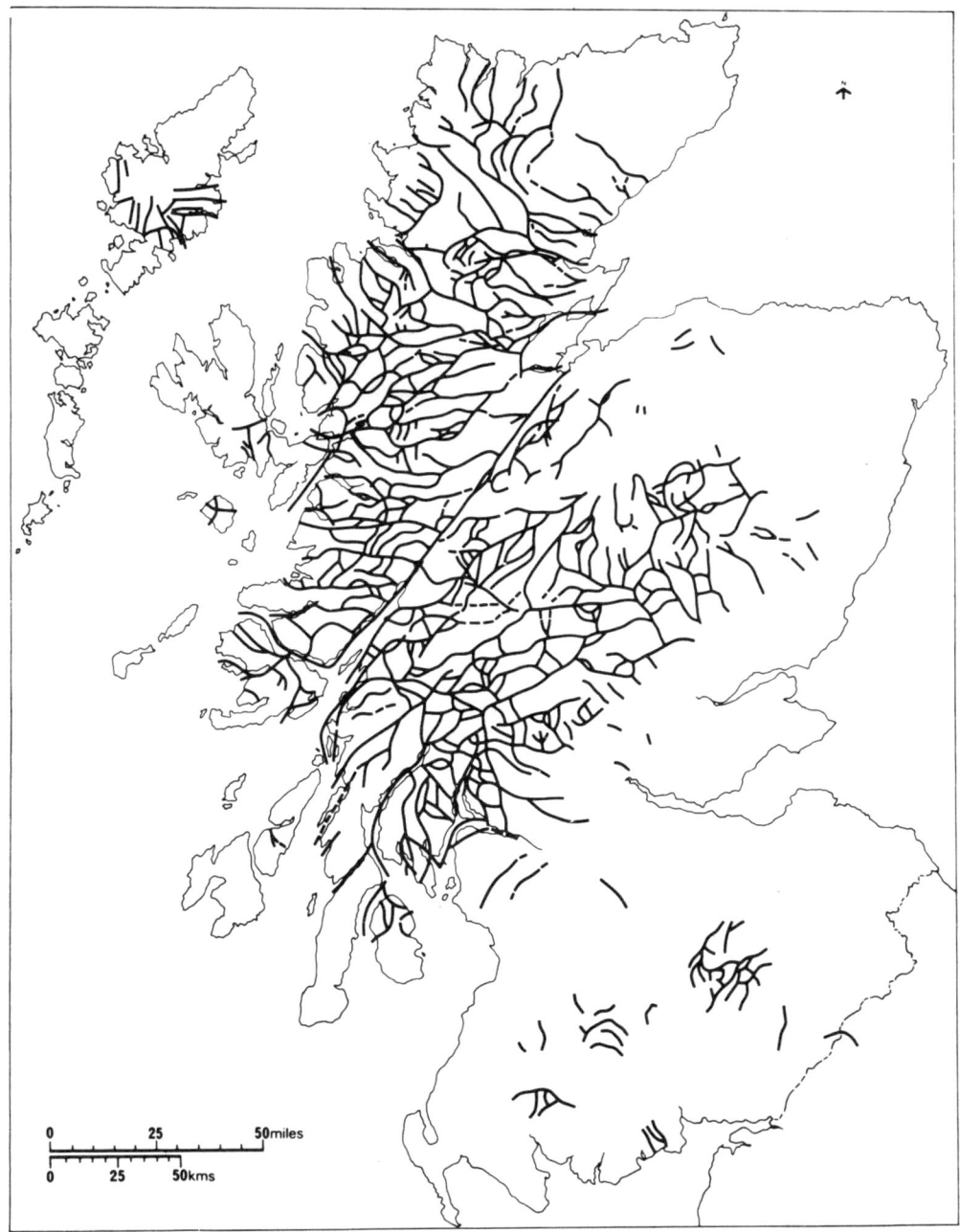

FIG. 5.30. GLACIAL TROUGHS IN SCOTLAND BY D. L. LINTON
(After Clayton, 1974).

with irregularities in the bed, the level within the glacier at which water pressures fall to the atmospheric value (the piezometric surface) expressing this tendency (Figure 5.31). In the Glacier d'Argentière in the western European Alps, for example, measurements have shown that the basal water pressure may be so high that effective normal stress is locally only half the value of the calculated total normal stress.

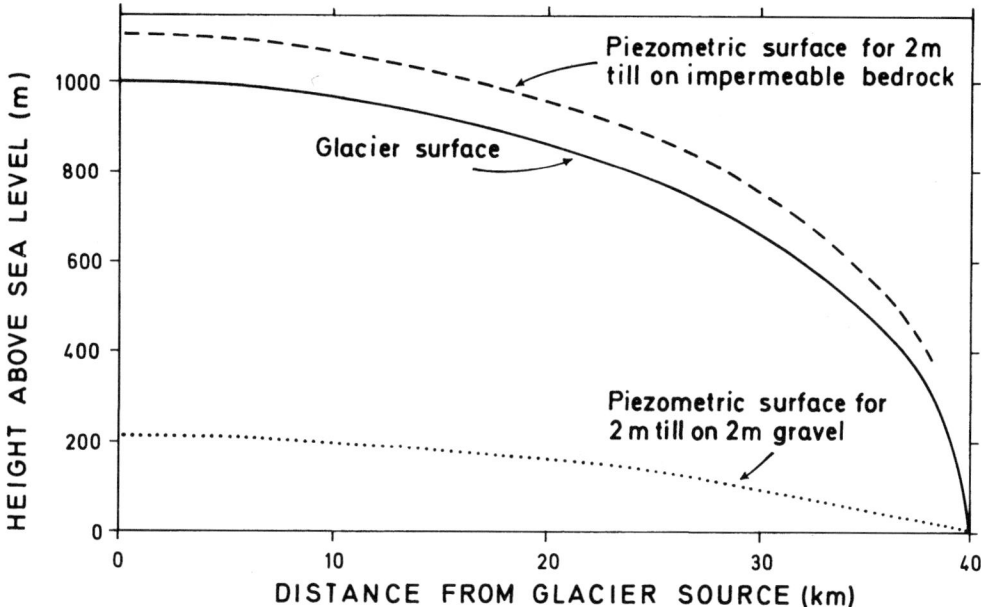

FIG. 5.31. GENERALIZED PROFILE OF THE GLACIER BREIDAMERKURJÖKULL, ASSUMING HORIZONTAL BED AND SHOWING CALCULATED PIEZOMETRIC SURFACES FOR 2 M TILL BED AND A 2 M GRAVEL BED
(After Boulton, Dent and Morris, 1974).

When a glacier rests on permeable rock and sediments, some of the water reaching the bed of the glacier will permeate into the bed. In such a situation, the water pressure at the ice-sediment interface is low and the effective stress is higher than in the case of an impermeable substrate. Where infiltration equals or exceeds the supply of basal water, the effective stress is raised to the value of the total normal stress (i.e. $\sigma_e = \rho g h$) which will be directly proportional to the ice thickness (h). Thus, variations in the permeability of the bed result in variations in effective stress and these, in turn, produce lateral variations in potential sliding velocity, abrasion potential and deposition (Section 5.6b).

With forward motion, effective normal stress fluctuates when the rock bed is irregular, values being higher than average on the upglacier side and lower than the mean on the leeward flank of a particular obstacle. Marked variations in normal stress are to be expected where bedrock obstructions are short and high and less marked when obstructions have low amplitude/length ratios. Spatial variations in normal stresses result in complementary variations in stress within the bedrock obstacle. This implies that shear strength (or the critical shear stress, i.e. the value of stress at which shear failure occurs) will vary in massive, non-jointed rock, being greatest where normal (i.e. containing) pressure is highest, and lowest where normal stresses are at a minimum. The shear strength of the bed is the sum of the cohesion and the product of the effective normal stress on the bed and the tangent of the angle of internal friction of the rock material (Coulomb's equation: $\tau = c' + \sigma_n \tan \phi$). In the lee of an obstacle where a small cave is formed by detachment of the glacier from its bed, the normal stress on the bed falls to zero, so reducing the value of the third term of the equation to zero ($\tau = c'$) (Figure 5.32).

Thus, when subglacial stresses reach the shear strength values of the rocks over which the glacier is moving, the elastic limit of the rock may be exceeded and failure will occur. The orientation of the resulting failure planes will tend to lie at an angle to the major principal stress axis of half the angle of internal friction. With internal friction angles of common rocks ranging from 35–45°, an aggregate of trapezium-shaped clasts may thus be produced from massive, isotropic rocks. Variable normal stresses around the bedrock obstacles also produce zones of compression and tension in rock, the readiest form of stress relief under tension being by the propagation of tensional cracks parallel to the rock surface. Such sheeting, or dilation jointing, has been described from Norway where the joints parallel the ice-rock interface in cirque basins and glacial troughs and cut across primary rock structures (Lewis, 1960).

Where basal fragments impinge on bedrock surfaces to produce very high stress con-

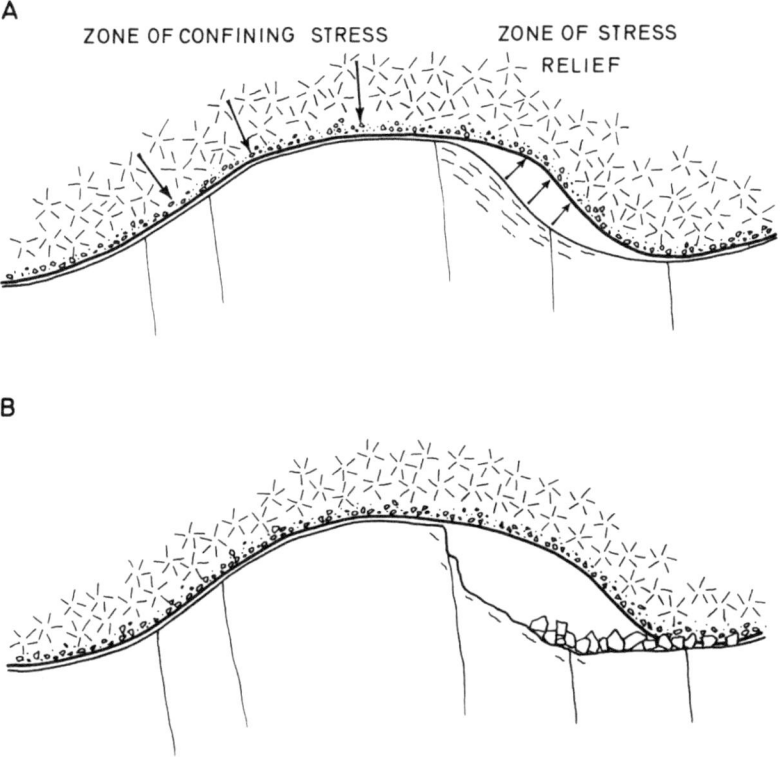

FIG. 5.32. FAILURE OF BEDROCK IN ZONE OF STRESS RELIEF CAUSED BY CAVE DEVELOPMENT IN LEE OF OBSTACLE AS THE GLACIER MOVES ACROSS IT FROM LEFT TO RIGHT
Stress relief joints (shown by discontinuous thin lines in A), together with rock joints (continuous, near vertical lines) are zones of weakness. Failure steepens down-glacier face of *roche moutonnée* and provides abrasive rock debris in basal zone of glacier (B).

centrations at a point, concentric cracks may develop. These occur on the beds of glacier streams and beneath glacier ice where they tend to lie transverse to the direction of glacier sliding. They have been explained as conoids of percussion produced by high tangential stress. Secondary fractures commonly intersect these to produce a variety of small erosional forms called lunate fractures and crescentic gouges (Figure 5.33).

Bedrock anisotropy arising from primary structures such as joints, bedding planes and faults result in critical strength values of a particular rock being strongly directional. The disposition of major structural features (the rock fabric) in relation to the major principal stress axis and to the rock surface form as affecting normal tensional stress, influences the rate of shear failure and the size and shape of the sheared blocks. For example, there is a tendency for shear block shape to range from tabular in phyllites and schists, to wedge, columnar and rhomboid in basalts, and to cuboid (equant) in granites, although the number of observations on freshly detached subglacial clasts is still very few.

Rock fragments detached from the bed by shear failure and dilation, together with those derived from uncemented subglacial sediments, become entrained in the basal zone of the glacier by the processes of regelation and plastic flow. The process of block removal by glaciers, in which the shear stress in the basal ice exceeds the frictional resistance of the rock fragment, is known as plucking.

Regelation is the process in which ice on the upstream side of an obstacle is differentially melted owing to the higher interfacial stress on that surface, only to re-freeze owing to the reduction of stress on the lee side. In the case of small obstacles (1–2 mm), some of the latent heat of fusion is conducted back through the obstacle to be utilized in further melting on the onset face. The utilization of heat derived from pressure melting will clearly be most efficient in the case of small obstacles and the velocity due to this process will be inversely proportional to the size of the obstacle, i.e. ice flows easily around small obstacles. The process gives rise to a distinctive type of basal ice with small crystal size, strongly marked preferred crystal orientation and few gas-filled cavities, known as *regelation ice*. This is relatively rich in water produced by frictional heating. As the strain rate for a given stress rises with the amount of liquid water within glacier ice, the thickness and water content of regelation ice are important factors in the creep of basal ice, in the rate of supply of water to the ice-rock interface and so in the total flow equation of a glacier. Typically, regelation ice is rich in debris incorporated in the regelation process and represents the *basal debris zone*.

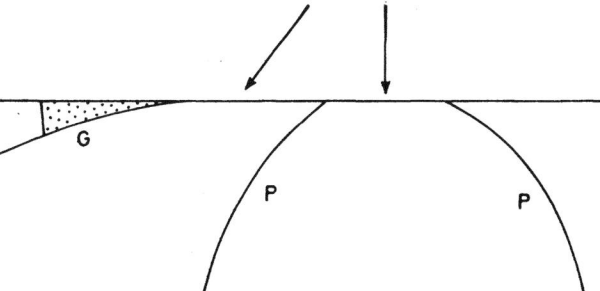

FIG. 5.33. CONOID OF PERCUSSION PRODUCED IN ROCK MASS IN RESPONSE TO VERTICAL IMPACT ON SURFACE, SHOWN BY RIGHTHAND ARROW
Impact oblique to surface produces conchoidal fracture. Combination of conoid and vertical fractures, as at G, releases a splinter of rock (stippled) to form a crescentic gouge. (After Gilbert, 1905).

In the case of large obstacles on the glacier bed (>1 m), the longitudinal stress and the strain rate in the ice is greater than the average for the lower part of the glacier as a whole. As the velocity is proportional to the product of strain rate and distance, larger obstacles enhance the zone of stress so that large obstacles generate zones of high basal velocity. This process is known as enhanced plastic flow. The size range of the obstacles intermediate in scale between that associated with enhanced plastic flow and that in which regelation is most efficient has been called the *controlling obstacle size* by J. Weertman (1961) in that it may be a major factor controlling the basal sliding velocity.

The basal sliding motion and the regelation process is clearly intermittent rather than continuous where it has been observed beneath the marginal parts of some temperate glaciers. The evidence for the spasmodic or 'jerky' nature of glacier sliding and regelation is to be seen beneath many glaciers including those in Iceland and Norway. Clusters and spicules of ice form at the moment of stress relief at the point where the thin meltwater film emerges between the glacier sole and the bed in the lee of an obstacle. Orderly series of such ice clusters mark successive slip-stick movements and may be used as a short-term calendar of sliding events.

Variation in the basal friction of glaciers may be a significant element in jerky glacier sliding at low to medium sliding velocities. The contribution to the basal friction of ice-rock adhesion varies with oscillation of the basal ice temperature around the pressure melting point. In addition, locally high salt concentrations may maintain water in the liquid phase at temperatures below the melting point. Local adhesion of ice to rock by freezing of basal water may result from stress relief as basal water pressures rise and fall rapidly over small areas of the bed. When the ice is not sliding, obstacles enter the sole owing to plastic deformation of the basal ice. The value of basal friction must then increase greatly for a cavity to form and, when this occurs, the friction value falls. It is also known that when a relatively hard clast is dragged over a softer surface, friction is high and the motion typically intermittent, whereas above a critical hardness value of substrate, friction drops to a low value. Thus, variations in rock surface hardness and in hardness differences between clasts and rock surfaces may favour jerky motion. At present however, a comprehensive theory to explain the magnitude and periodicity of intermittent basal sliding in different types of glacier remains to be formulated.

The development of a thin meltwater film on the upglacier (higher stress) side of small fragments and its re-freezing on their lee sides where basal stress is relieved by a periodic sliding event provides an explanation of the mechanism of debris incorporation in basal ice. In the case of larger fragments, basal ice impinging on them deforms plastically until they are incorporated and begin to move under the shear stress imparted by the moving glacier. As frictional resistance in the debris is raised by the high normal stresses where ice is in contact with the bed, removal by plucking is most likely at times when caves open and close with variations in ice thickness and ice velocity.

The rate of transportation of fragments in the process of plucking depends on the glacier's traction force and the value of friction at the clast-rock interface, the latter varying with a range of factors notably with size and shape of the clast (as affecting the area of contact between bed and clast) and the roughness of the bed. The limiting condition for particle movement on the bed may be expressed in the form

$$\tau = (\rho g h - P_w)F \tag{5.4}$$

where τ is the shear stress imposed by the ice on the particle and F is the coefficient of friction between glacier sole and bed. Large boulders under traction beneath some temperate glaciers travel at a rate of between one and two orders of magnitude slower than the ice, as is evidenced by the development of narrow, elongate cavities in the basal ice in their lee (Figure 5.34). Finer-grained fragments (<8 mm) also appear to move slowly as a

FIG. 5.34. TUBULAR CAVITY AT BASE OF BLÅISEN, SOUTHERN NORWAY
Produced by deformation of basal ice around large boulder (diameter 1.25 m) resting on a rock bed. Note concentration of debris in lowermost 10 cm of glacier. View up-glacier. (From photograph: E. Derbyshire).

result of the efficiency of the regelation process. Thus, the initial velocity of plucked clasts under traction is likely to be greatest in the size range 8–150 mm (Boulton, 1974).

Entrainment of loosened rock by the glacier is controlled by the temperature regime at the glacier sole. Assuming a planar bed, Weertman (1961) has set out three limiting conditions based on the ratio between heat produced by glacier sliding plus the geothermal heat flux and the product of the englacial temperature gradient just above the sole and the thermal conductivity of ice.

When the ratio is greater than unity, heat produced at the base being greater than the amount which can be conducted through the glacier, excess basal melting and basal sliding both occur. When the ratio is less than unity, water in the rock beneath the glacier is frozen, the glacier is frozen to its bed, no basal meltwater is produced and no basal sliding occurs. In the third case, when the two terms are approximately balanced, the glacier just slides over the bed in the absence of either excess melting or freezing.

A condition of excess basal melting is not conducive to the plucking mechanism which requires that meltwater produced at points of melting refreezes to the glacier sole and in so doing causes loosened rock to adhere to the sole of the glacier. Temporal and spatial variations in the incidence of melting and refreezing zones under a single glacier are to be expected with variations in climate, mass balance and bed morphology. In relatively thin polar glaciers, the frozen substrate will be frozen to the glacier sole, no basal meltwater phase will exist and basal sliding will not occur. Glacial erosion in such conditions may be absent as is suggested by the appearance of undeformed tundra polygons from beneath retreating glaciers in Antarctica. Such glaciers may be protective in the sense of eliminating subaerial weathering and erosion of the surface. Where the substrate is frozen only to a shallow depth however, and where structural planes coincide with the top of the thawed zone especially if porewater pressures are high (so reducing the frictional strength at the potential failure plane), large blocks of subjacent frozen rock or sediments may be plucked. This process has been put forward as an explanation of the giant erratics of chalk found within till in the cliffs of Norfolk (Boulton, 1972; Figure 5.35).

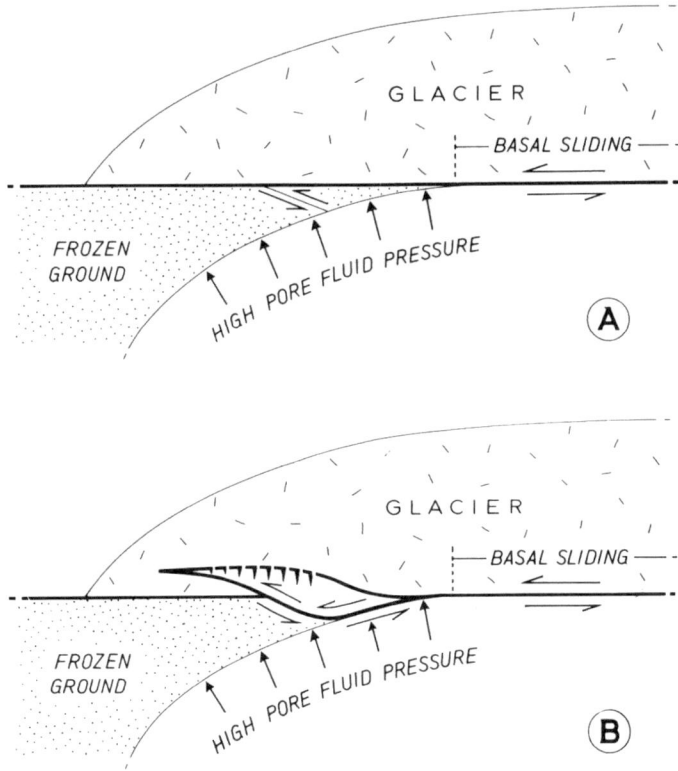

FIG. 5.35. SUBGLACIAL ENTRAINMENT OF WEDGE OF BEDROCK IN AREA OF THIN PERMAFROST (after Boulton, 1972), FOLLOWED BY FAILURE RESULTING FROM TENSION IN ENTRAINED BLOCK (VERTICAL FRACTURES IN 'B').

Interaction between rock fragments in the debris-rich basal surface of a glacier may result in the processes of grinding and crushing (comminution). Locally high loading stresses within fragmented debris at the rock head and between that debris and debris in the basal ice may produce high stress concentrations at particle contacts so that the compressive strength of the rock particles is exceeded and fracturing results. This may occur at relatively low values of normal and shear stress in poorly sorted debris such as till because the wider range of particle sizes increases the number of inter-particle contact points (*cf.* Billam, 1972). The process is clearly dependent on variations in ice thickness, ice velocity and bed roughness as they affect effective normal stress, shear stress and the rate of plastic flow of basal ice around clasts held in the sole, as well as on rock hardness and the frequency of cleavage planes and other planes of weakness. The proportion of material reduced to the finer particle sizes increases with distance from the source outcrop. The smallest grain size attainable by this process (the 'terminal grade': see Section 5.6a) varies with the rock type.

Comminution is the result not only of crushing but of abrasion. Clean ice will not abrade bedrock: abrasion is effected by sliding of the debris-rich basal layer of the glacier. Basal pressure melting in wet-based zones of glaciers and accommodation of the basal ice to large obstacles by plastic flow constantly renew the basal debris. The meltwater film found at the sole of temperate glaciers, and locally beneath sub-polar and some polar glaciers, reduces the value of the basal shear stress (p. 195) where bed roughness and basal debris sizes are small. In this case, the relatively higher stresses are borne by the larger obstacles and the coarser basal fragments. If the effective stress remains below a certain critical value (as beneath relatively thin, rapidly moving glaciers, surging glaciers, or as a result of the maintenance of high basal water pressures in the case of very thick glaciers), the adhesion between glacier and fragment may exceed the value of the internal frictional force in the subjacent fragmental debris. This will cause the debris to roll and slide along the glacier bed. Thus, the rate of abrasion will increase with the sliding velocity and with any increase in the effective normal stress (and so with ice thickness). This is true only up to a limiting value of effective normal stress however, because high values of effective normal stress result in plastic deformation of ice around the obstacle. As the rate of plastic deformation increases, the abrasion rate declines rapidly for any given sliding velocity relative to the rate of sliding of the basal fragments. Finally, the frictional force between fragment and bed exceeds the force exerted on the particle by the glacier and traction and abrasion cease entirely (Figure 5.36). In the case of a glacier on a freely permeable bed, this threshold value will be reached under smaller thicknesses of ice as the basal water pressures will be zero.

The effectiveness of abrasion is influenced by the weight of a basal particle and by the area of the particle in contact with the bed. According to G. S. Boulton, fragments between 8 and 150 mm in diameter have the highest transport velocities (see above) and so play a major role in abrasion. While transport velocities decline for basal fragments with diameters greater than *c*. 150 mm this is partly compensated for by their great weight. Under consistent subglacial hydrostatic pressure gradients, the finer debris may be mobilized and preferentially washed out of the subglacial debris.

Abrasion rates vary with rock hardness and particle shape. Experiments under the Glacier d'Argentière by Boulton and Vivian (1973) suggested that marble abrades three times more rapidly than basalt. Evidence of abrasion of rock surfaces consists of a variety of small erosional markings including polished surfaces, striations and grooves. Sichel-wannen, cavettos and some grooves are viewed as products of glacial erosion by some authors and as glacifluvial erosional forms by others (see Section 5.7a).

Glacially polished rock surfaces are made up of a myriad of fine scratches not all of which are visible to the naked eye. These are produced by the finest products of comminution (rock flour; silt and clay size material in which quartz is commonly dominant) which form a concentrated but very thin layer on the soles of many glaciers. Here, in the presence of pressure melting, it acts as a layer of low strength. It is constantly remobilized as the basal stress varies and it can sometimes be seen to extrude into minor irregularities between glacier and rock bed and in shear planes where upper, active ice is thrusting over lower, inert ice. Scorings on rock surfaces produced by ice-borne debris of sand grade and larger are known as *striae*. Individual striae are rarely of great length owing to the abrasion of the particle itself and its tendency to respond to frictional stress at the bed by rotation and enhancement of the plastic deformation of the glacier surrounding it. Beneath a glacier sole rich in debris, however, extensive and highly consistent striation may occur. Parallelism of the scratches decreases with increased bed roughness and with decreased sliding velocity.

It has been shown that the effective normal stress exerted on a rock surface varies with the form and amplitude of any undulations in the bed. At relatively low values of effective normal stress adhesion between fragment and glacier is conducive to rapid abrasion rates. At the same time, the shear strength of the rock bed is most likely to be exceeded in the lee

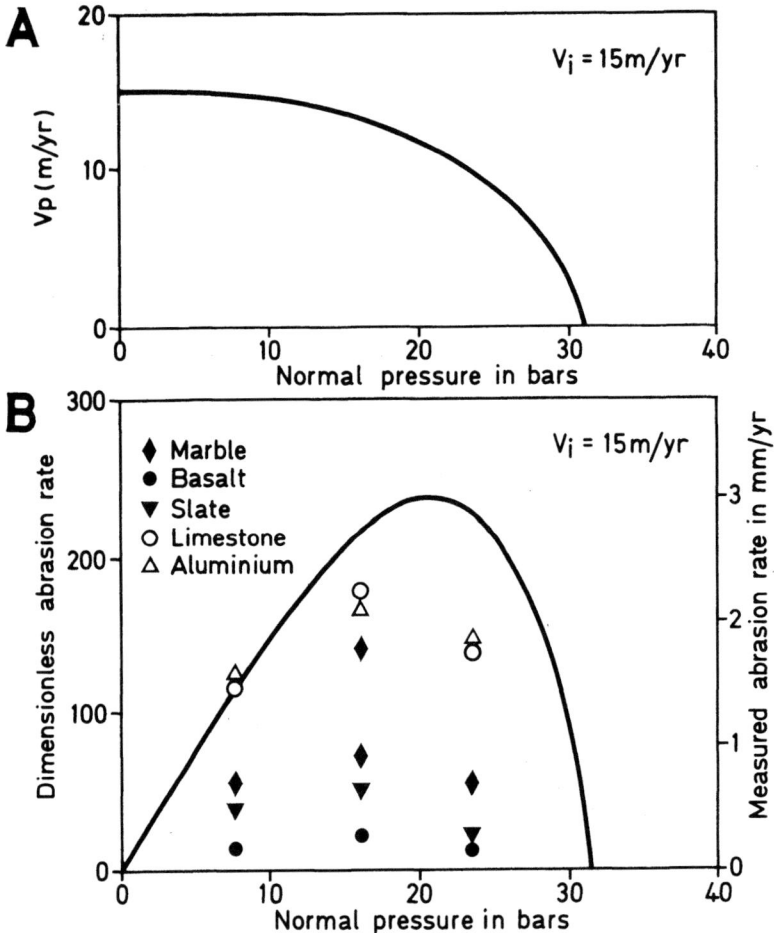

FIG. 5.36. THEORETICAL ABRASION RATES V. EFFECTIVE NORMAL PRESSURE AT GLACIER BASE, FOR DIFFERENT ICE VELOCITIES
A – velocity of a particle over the bed of a glacier (Breidamerkurjökull, southeast Iceland) with a sliding velocity (v_i) of 15 m/yr, as a function of effective normal pressure
B – theoretical glacial abrasion rate for Breidamerkurjökull at a point where v_i = 15 m/yr. Abraded plates were placed at points on the sublacial bed where v_i was constant and effective pressure varied. Data have been normalized to the value for limestone adjusted to fall on the theoretical curve
(Diagrams kindly supplied by Dr G. S. Boulton prior to publication in *J. Glaciol.*)

of a rock convexity and the loosened mass removed by plucking in conditions of limited leeside cave development. This combination of conditions produces the characteristic asymmetry of the typical *roche moutonnée* with polished and striated gentle surfaces dipping upglacier and sharp, steep angular plucked surfaces facing downglacier (Figure 5.37).

FIG. 5.37. GLACIALLY POLISHED AND STRIATED SCHIST BEDROCK SHOWING ASYMMETRICAL FORM TYPICAL OF ROCHE MOUTONNÉE
Franz Josef Glacier, New Zealand, moved from right to left. (Photograph: E. Derbyshire).

With increase in normal stresses on the onset face of the rock obstacle, abrasion rates may decline to such an extent that abrasion ceases altogether and local deposition by lodgement occurs. Normal stress levels on the summit and lee slope, being lower than on the onset face, allow the maintenance of abrasion on these surfaces, so that the feature takes on a streamlined form which slopes steeply upglacier, the gentler polished face pointing downglacier, the converse of *roche moutonnée* asymmetry. Such forms have been called rock drumlins.

The characteristic valley form associated with glacial erosion is the parabolic cross profile. The longitudinal form is a single or a composite concavity in which 'overdeepening' (or the development of reverse, upglacier-facing bedrock slopes) is a characteristic though not a universal property. The sweeping concavities of glacial erosion contrast with the broad convexities of gelifluction slopes and with the rectilinear or limited convexo-concave forms of fluvial landscapes. The contrast between the broad V-shape of river valleys and the rough U-shape of glaciated valleys has long been recognized. Much of the fluvial valley surface is of subaerial origin whereas the parabolic glacial valley is a subglacial form (below the upper break of slope or 'shoulder'. Figure 5.40) and is to be compared with the river channel rather than the river valley.

Factors involved in the modification of the transverse form of a river valley as a result of its invasion by a valley glacier have been set out by G. S. Boulton. Effective normal stress on the valley walls can be expected to rise with depth below the glacier surface so that for any given velocity, the abrasion rates will rise to an optimum level (Figure 5.36), after which they will decline until abrasion ceases and deposition occurs. As the attainment of optimum abrasion rates is dependent principally on the ice thickness, the most rapid abrasion may take place at any level on the valley sides. If it occurs near the base then a parabolic form will develop in bedrock. If effective pressures are high such that optimum abrasion rates are reached well above the valley bottom and abrasion gives way to deposition at the greatest depths, the parabolic cross profile assumed by the valley may be made up of rock wall concavity and a surface of deposition along the valley axis. Subsequent (postglacier) excavation of the till by stream action will produce a valley-in-valley form indistinguishable from similar forms produced by subglacial fluvial erosion and postglacier river incision.

As the parabolic cross profile develops, shear failure in the bedrock and its removal by plucking will vary according to the value of the normal stress. As shear failure in an uneven rock bed is optimized in conditions where lee-side normal stress is minimized, as when a small lee-side cavity develops between glacier and bedrock, the optimum depth for erosion by block removal will again be a function of ice thickness. Optimum conditions will prevail where ice is thick but not so thick as to prevent the development of basal cavities. Under a thick valley glacier therefore, maximum erosion by block removal is likely to be concentrated on the lower half of the valley wall rather than on the upper slopes beneath thin ice or at great depth where the confining pressure of the ice prevents cavities forming. The process of dilation (p. 36) may also contribute to this process such that dilation-jointing develops on a small scale owing to periodic stress relief associated with jerky basal sliding, and on a large scale as progressive valley erosion unloads the rock to such an extent that the rock yields by tensile fracture, the load represented by the glacier being progressively smaller in comparison to the original rock load at any specified depth as rock erosion proceeds. The process is probably accelerated by a period of high erosion rates followed by a period of rapid glacier thinning.

The manner in which a rectilinear longitudinal profile may be modified to a composite concave form has been explained in terms of slip-line fields by J. F. Nye (1952). Assuming a thick glacier everywhere in contact with the bed upglacier of point A (Figure 5.8a,b) the bed is an envelope of slip lines but the line coinciding with A meets the bed tangentially again at point B, so that the slip line A–B is the boundary of the plastically deforming region. Ice and debris below the line will be stationary and erosion will not occur. Erosion will be concentrated where the slip lines or slip line envelopes coincide with the bed, i.e. above A and between B and C, and also below C where the surface coincides with an envelope of slip lines radiating from C. In time, therefore, erosion above A and between B and C will lower that part of the bed until the whole reach is a slip-line envelope (above E,

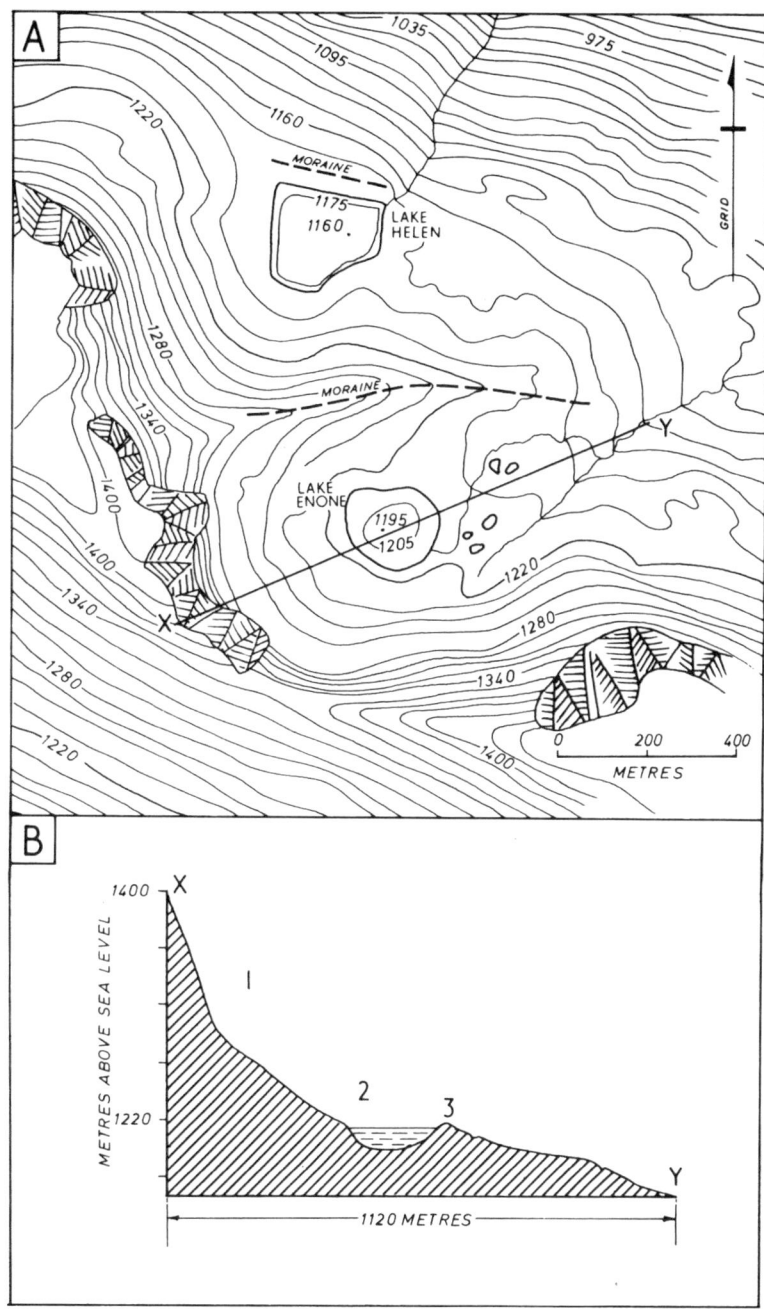

FIG. 5.38. A – MAP OF TWO CIRQUES ON MOUNT OLYMPUS, CENTRAL TASMANIA
B – profile across Lake Enone cirque showing (1) steep headwall, (2) rock basin floor modified by accumulation at foot of the headwall, and (3) threshold of bedrock covered by thin layer of drift.

F–G and G–H) or coincides with a slip line (as in E–F). This sequence assumes compressing flow and can be matched by one for the extending flow condition (Figure 5.8c,d). The radius of curvature of the slip lines is influenced by the bed slope and the ice thickness and appears to have a minimum value of about 100 m. This implies that, beneath thick ice, concavities in the bed with a radius of curvature smaller than this value will not develop and any present initially will be eliminated by erosion of their bounding convexities, there being no minimum radius of curvature for bed convexities. Once the rock-step and basin form is initiated it is perpetuated and probably achieves an equilibrium form as suggested by slip-line field theory (see Section 5.2).

Closed longitudinal concavities in glacial erosional features (rock basins) tend to have an upglacier asymmetry in the absence of marked bedrock structural control, i.e. their steeper shorter slopes face upglacier. In a simple glacier-filled basin, ice thickness and ice velocities increase downslope, reach a maximum at the equilibrium line and decline thereafter. The pattern is one of downward directed flowlines in the accumulation zone and upward-directed flowlines in the accumulation zone. Given the general linear relationship between ice thickness and abrasion and block removal rates, maximum erosion is expectable beneath the equilibrium zone. Lowering of this zone of the valley floor proceeds relatively rapidly, therefore, and continues until a reverse slope is created. In the relationship

$$\tau_b = \rho g h \, \sin \alpha \qquad \qquad (5.5)$$

the term $h \sin \alpha$ is essentially constant. It follows that, in the terminal zone, the basal shear stress (τ_b) is determined by the ice-surface slope (α) and as a result, ice flows in the direction of its steepest surface gradient whatever the direction of bedrock slope. The development of overdeepened rock basins, such as are typical of mountain-foot situations in glaciated uplands, may be explained in these terms. Overdeepening of a glacier basin will ultimately be limited by high normal stresses in the manner outlined on p. 237, so that some equilibrium may be characteristic of stable, long-lived ice bodies. The simplest illustration is probably provided by the small glacier which develops a basin in a fluvial valley head or valley side alcove in rock but which, because of its proximity to the equilibrium line, does not grow so large as to extend for any great distance downvalley. A steep bank of firn may give rise to such a steep, equidimensional glacier, so that erosion is concentrated over a relatively small area. With the development of a rock basin, a lip or threshold is formed and glacial and cryonival processes combine to maintain a steep rock face at the head of the glacier, known as a *headwall*. The resulting erosional landform is a *glacial cirque* (Figure 5.38). Strong evidence of the dominance of basal slip in many cirque glaciers led to the view that cirque glaciers move as a unit by rotational sliding. This was held to be the cause of rock basin development both in cirques and in some glacial valleys where rotational movement was concentrated below rock bars. There is a danger here of confusing cause with effect, however, for it can be argued that the development of a smoothly concave basin will enhance rotational sliding rather than be the product of it, the real course of events.

5.5 DEBRIS ENTRAINMENT BY GLACIERS

Rock debris is transported upon, within and at the base of glaciers. Debris may become *englacial* as a result of burial by accumulating snow and subsequent firnification above the equilibrium line of a glacier. When debris falls and slides on to the glacier surface in the accumulation zone, interbedding with snow occurs and such debris may ultimately extend throughout a glacier's thickness and become concentrated in subvertical zones or 'septa' with transverse compression in the ice (Figure 5.39). When two tributary glaciers con-

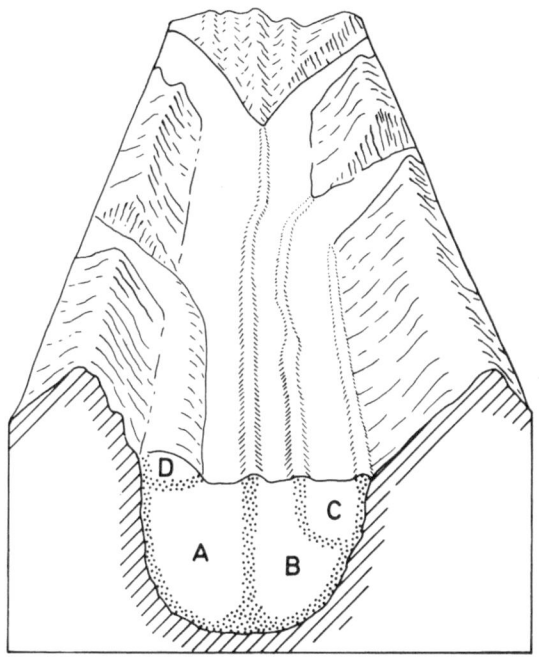

FIG. 5.39. MEDIAL MORAINES PENETRATING TO GREAT DEPTH WITHIN GLACIER AS A RESULT OF SEVERAL TYPES OF GLACIER CONVERGENCE
These include (A and B) side-by-side relationship, (C) an inset glacier, and (D) superimposed glacier. (After Sharp, 1948).

verge, debris on their margins, or in the crevasses at the confluence of two ice-falls, becomes incorporated in the trunk glacier. In this case, also, debris may penetrate to the glacier sole. Englacial debris may also become concentrated in debris bands as a result of the development of basal freezing-on of debris at the glacier sole and as a consequence of shear (thrust) zone development in areas of compressive flow. The *superglacial* debris, derived from a variety of subaerial processes, may not become incorporated in the glacier if accumulation occurs within the ablation zone; such debris may suffer severe frost shattering, washing by meltwaters and sorting by wind action while still in transport on the ice surface (Figure 5.40). Debris may also appear on the glacier surface as a result of concentration by downmelting of the ice.

Below the equilibrium line, melting out of englacial septa or constant addition of rock debris from outcrops will reduce ablation in these zones of the glacier and so give rise to

FIG. 5.40. SUPERGLACIAL DEBRIS ON MUELLER GLACIER (right foreground), NEW ZEALAND WITH LARGE LATERAL MORAINE (centre foreground) LEFT BEHIND AS GLACIER HAS DOWN-MELTED
Note glacial shoulder on valley side in left middle distance. (Photograph: E. Derbyshire).

ice-cored ridges called medial moraines*. Medial moraines tend to become broader with distance down the glacier forming salients along the ice front where locally thick sediment accumulation thus occurs. Zones of debris concentration, such as shear zones within the ice, melt out more slowly because of the insulating effect of the debris layer. Debris making up the *basal* load is concentrated in bands roughly paralleling the glacier sole and may constitute up to 40 per cent by volume in particular zones. It is here that most of the rock flour found in various glacial deposits is produced. High concentrations occurring as a very thin layer between glacier and bed serve as a layer of low strength facilitating glacier sliding. Such material tends to be extruded into rock and ice cavities in response to basal pressure gradients. Within temperate glaciers, englacial and subglacial water systems also transport debris between the surface and the bed. Glacial sediments may thus be recycled along a variety of routes (Figure 5.41).

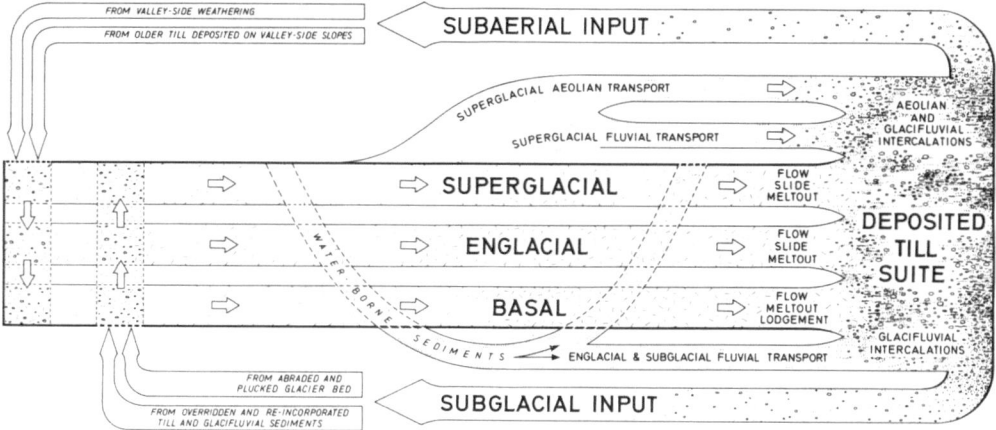

FIG. 5.41. SCHEMATIC DIAGRAM REPRESENTING GLACIAL SEDIMENT SYSTEM

The relative amounts of superglacial, englacial and basal debris vary with the glaciological regime and hence with climate and ice thickness. In a temperate glacier, pressure-melting upglacier of small obstacles is balanced by regelation in their lee. Basal debris becomes elevated above the base of a layer of regelation ice. Individual layers of debris are destroyed by pressure melting on one side of small obstacles but reform downstream following regelation. The range of small obstacles and basal fragment sizes is represented in the base of the glacier as a fine alternation of thin debris-rich laminae and clean, regelation ice layers. As regelation and pressure melting are limited to small obstacles, the debris-rich basal ice of temperate glaciers is rarely more than 20 cm thick. Nevertheless, with the enhanced plastic flow induced by obstacles of 1 m or greater diameter, compressive stresses may be greater between two obstacles than the local vertical compressive stresses, so that debris-rich basal ice may converge into the lee-zone of the obstacles and so produce a locally thick basal debris concentration. In general, however, the greatest volume of the load of many temperate glaciers is to be found on the glacier surface as a result of rockslide, rockfall, avalanching and similar processes (Figure 5.40); only a small

*The term moraine refers to landforms composed of glacigenic sediments. It is sometimes used as a synonym for 'till' but this usage may cause confusion and is not recommended.

percentage is englacial material, perhaps derived from the cirque headwall region and *in transit* along flowlines, with the basal debris being highly concentrated but relatively thin.

In thin polar glaciers, on the other hand, movement is entirely by internal deformation because the glacier is frozen to its bed. Glaciers of this type usually have extremely low debris concentrations. Incorporation of debris from the bed is insignificant and input of superglacial debris from subaerial sources may be limited to blown sand and silt interbedded with snow. As a result, the sparse, predominantly fine debris load is often entirely englacial and located in the planar flow lines.

Incorporation of subglacial debris and its transportation in suspension as basal and englacial load or its rise to the surface to become superglacial debris is favoured by the presence of a zone of basal sliding upglacier of a zone in which the glacier is frozen to its bed. Such a situation is found at a variety of spatial and temporal scales. It occurs, for example, in some parts of large polar ice sheets where ice is locally thick enough to raise basal temperatures above the pressure melting point. This condition is perennial rather than seasonal in the case of thick polar ice sheets. It is also found in subpolar glaciers and ice sheets and, on a smaller scale, in some temperate ice bodies. In temperate glaciers and in some sub-polar glaciers, freezing of the outer zone of the glacier bed is a seasonal feature, the width of the zone varying from a few metres in the former to tens or hundreds of metres in the latter. Other sub-polar glaciers maintain in their outer parts a condition of basal freezing throughout the year.

The importance of this relationship is two-fold. In the first case, basal debris in the inner zone of glacier sliding may become frozen into the basal layer as it crosses the freezing isotherm, thereafter being carried along internal flow lines so that it rises into the glacier (Figure 5.42). In addition, when the outer parts of a glacier are frozen to the bed, the plane

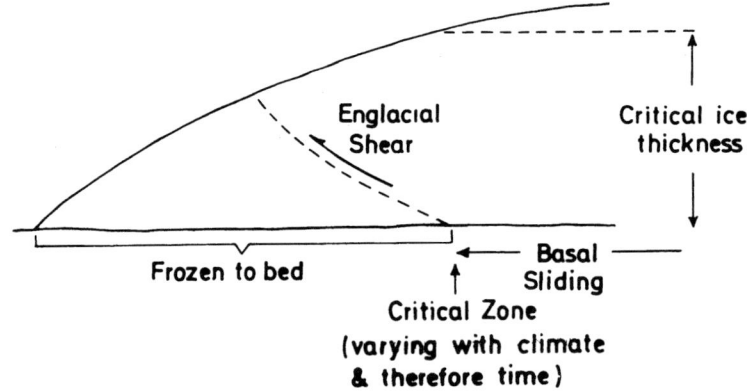

FIG. 5.42. DEVELOPMENT OF SHEAR ZONES IN GLACIER AS RESULT OF FREEZING OF OUTER ZONE TO BED

of greatest weakness occurs at a depth coinciding with the base of the perennially frozen layer. In this case, stress relief occurs below the interface of glacier and bed, and frozen bedrock or previously deposited till may become incorporated in the glacier (Figure 5.35). The development of very high pore fluid pressures below the essentially impervious permafrost-glacier body plays a major role in this process of detachment. Blocks of rock or frozen drift dislodged in this way, together with the adjacent glacial ice, may be subjected to tension during transport in the basal zone of the glacier as the adhesive strength of the bed/glacier contact is at least six times greater than the 'yield stress' of ice (Weertman, 1961). Examples are provided by tensional failure joints in sheared masses of till beneath some sub-polar glaciers. Successive incorporation of rock and debris by this process probably explains the great thicknesses of debris-enriched ice found in some sub-polar glaciers.

The second case involves the generation of high compressive stresses in the glacier in the zone of transition between basal sliding and basal adhesion. The resulting strain energy may be released by brittle failure in the ice mass to produce shear or thrust planes along which active ice slides over immobile ice (Figure 5.42). Once established, the process may be perpetuated because the strength of the debris incorporated within the shear plane is lower than that of clean, unsheared ice. Evidence of this process occurring each winter can be seen at the snouts of many mid-latitude glaciers. At Blåisen in southern Norway, for example, the shear plane generated during the winter, when the outer 10–20 m of the glacier freezes to its bed, melts out into a narrow, planar cave with clean ice above and inert ice coated with shear plane debris below. In large sub-polar glaciers, such as those in Spitsbergen, this process adds to the already high debris concentrations in the marginal ice derived from freezing-on (Figure 5.64c). Clearly, the thermal conditions at the boundary between glacier and bed are a critical factor in debris entrainment. As they are affected by climatic regime, ice thickness and the morphology of the bed, they may be expected to vary quite widely in both space and time.

5.6 GLACIAL DEPOSITIONAL PROCESSES

The debris transported by glaciers is ultimately deposited as a group of sediments referred to as *glacial drift*. It is important to consider the processes by which glacial deposition occurs for two reasons: first, because of their influence on the nature and properties of the sediments laid down, and second, because of their relationship to the landforms produced. It is useful to distinguish two families of deposits: those laid down directly by glacier ice and those derived from glaciers but deposited in water bodies. These are known, respectively, as *glacigenic* and *glaci-aquatic* sediments.

5.6a Glacigenic sediments

Glacigenic sediments are deposited by the direct agency of glacier ice, often in the presence of meltwater but without significant modification by flowing water or by differential settling of its component grains in still water. Glacigenic sediments are known by the collective term *till*.

Tills vary systematically in their particle size composition, lithology, mineraology and fabric as a result of:
  (i) derivation from different bedrock types, and
  (ii) the process history of the sediments during entrainment, transportation, deposition and post-depositional events.
Given that the two major processes of bedrock erosion by ice are abrasion and quarrying, a

broad correlation is to be expected between till characteristics and bedrock properties, specifically abrasion hardness, and strength as influenced by the frequency and disposition of rock joints. At one end of the scale, a predominance of abrasion of fine-grained rocks of low hardness (clays, soft limestones and some mudstones) produces massive clay tills with sporadic stones (matrix dominated tills). At the other extreme, plucking of hard rocks which are well jointed (granites, gneisses and some quartzites) yields abundant clasts, abrasion producing only minor amounts of rock flour matrix. The expected result is clast-dominated tills (Table 5.2). Although the relationship between bedrock type and till

Table 5.2
GRADATIONAL SERIES OF TILL TEXTURES AND GENERALIZED ROCK TYPE CORRELATIVES

	Major rock types	Till textures (Derbyshire, McGown and Radwan, 1976)
Increasing rock hardness / Increasing frequency of orthogonal jointing systems	Gneisses/granites Quartzites Fine-grained igneous Sandstones	Granular till (Clast dominated)
	Hard limestones Slates/greywackes Mudstones and shales	Well-graded till ('Typical' till)
	Soft limestones Clays	Stony clay-silts (Matrix dominated till)

grain size is not a simple one because of the glacial process factors discussed below, the generalized pattern of till types deduced from the distribution of rock types in Britain (Figure 5.43) accords with field descriptions of the tills.

The particle size distribution also varies with the distance of glacier transport. Tills made up of short-travelled debris, such as is common in many alpine type glacier basins, tend to be clast-dominated. With greater distances of transport and hence greater likelihood of sediment processing, the till matrix becomes increasingly dominant (Figure 5.44). With prolonged transport, a notable proportion of the rock debris may be comminuted, the initial size and lithology of the clasts being an important influence on the final product. Ultimately, the modal size of the matrix particles reflects the fact that there is a size below which a particular mineral will not be reduced by comminution. A. Dreimanis has called this the 'terminal grade' (Figure 5.45). Thus, far travelled material may appear rather better sorted than tills made up of local materials.

The bedrock source of the till and its processing during glacier transport influences the shape of the clasts in till. Superglacial debris is distinctly more angular than material transported by traction at the glacier bed which characteristically shows one or more flattened facets with striations as well as some edge rounding. There is some evidence to suggest that the shape of glacial stones changes progressively the greater the distance they are transported. In any one location, however, there may be wide diversity in pebble shapes because of a complex history of deposition, re-incorporation in the glacier and entrainment by glacial meltwater streams.

Several processes are involved in the release of debris from glacier ice and its deposition as till. The factor common to all processes is the melting of the interstitial ice component of an ice-debris mixture. This is effected at the glacier base by pressure melting. It also occurs

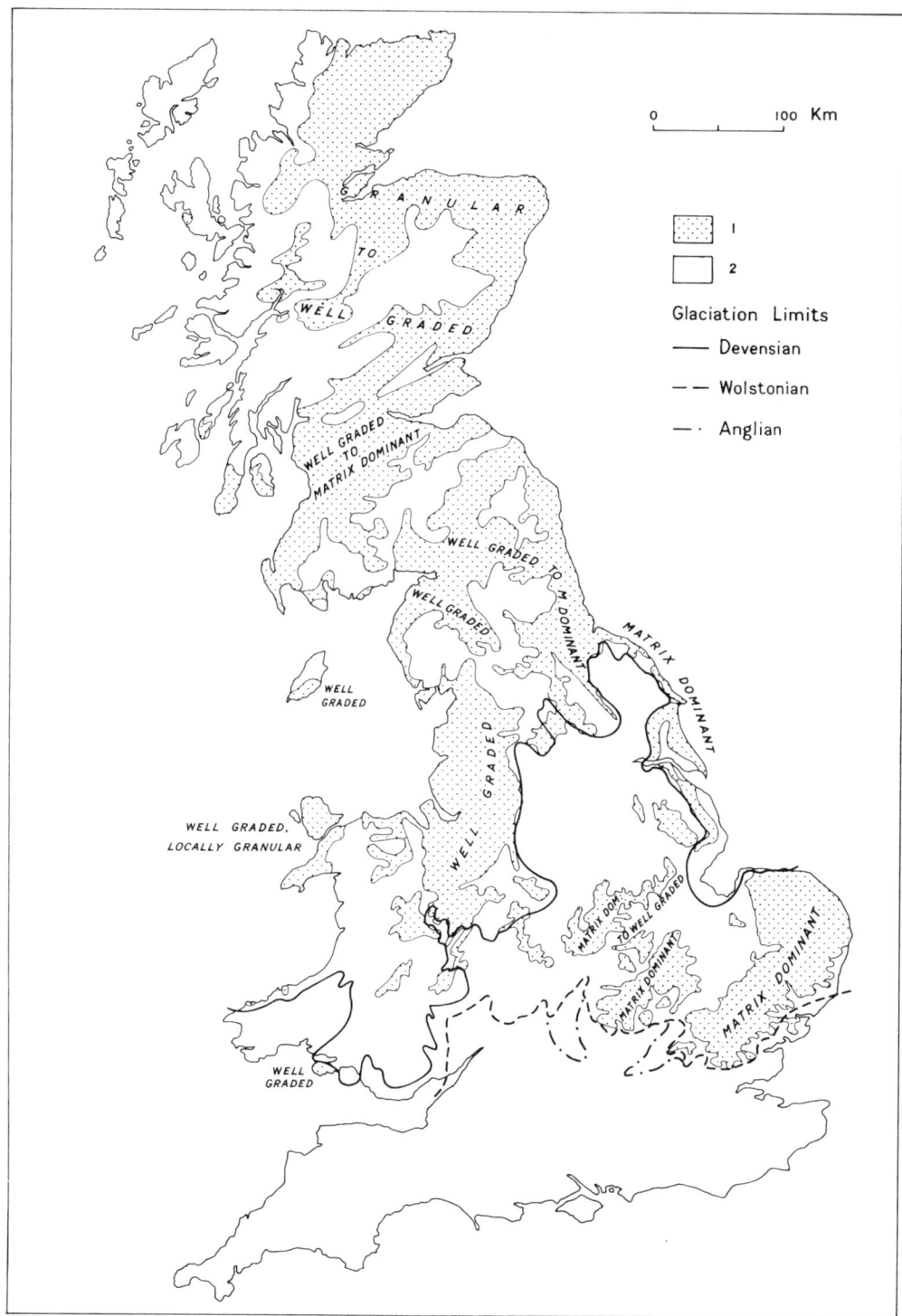

FIG. 5.43. MAP OF GENERALIZED TILL TEXTURES IN GREAT BRITAIN
(From Derbyshire, 1975).

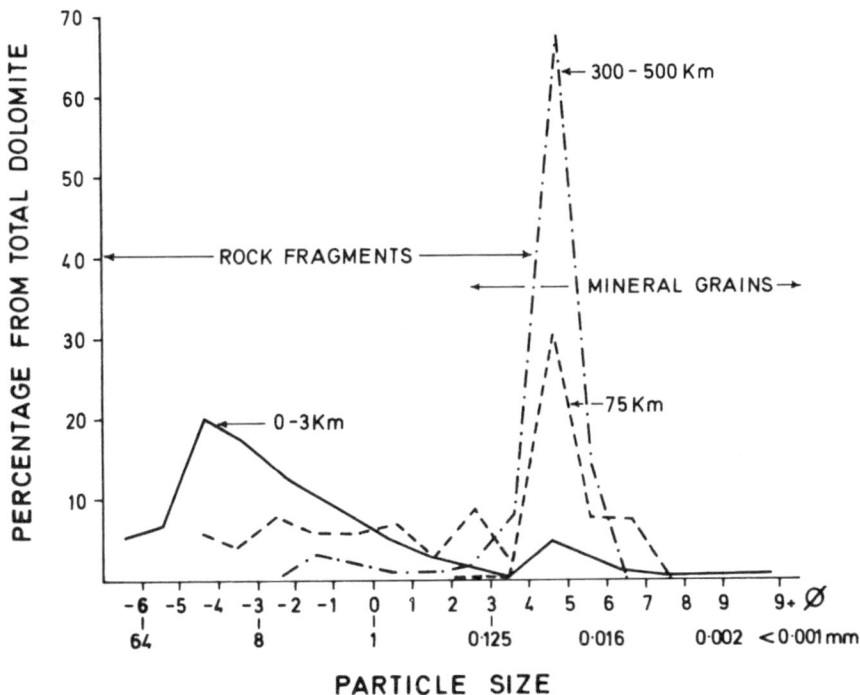

FIG. 5.44. FREQUENCY DISTRIBUTION OF DOLOMITE IN THREE TILL SAMPLES FROM HAMILTON-NIAGARA AREA
(After Dreimanis and Vagners, 1971).

Table 5.3
PROCESSES OF DEBRIS RELEASE FROM GLACIER ICE

	Generic terms	Specific terms		
Glacial debris in transport	Glacial debris	By transport location		
		Superglacial debris		
		Englacial debris		
		Basal debris		
		By depositional process		Qualified by location of primary deposition
Glacial debris deposited	Till		Meltout till	Superglacial
				Subglacial
		Ablation till	Flow till	Superglacial
				Subglacial
			Lodgement till	(Subglacial by definition)

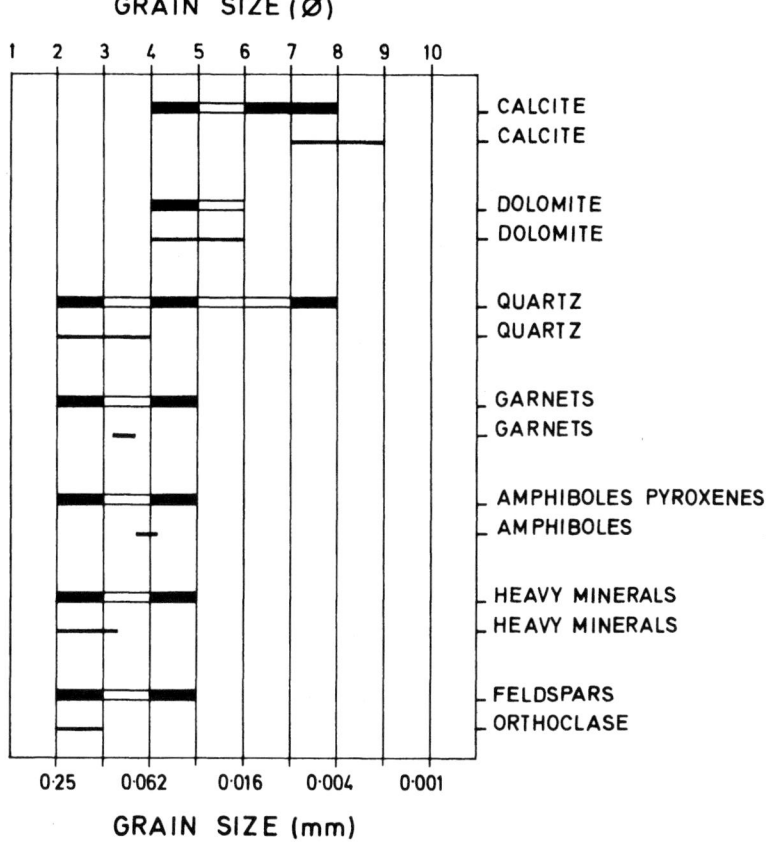

TERMINAL MODES IN BASAL TILL

FIG. 5.45. 'TERMINAL GRADES' OF SELECTED MINERALS IN BASAL TILL OF PLEISTOCENE AGE FROM SOUTHERN ONTARIO, SHOWING MEANS, STANDARD DEVIATIONS AND MODAL CLASSES
(After Dreimanis, A. and Vagners, U. J. in Yatsu, E. and Falconer, A. (eds) *Research Methods in Pleistocene Geomorphology* Norwich, Geo Abstracts Ltd., (1972) 74. By permission of A. Dreimanis.)

as the static process of ablation (melting and sublimation) beneath immobile ice and on the surface of glaciers. The two primary processes of debris release from glaciers are, therefore, lodgement and meltout (Table 5.3).

5.6b Lodgement

The subsequent behaviour of the material laid down by these primary processes will vary according to the form and nature of the underlying surface (permeable rock or debris, impermeable rock or debris, or ice), and the moisture content, pore fluid pressure and particle size distribution of the tills. In subglacial situations where some subglacial melting occurs, debris-enriched basal ice releases its debris as a result of transfer of geothermal heat from below and in response to the frictional heat generated in glacier sliding. This process is sometimes called pressure melting deposition. When the frictional force exerted by the bed on a rock fragment in the glacier sole becomes greater than the force exerted on it by the glacier, it will come to rest. This force *(F)* is a product of ice density (ρ), ice thickness *(h)*, the gravity force *(g)*, the surface area of the particle in contact with the bed (S_1) and the coefficient of dynamic friction (μ), thus

$$F = \rho gh\, S_1\mu \qquad (5.6)$$

The value of *F* will be reduced by positive water pressure (P_w) at the bed (when $F = (\rho gh - P_w)\, S_1\mu$ and variations will also arise as a result of variations in the size and shape of the particle itself.

The difference between the velocity of a particle at the bed (V_p) and that of the ice around it (V_i) is termed the relative velocity (V_r). This is greatest for particles of less than *c.* 8 mm diameter (because of efficient regelation) and greater than 150 mm (because of efficient plastic deformation around them). The value of this relative particle velocity has been expressed by Boulton (1975) in the following relationship:

$$V_r = A \left(\frac{F}{S_2}\right)^3 \; 1+\frac{CKF}{L\rho lS_2} \qquad (5.7)$$

where S_2 is the cross-sectional area of the particle transverse to flow, *1* is the length of the upstream side of the particle, *K* is the thermal conductivity of the particle, *L* is the latent heat of fusion of ice, *C* is a constant relating pressure and the resultant lowering of the freezing point and *A* is the temperature-dependent constant in Glen's flow law of ice (see Section 5.2). Because of the influence of S_1 on *F* and of S_2 on V_r, shape (expressed as the ratio $S_1:S_2$) is an important source of variation in bed forces and hence in depositional rate. Platy clasts at the glacier sole will have high values of S_1 and low values of S_2, so that both the frictional drag *(F)* and the relative ice velocity (V_r) will be high favouring lodging of the particle. Wedge-shaped clasts (those with triangular cross-section form) may have either high or low values of S_1 depending on the surface reaching the bed (Figure 5.46) but the value of S_2 will always be relatively high. In its most stable position with one face on the bed, the value of *F* will be at the maximum and that of V_r at the minimum, a situation favouring maximum effective abrasion of the particle. This may explain the high frequency of 'flat iron' shaped clasts with one or two smoothly abraded facets considered to be the typical till stone shape. The degree to which the glacier bed is deformable will also have a strong influence on whether or not clasts become lodged.

As basal glacier debris characteristically contains particles of widely divergent sizes and shapes, particular values of effective normal stress at the bed and ice velocity will favour deposition of some particles and continued transport of others. The deposition of clay-rich till is often termed 'plastering on' but the essential control, namely the ratio of frictional drag and tractive force, is the same for all scales of operation. Where the frozen basal debris is very rich in fine-grade particles, it may be deposited as relatively rigid units, and aggregates of clay rather than individual clay particles may behave as frictional units. Accordingly, lodgement is the preferred term for this process. It is used in this sense to describe tills released at the glacier bed, the so-called *lodgement tills*. Comminution products are present in most tills but they tend to be the dominant element in tills of lodgement origin. Admixtures of debris of other origins, such as meltwater deposits or valley-side scree, modify the grading curve. Several other properties of lodgement till throw light on processes operating both during and following its deposition. Layering which varies in thickness from a centimetre or two to a metre is thought to be the product of successive accretion of till by lodgement. Such till is fissile (readily split), the parting planes being marked by sand and coarse silt concentrations only one or two grains thick between otherwise massive till units.

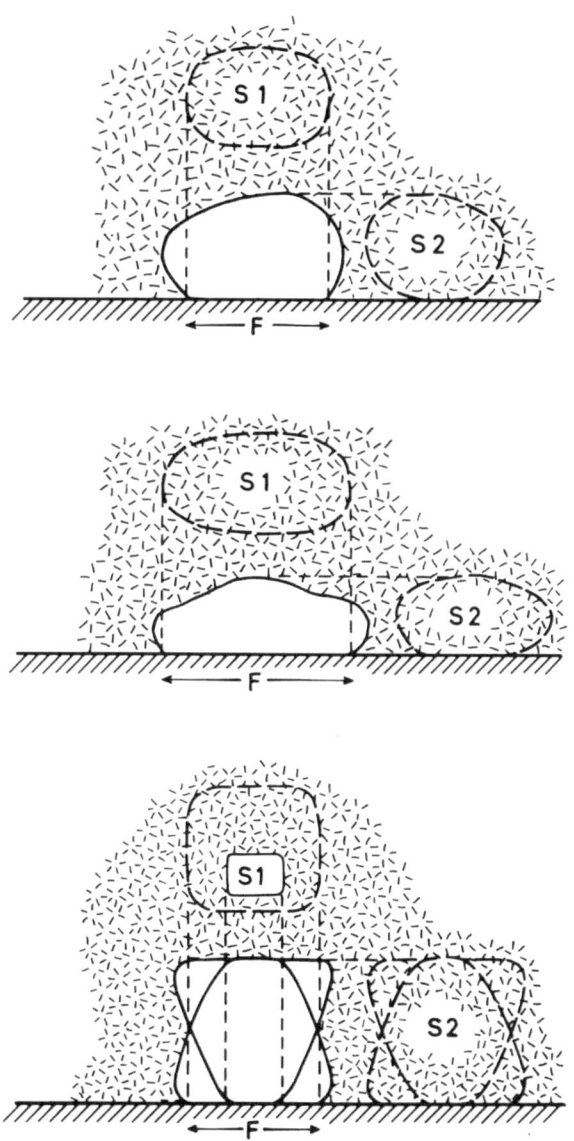

FIG. 5.46. SCHEMATIC DIAGRAMS SHOWING AREA OF PEBBLE ENCLOSED IN BASAL ICE
Pebble area in contact with rock bed beneath (S1), cross-sectional area of clast facing up-glacier (S2), and area over which frictional drag occurs (F), for several shapes of clast. For further explanation see text.

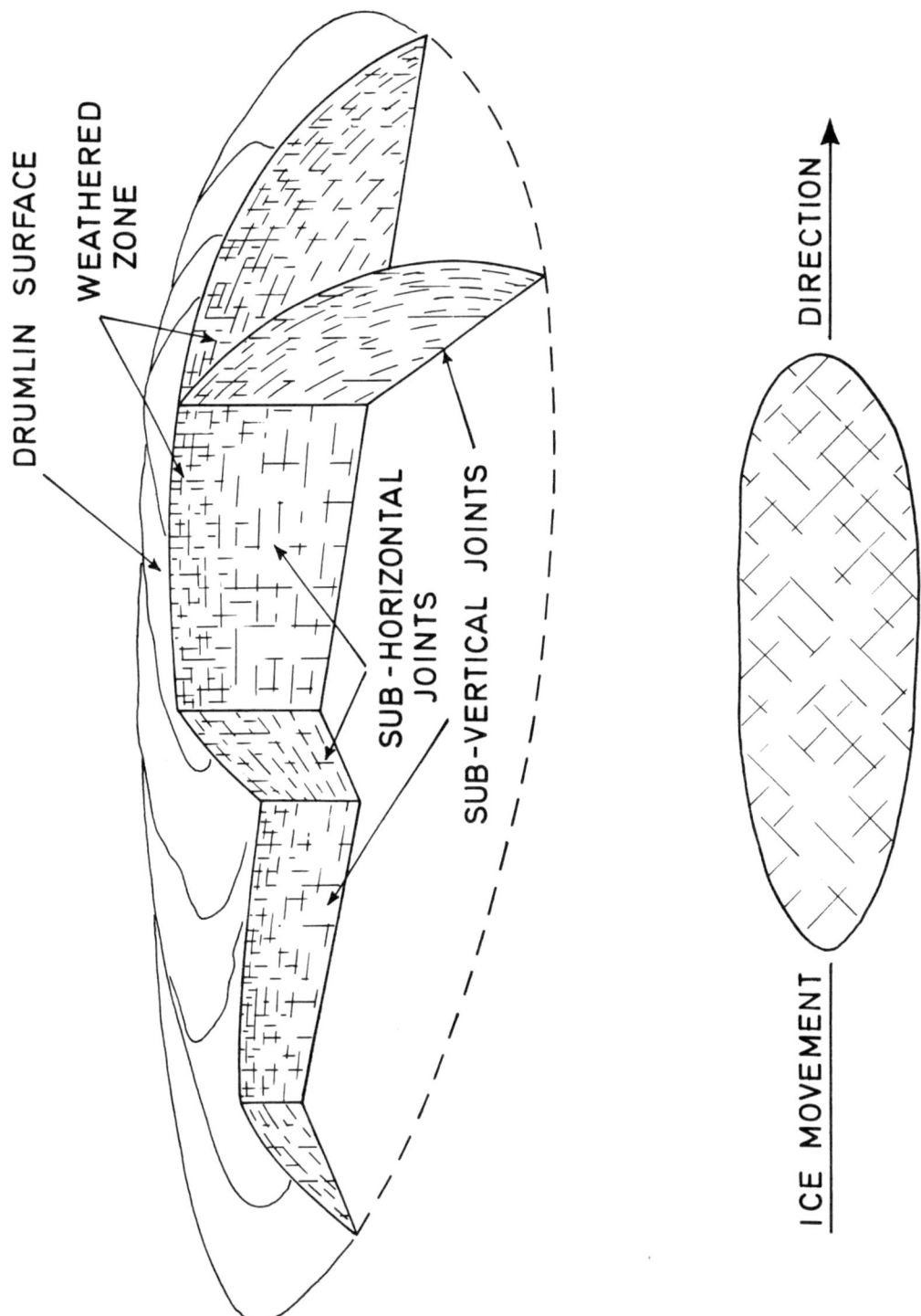

FIG. 5.47. PERSPECTIVE SKETCH OF DRUMLIN, SHOWING HOW SETS OF JOINTS RELATE TO SURFACE OF FEATURE AND TO DIRECTION OF GLACIER FLOW
(Based on an original drawing by A. McGown). Lower diagram is a plan view of drumlin showing sub-vertical joint sets disposed conjugately about ice movement direction.

Similar planar features which, however, lack the particle size differentiation produced by accretion are well developed in some lodgement tills and clays of other origins. Their disposition parallel to the ground surface suggests that they are the result of relief of stress following removal of overriding ice. Other features of lodgement till ascribed to stress relief include almost vertical joints disposed normally about the former ice movement direction (Figure 5.47). The stress in this case was applied by periodic glacier sliding, alternating episodes of stressing and stress relief producing a systematic brittle failure in compact lodgement till. Another form of brittle-type fracture in consolidated till to be seen beneath some modern glaciers is expressed as low-angle shear planes, the faces of the sheared blocks being characteristically *slickensided* (grooved and polished). Compressional shearing of frozen till and bedrock occurred on a large scale during the Pleistocene glaciations.

It has been known for decades that the average direction of the longest axes of a randomly selected population of stones in lodgement till lies parallel to the direction of movement of the glacier which deposited the till. In addition, the longest axes tend to dip upglacier, i.e. against the flow. As a result, the orientation of the *a* axes of clasts has been the most widespread sub-fabric element studied, primarily for the purpose of establishing the direction of movement of parts of the great Pleistocene ice sheets. The use of till fabrics as an analytical tool to aid understanding of the processes of till deposition is a relatively recent development, however. Both field evidence and theory suggest that alignment of particles parallel to flow is to be expected during extending flow, with a transverse alignment occurring during compressing flow. Alignments of long axes are parallel to flow in thrust planes and shear zones (Figure 5.48) and where high pore-fluid pressures induce subglacial flow of till. As the matrix is the stress-transfer medium, flow is induced in the matrix. Elongate clasts present their smallest cross-sectional area to the flowing matrix; this minimizes their resistance to flow and gives rise to long axis orientations parallel to flow. As in the case of zones of compressing flow in the glacier, folding of till produces a long axis orientation maximum transverse to the ice flow direction.

The process of lodgement produces a depositional surface which is irregular in detail because of the bimodal particle size composition of most lodgement tills. Particle collision at the depositional surface results in a tendency for the clasts to become clustered together. Individual boulders will plough into the till matrix unless it is very compact and additional clasts will lodge upglacier of such obstacles. Thus aggregations of boulders and cobbles may form a locus of till deposition. In a dynamically active glacier in which the frontal margin is maintained in approximately the same position over a period of time, deposition of lodgement till from englacial debris bands may result in a thickening of the till by successive lodgement. With retreat of the ice front, the thickened zone of till will appear as a ridge known as a *lodge moraine*. Unmodified frontal ridges or *end moraines* of this origin tend to be asymmetrical, the gentler flank being the proximal one, i.e. facing the glacier (Figure 5.49). Tabular clasts exhibit consistent upglacier dips, elongate stones being parallel to ice flow in response to matrix remoulding beneath the ice edge (Figure 5.50a) but also occurring transverse to flow as a result of rolling and particle collision (Figure 5.48).

End moraines are also produced by the 'bulldozing' action of the glacier as it pushes into till surfaces. The process is most clearly seen where thin till overlies bedrock. This is a

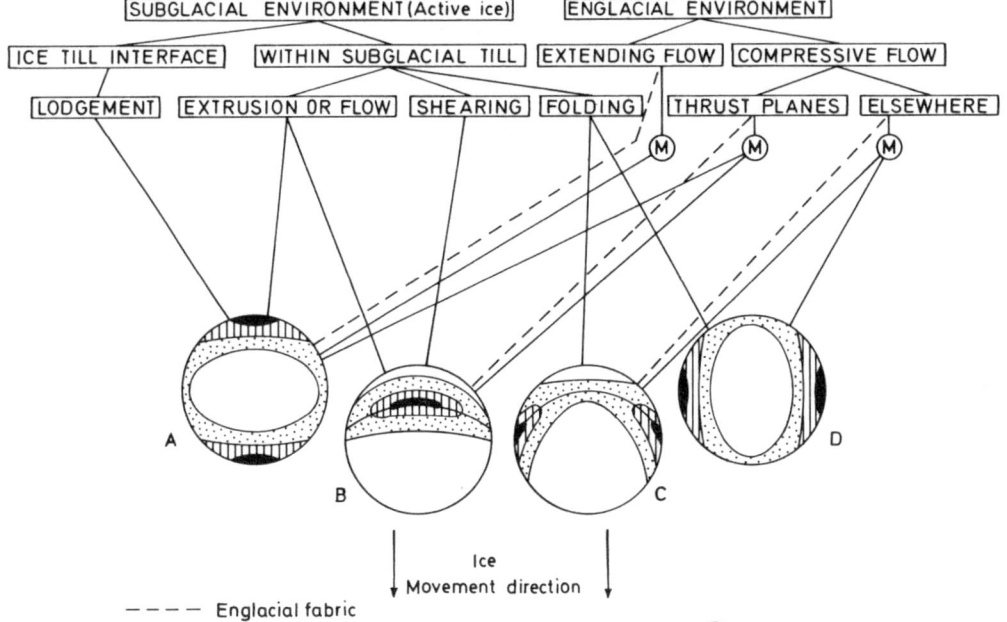

FIG. 5.48. PROPOSED RELATIONSHIPS BETWEEN FABRIC-FORMING PROCESSES AND HYPOTHETICAL FABRIC TYPES IN LODGEMENT AND MELTOUT TILLS

Pebble longitudinal *(a)* axis orientations of type A and B lie parallel to ice movement while those of types C and D are transverse; cf. Figure 1.7B. (After Mark, 1974, with permission of Geological Society of America).

FIG. 5.49. END MORAINES OF LODGE TYPE IN FRONT OF HOOKER GLACIER, NEW ZEALAND

Hummocky moraine produced by deposition of superglacial debris around dead ice blocks can be seen adjacent to small lake. (Photograph: E. Derbyshire).

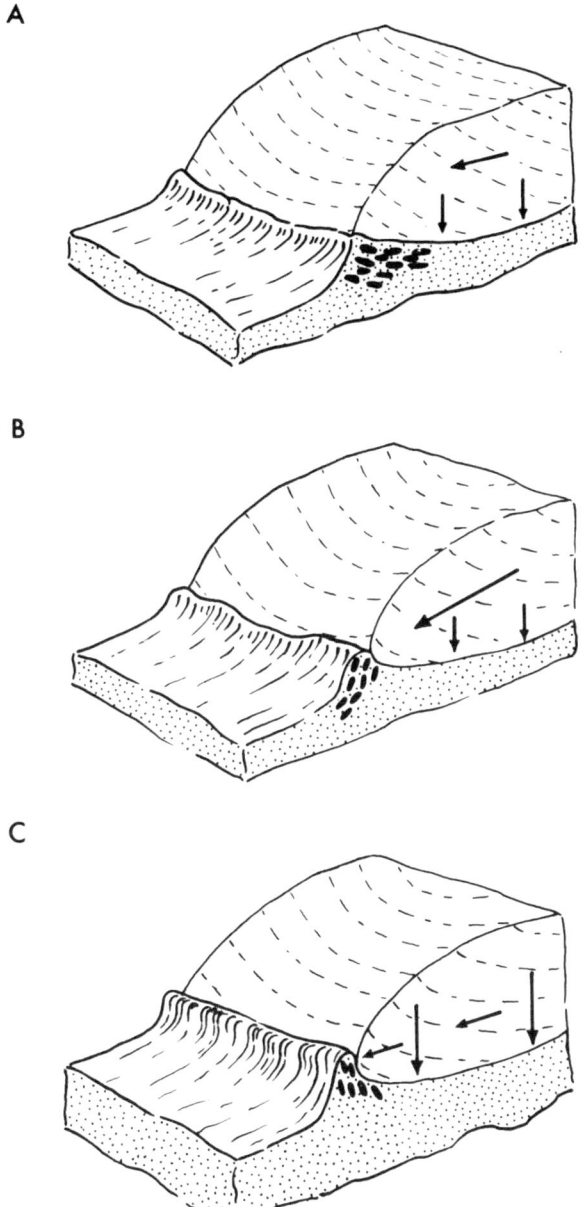

FIG. 5.50. END MORAINES OF SIMILAR SIZE AND MORPHOLOGY PRODUCED BY DIFFERENT PROCESSES AND HAVING DIFFERENT STONE FABRICS

(a) – stresses arising from forward motion and normal stresses approximately in balance and lodgement process produces stone fabric which is imbricated *up-glacier*. (b) – stresses arising from relatively rapid forward motion are more important than normal stresses hence the glacier 'bulldozes' till surface, and platy clasts tend to line up normal to bulldozing stress, producing steep *down-glacier* dips in pebbles. (c) – when stable ice front rests on thick saturated till, normal stresses may be dominant factor in end-moraine formation, resulting in 'squeeze-up' of till at ice margin. In this case stones tend to dip very steeply *up-glacier*.

regular occurrence in winter when short-distance advances cause the ice front to plough into the till released in the previous summer. In their fresh, unmodified form, these *push moraines* are asymmetrical with the proximal slope being the steeper of the two. The *a/b* planes of tabular particles tend to dip *down*-glacier as do particle long axes but the tendency toward a transverse maximum (i.e. parallel to the moraine crest) is also strong. These fabric properties are a result of thrusting of water-soaked till, *a/b* planes of particles tending to align themselves normal to the direction of applied stress (Figure 5.50b).

The normal stresses acting on saturated till as a result of loading by an ice front may cause the till to extrude from beneath the ice. This produces small end moraines which preserve in their plan shape the precise form of the ice edge. Such *'squeeze-up' moraines* have a morphology identical to small push moraines but their fabric is quite different. The stress applied by the ice front is transmitted by the wet matrix and the clasts tend to line up parallel to the flow of the matrix, so producing very steep upglacier dips. This process requires a thick layer of till (Figure 5.50c), but both squeeze-up and push moraines are known to occur along the same ice margin, and both processes probably operate in producing end moraines.

The tendency for lodgement of large clasts at the depositional surface of till may result in linear zones of boulders aligned parallel to the ice flow. Such aggregations constitute an obstacle which, although it may continue to be moved by slow traction, will travel much less rapidly than the surrounding ice. Plastic deformation around the obstacle thus tends to produce a vault or tunnel which is controlled in its diameter by the size of the obstacle and which is elongated leeward of it by an amount proportional to the ice thickness and the velocity difference between the glacier and the obstacle (Figure 5.51). Depending on the degree of melting of the bridging part of the glacier sole, the debris content of the sole and the pressure of any water in such cavities, deposition of till may be concentrated there. Two

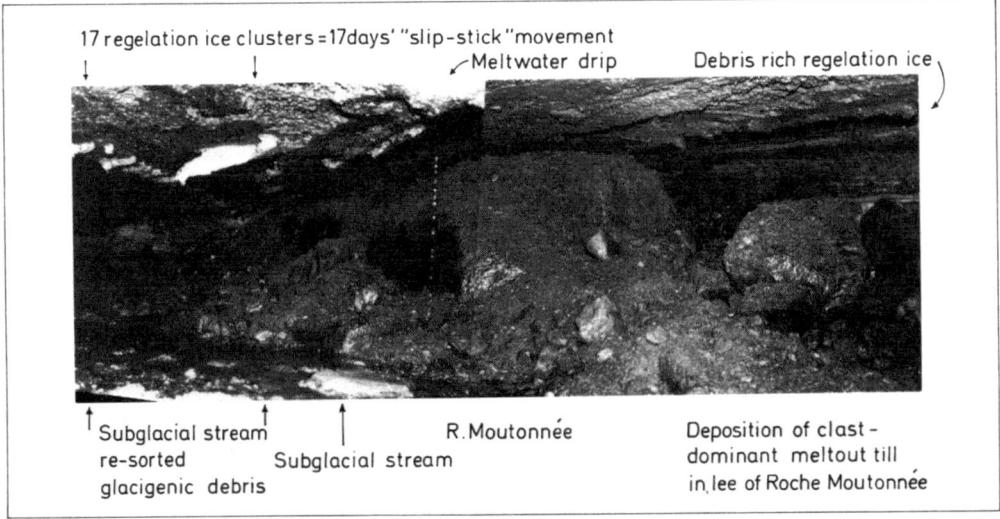

FIG. 5.51. SOLE OF BLÅISEN, SOUTHERN NORWAY, AT A POINT 40M INSIDE ICE EDGE WHERE IT OVERRIDES A ROCHE MOUTONNÉE
(Photograph: E. Derbyshire).

important depositional processes involved are free fall of basal debris from the tunnel roof and extrusion of basal till immediately in the lee of the obstacle where confining pressures suddenly decrease. Sedimentation of till by meltout, flow and lodgement may ultimately fill the vault and a streamlined equilibrium form, with respect to prevailing glacier conditions, may result.

Streamlined depositional landforms produced in this way are linear in form and rarely more than two or three metres in height around modern glaciers. They are known as *flutes* or *fluted moraines*. Till may be fluted with great regularity and straightness. The flutes are flanked by grooves. Examination of subglacial cavities reveals that plastic deformation of the basal ice around obstacles produces a vault in the glacier sole and that pendants or ribs of ice, in which regelation ice is dominant, are produced by enhanced plastic flow at low points between such obstacles (Figure 5.52). Lateral loading of wet till by the ice produces

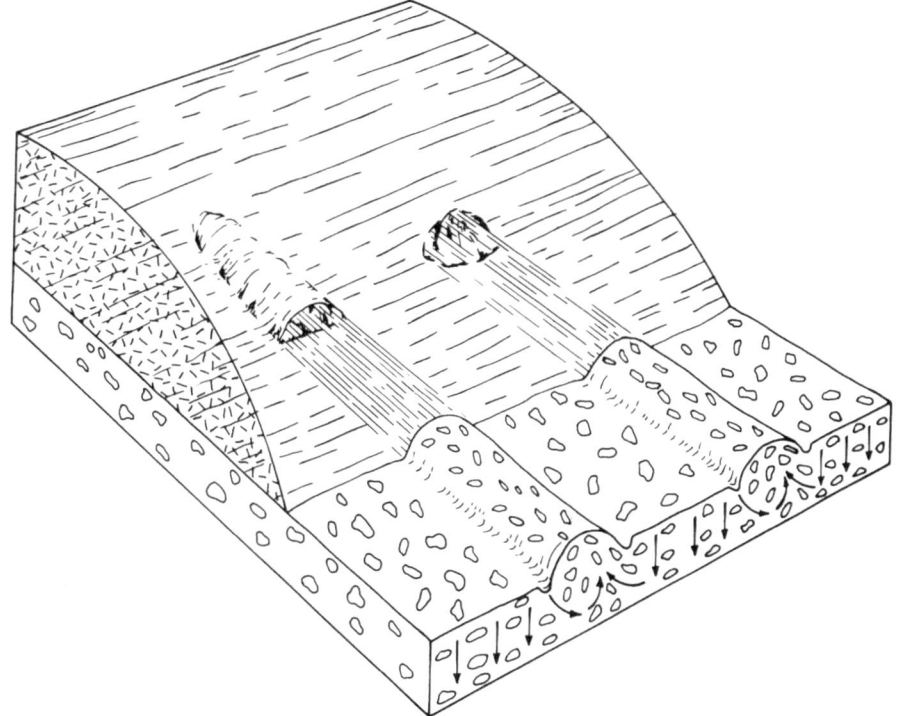

FIG. 5.52. SCHEMATIC BLOCK DIAGRAM SHOWING DEFORMATION OF GLACIER ICE ACROSS ROCHE MOUTONNÉE (LEFT) AND BOULDER BEING DRAGGED ALONG SOLE (RIGHT)
Linear vaults produced in base of glacier by plastic deformation around these obstacles provide zones of low stress into which saturated till is squeezed from adjacent areas of high stress. Such alternations of high and low stress linear zones produce till flutes, the generalized movement of mobilized till being shown by arrows, together with distinctive modification of orientation of stones in till.

grooves, and the till moves away from such zones of high confining pressure into the zone of low pressure provided by the subglacial vault. The expected flow lines will be transverse to the line of the cavity but, as the ice is moving, longitudinal shear forces can be expected to impose a down-glacier element into the pattern. This hypothesis is confirmed broadly by stone orientation analysis of modern flutes. This ideally symmetrical pattern is modified in natural conditions however, by factors such as the presence of very large clasts in till, variations in the strength of the till and in the strain-rate vectors at the till-ice interface.

Much larger streamlined features which are of typically ovoid or of 'half-egg' form rather than ridges, occur in many formerly glaciated regions and are known as *drumlins*. They may be as much as 50 m high and several kilometres long, although transitional features sometimes called megaflutes may reach 20 km in length. Thus flutes, drumlins and megaflutes are members of a continuum of streamlined forms. They may also constitute a hierarchy of bedforms by analogy with much smaller features on stream beds (plane bed, ripples, dunes, plane bed, antidunes). Actively developing streamlined depositional land-forms the size of most Pleistocene drumlins are not accessible for direct study beneath modern glaciers. It might be argued that, by analogy with flute formation, the process of shearing and flow of previously deposited sediments from points of high confining pressure into low pressure zones is sufficient to produce drumlins. Mechanical difficulties arise, however, in the application of this process model to streamlined depositional forms of more than two or three metres in height. It is not enough merely to 'scale up' the flute mechanism by increasing the obstacle size, the glacier thickness and the ice velocity. The simple relationship between size of obstacle and size of basal cavity breaks down because increase in plasticity with ice thickness inhibits development of cavities of more than a few metres in height. In other words, drumlins are not primarily the product of the squeezing of remoulded till into existing basal cavities in the ice, and recourse must be made to other processes to explain their origin.

Bulk expansion, or dilation, of wet subglacial till is one important mechanism. The amount of dilation is proportional to the angle of internal friction of the till and the degree to which the clasts interlock; zones of dilatant till provide a source of variation in the form and strength of a mobile glacier bed. Smalley and Unwin (1968) described deformation in terms of three stress ranges (Figure 5.53). With stresses above point A, deformation is continuous. Below point B, deformation does not occur. With shifts in stress levels between A and B, which are to be expected in zones outside the areas of thickest ice but well inside the glacier margin, basal debris masses which fall below stress level B will form resistant mounds (Figure 5.53). The till between such zones and the glacier sole will continue to be sheared with passage of ice across it. A preferred stone long-axis orientation pattern similar to that found at the surface of small flutes will be induced, platy particles generally lying parallel to the drumlin surface. Layering in the surface till of drumlins with sand or coarse silt present in the parting planes suggests accretion of units by lodgement. Sub-horizontal stress-relief joints and conjugate sets of near-vertical joints so common in lodgement tills are also present (Figure 5.47). Drumlins may consist of till, glacifluvial and glacilacustrine deposits and a variable proportion of bedrock.

Fluctuations in the condition of the ice sheet, notably in its thickness and sliding velocity, induce variations in stresses at the bed. These, together with differences in shear strength of the bed material, are reflected in the amplitude of the bedforms. Under relatively constant stress conditions at and within the bed, the bedforms will tend towards a quasi-equilibrium in both form and scale. Many drumlin fields of Pleistocene age exhibit an impressive consistency in these properties, while others contain drumlin populations of different scale and form. The two may, in fact, be superimposed one on the other. A

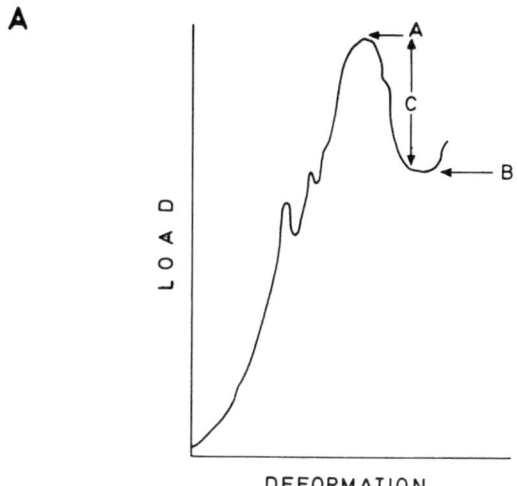

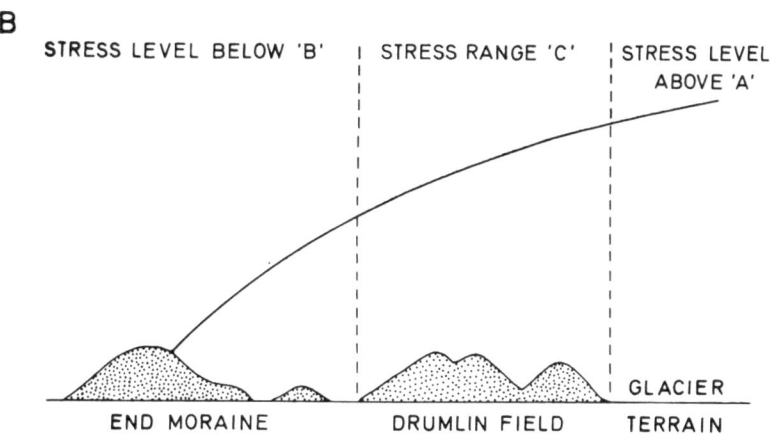

FIG. 5.53. A – GENERALIZED LOAD-DEFORMATION CURVE FOR TILL. B – THE CRITICAL STRESS RANGE REGIONS NEAR THE EDGE OF ICE SHEET SHOWING RELATIONSHIP OF STRESS RANGE 'C' TO ZONE OF DRUMLIN DEPOSITION. (After Smalley and Unwin, 1968).

schematic summary of the relationship between basal ice conditions, bed material and resultant (drumlin) bedforms is shown in Figure 5.54.

Many surfaces glaciated during the Pleistocene era display an intimate intermingling of flutings, drumlins and transverse moraine ridges (Figure 5.55). This association, together with the remarkably regular spacing suggests that the transverse ridges are not of end moraine origin. It seems more likely that, in common with drumlins and flutes, they are subglacial bedforms around which complex flow patterns occurred in the basal ice (Aario, 1977). Small (5–15 m high) regularly spaced transverse ridges (called De Geer moraines) containing abundant bedded sediments, are interpreted as the product of deposition at the

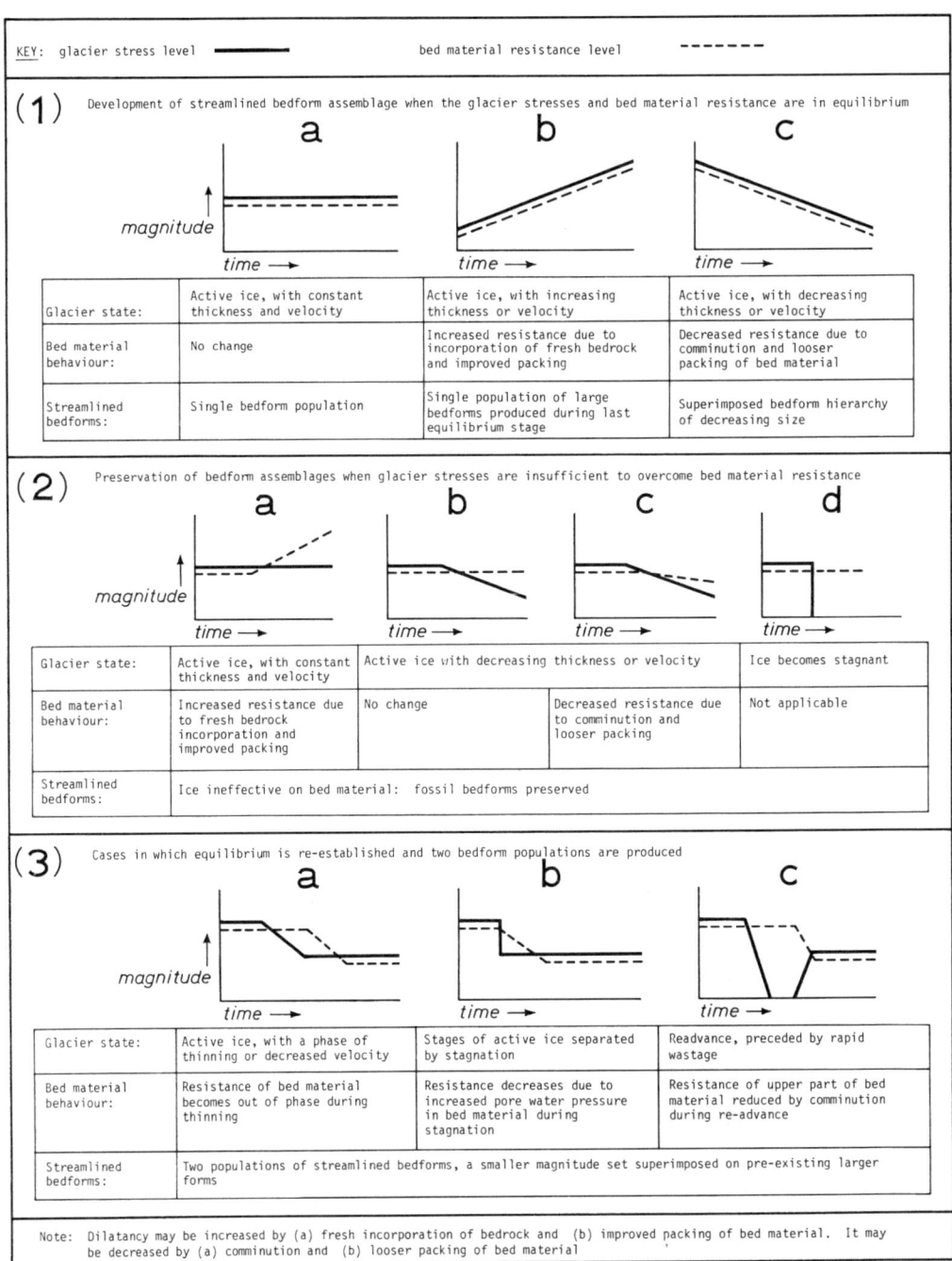

FIG. 5.54. HYPOTHETICAL CONDITIONS AT BASE OF GLACIER LEADING TO FORMATION OF STREAMLINED BEDFORMS (After Rose and Letzer, 1977).

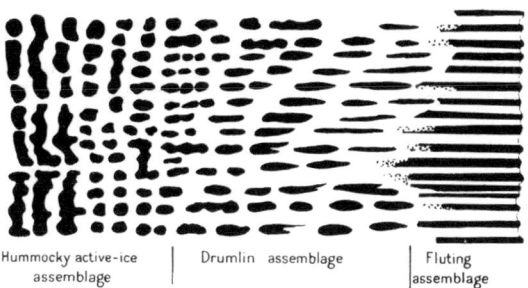

Hummocky active-ice assemblage | Drumlin assemblage | Fluting assemblage

FIG. 5.55. TRANSITIONAL SERIES OF BEDFORM ASSEMBLAGES ASSOCIATED WITH ACTIVE ICE
Flow from left to right. (After Aario, 1977).

grounding line of a thick ice sheet partly floating in deep water. Their regularity of spacing may be a reflection of the spacing of tension crevasses in the ice sheet. De Geer moraines provide a warning reminder that some important moraine-forming processes are not now available for scientific study and that those processes which can be studied today may have been of only relatively minor importance in the generation of Pleistocene depositional landforms.

5.6c Meltout, flow and sliding

At the base of a glacier with some basal melting, rock-surface irregularities and variations in ice thickness give rise to hydraulic gradients at the bed, which induce subglacially-released till to flow as a slurry. Till may thus be moved from zones of high pressure to be deposited finally in areas of lower pressure. This may be a recurrent phenomenon, temporarily deposited till being reworked by kneading (remoulding) and redeposited. Such cyclic consolidation and dilation is common where glaciers override previously deposited sediments. Dilation of lodgement tills may loosen the fabric, increase the void ratio and so reduce the resistance to shear. Dilated lodgement till may thus have a shear strength which is lower than the value when first deposited by the glacier.

Another process of subglacial till deposition, and one which has been observed in operation, is the release of basal debris in caves formed as the glacier sole arches over obstacles in the bed such as *roches moutonnées*. Sheets and blocks of debris with interstitial ice are released by melting as a result of geothermal heat flow and of circulation of air within cavities, following which the blocks slowly melt out. Frozen sheets and blocks of debris may reach considerable thicknesses within stagnant ice bodies. Beneath moving glaciers, they may be released by shearing as well as by melting and free fall. In all cases, the product is *subglacial meltout till*. Given suitable depositional surface slopes and moisture contents, this meltout till may flow under its own weight within subglacial caves and so be remoulded into *subglacial flow till*.

In the marginal areas of glaciers and ice sheets, englacial debris tends to become concentrated on the ice surface as a result of thermal conditions favouring persistent accretion by regelation or englacial shearing (Figure 5.42). As the glacier wastes down, so the debris layer thickens as a result of surface ablation. The melted-out debris layer may reach several metres in thickness on some temperate and sub-polar glaciers. However, considerable thicknesses are most usually the result of differential melting of the buried ice with periodic sliding and flow of the superglacial debris towards low points and the exposure of clean ice below. Having been stripped of its insulating cover of debris, the bare ice suffers accelerated melting. It follows that *superglacial meltout tills* vary greatly in thickness, surface form and density (Figure 5.56).

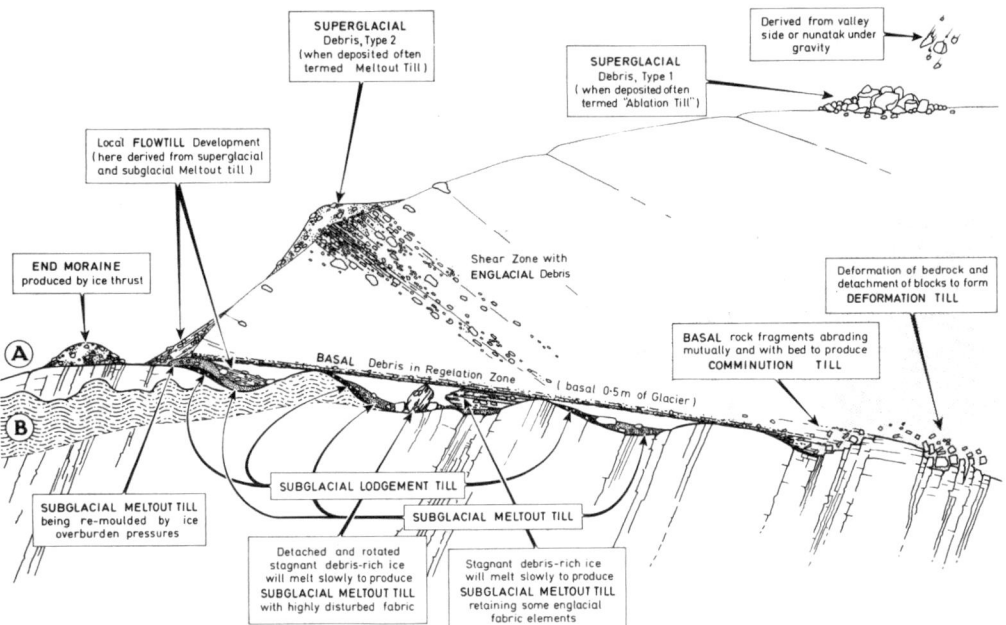

FIG. 5.56. GENERALIZED RELATIONSHIPS OF ICE AND DEBRIS IN TEMPERATE GLACIER–VERTICAL SCALE GREATLY EXAGGERATED
Two cases of end moraine formation shown. A – glacier piles up ridge as it slides over bedrock. B – glacier rests on thick saturated till and produces 'squeeze-up' moraines. (After McGown and Derbyshire, 1977).

In extremely arid, polar environments such as the 'Dry Valley' area of eastern Antarctica, sublimation is an important element in ablation, with melting being relatively less important than in more maritime conditions. Loss of the ice in an ice-debris mixture by this process minimizes the action of meltwater and reduces the likelihood of mass flowage. It has been suggested by Shaw (1977) that this is the situation in which preservation of former englacial relationships of the sediments, including foliation, thrust and fold features, is most likely to occur. He has proposed that such *sublimation till* can be distinguished from the meltout till of more humid areas.

The release of meltout debris on sloping ice surfaces may be followed by sliding and flow. In debris rich in clay and silt, the type of movement is dictated by the moisture content, behaviour passing from that of a liquid to a plastic and finally to a solid as moisture content falls. Thus, saturated sheets of meltout debris rich in fines may be mobilized into superglacial slides and flows in response to a rise in pore water pressures owing to the persistence of melting of the ice beneath them. Debris continues to be supplied from below and movement often involves some basal sliding at the debris-ice interface as well as viscous flow. The resultant subaerial (superglacial) flow tills are common in the marginal zones of subpolar glaciers but they also occur on a smaller scale on temperate glaciers and, in certain conditions, on polar ice sheets.

Thick layers of meltout debris on glacier surfaces may move short distances by settling, internal deformation and basal sliding. Apart from occasional flow structures, such massive flow tills are difficult to distinguish from meltout tills. In contrast, flows of glacial debris with a very high moisture content may extend for considerable distances especially when the ice surface is steep. They may overrun thick flow, meltout and lodgement tills in front of the glacier. These highly mobile flows are usually thin (up to 30 cm thick) and display particle-size segregation (sorting) with distance downslope. Large clasts can be seen to slide and roll under traction and, in the thicker flows, to concentrate at the base of the flow. There is constant segregation of the fines, with silt and clay-rich suspensions at the surface which merge, at the snout of the flow, into turbid meltwater rills. As the flow ceases, dewatering occurs bringing a layer of fine silts and clays to the surface. Thin, mobile flow tills thus possess many characteristics, including occurrence as a superposed series of flows and common interbedding with thin beds of sediment of meltwater origin.

Stone orientation in flow tills often equals in its consistency that typical of lodgement tills. In flow tills, however, the primary *a* axis mode lies parallel to the local flow direction rather than the ice movement direction, and the *a/b* plane mode lies parallel to the surface of the flow and imbricated upflow. Thus, the clast fabric properties are controlled by local factors, notably meltwater supply and slope direction. Flow till units frequently contain both types described above, the thick, massive types making up the lower unit and the thin, multi-layered flow tills being the upper element.

Lodgement, meltout and flow tills may be remoulded by thrusting and ice-loading during glacier advances. This happens on a small scale during the minor winter advances of some temperate glaciers as they thrust the tills which they deposited during the preceding summer. More extensive and rapid overriding, such as may occur during surges, may result in shearing, faulting and folding of previously deposited but subsequently frozen sediments. Consolidation and dilation of tills may thus be a cyclic phenomenon.

The relative importance of debris sliding, fall, and flow varies with glacier type and glacier regime and the particle size composition of the debris; the form and fabric of the resultant *hummocky moraine* varies in consequence. On the lower Mueller Glacier in New Zealand, for example, the dominant depositional process affecting superglacial debris is sliding. Here, the debris is commonly very coarse-grained and lacking in cohesive fines as a result of loss by wind action and substantial removal of particles of sand size and finer by the abundant meltwater produced by the very high summer temperatures within deep, sheltered valleys. Subglacial thermal erosion and the development and enlargement of a network of subglacial and englacial tubes and caverns leads to widespread collapse analogous to periglacial thermokarst (Figure 5.40). Where a preferred orientation of clasts is evident, local factors such as ice slope and rock and boulder surface slopes are important influences.

By way of contrast, meltout and flow are the dominant depositional modes in the case of superglacial concentrations of clay- and silt-rich (matrix dominated) debris derived from the basal zones of subpolar glaciers. Debris-rich ice-folds and shear zones often give rise to ice-cored ridges or irregular ice-cored hummocks lying parallel to the ice front (Figure 5.64c(iii)). Meltwaters are channelled along the depressions between the hummocks and ridges, and ponding also occurs there. As the superglacial debris flows from time to time into the adjacent depressions, an intimate interstratification of tills, glacifluvial sands and gravels and lake clays may be laid down. The final result may be an inverted relief with depressions floored with meltout and flow tills and hummocks and ridges made up substantially of glacifluvial sands and gravels. Such ridges and hummocks have a morphology which is indistinguishable from similar forms in till produced by quite different processes. For example, hummocky moraine has also been explained as a product of the squeezing up of thick, saturated till into cavities and crevasses at the base of thick, immobile ice during regional stagnation of ice sheets. There is some till fabric evidence in support of this subglacial hypothesis. Also, ice-front streams may dissect till deposits which then become smoothed in form by mass movement. Hummocky moraine produced by subaerial dissection may be difficult to distinguish from forms produced subglacially unless the constituent sediments are well exposed. Hummocky moraine is a useful reminder that similar geomorphic forms may be produced by diverse suites of processes.

Considerable amounts of debris frequently accumulate on the lateral margins of glaciers. Subglacial and englacial debris rarely constitutes a major source of such material, the bulk of it being derived from mass wasting of adjacent unglacierized rock and debris-covered slopes. The input from this source may include previously deposited till from higher elevations (Figure 5.41). Many of these lateral moraines are made up of coarse, angular talus in a fine matrix. A tendency for clasts to dip away from the glacier, together with a crude but clearly visible bedding dipping in the same direction are indicative of the importance of deposition by sliding and rolling (with some flow of fines) in the growth of lateral moraines (Figure 5.40). In many areas of alpine type glaciation the largest moraines are those of lateral type.

Many glaciers and ice sheets terminate in part in glacial lakes and in the sea, but little is known about the *waterlain tills* laid down in such conditions. In shallow waters in front of a temperate glacier, deposition by lodgement and subglacial meltout may occur in much the same way as elsewhere. In deeper water bodies, however, the ice tongue will tend to float and a continuous 'rain' of both clasts and matrix may fall on the bed without significant water-sorting. The sediment produced may be very similar to a till laid down in subaerial conditions, though perhaps with relatively higher void ratios. Elongate particles will tend to settle parallel to the bed or, in falling through some depth of water, become embedded in a sub-vertical position as a 'dropstone'. Thus stones in waterlain tills may have two distinct preferred orientations – subhorizontal and subvertical. Flow tills may also be produced at the bottom of glacial lakes by movement of waterlain tills down steep slopes and by the release of basal debris as *turbidity currents* (Highly charged suspensions which are much denser than the surrounding lake waters). Locally concentrated deposition and loading structures in tills may result from the grounding of icebergs especially in lakes with widely fluctuating depths resulting from periodic rapid drainage. The incidence and extent of types of till in different glaciological regimes is summarized in Table 5.4.

Table 5.4
INCIDENCE AND EXTENT OF TILL IN DIFFERENT GLACIOLOGICAL REGIMES

Glaciological regime	Meltout till	Lodgement till	Flowtill
Temperate	Common to rare Localized	Common Widespread to localized Thin	Rare Localized
Sub-temperate	Common Localized	Common Widespread Thin	Common Localized
Sub-polar	Abundant Widespread	Abundant Widespread Thick	Abundant Widespread
High polar (Non-surging)	Common Localized (Sublimation till may be widespread)	Presence or absence is dependent on ice thickness	Rare to common Localized

5.7 MELTWATER EROSION AND GLACI-AQUATIC SEDIMENTATION

One of the most impressive features of the surfaces, soles and margins of temperate glaciers is the volume of water present in thin films, streams and ponds. This is particularly true in summer, of course, but it is also the case in winter in the lower reaches of many temperate glaciers with high values of mass flux such as those in coastal mid-latitude mountains.

Meltwater may be present in appreciable amounts in the upper parts of some subpolar glacier systems. On polar glaciers meltwater may be present at the base only where the ice is of great thickness (see p. 235). Surface meltwater streams are rare on high-polar glaciers, except where re-radiation from rock surfaces concentrates melting. The presence of meltwater is important because of its influence on mode and rate of glacier flow (Section 5.2). It is produced beneath and within glaciers by the flow of earth heat from the rock head. Heat for meltwater production is also released during recrystallization associated with plastic creep and in overcoming the basal sliding friction. Another, locally important, source of basal meltwater is groundwater moving out of subglacial aquifers into the basal zone.

Meltwater produced by the surface melting of ice and snow is added to that derived from glacier flow especially in summer. The volume of seasonal meltwater varies along climatological gradients (both latitudinal and altitudinal), being high in the lower reaches of maritime temperate glaciers, but declining systematically towards continental interiors. Summer precipitation in the form of rain enhances the superglacial and englacial discharge of meltwater and, together with the runoff from adjacent unglaciated rock slopes, it may cause rapid peaking of discharge in proglacial streams fed from subglacial and marginal sources.

Summer melting of the previous winter's snow cover, and ultimately of the bare surface of glacier ice, produces runoff on the glacier surface. Water sometimes moves as thin films on smooth ice but flow in channels ranging in scale and form from rills to well-developed meandering channel systems is more common. Meanders may become incised largely as a result of thermal rather than mechanical erosion.

Water moves through and beneath glaciers in several ways. Structural discontinuities such as crevasses produced by tensional stress in the ice commonly provide sinks for superglacial stream discharge. Where meltwater maintains a vertical avenue into the glacier in conditions favouring crevasse closure, a glacier mill or *moulin* is formed. There is a strong correlation between moulin distribution and crevasse location on many glaciers (Figure 5.57). The water stream may advect warm air into the moulin which enhances melting of the ice walls and so helps maintain the passage. There is a lack of precise knowledge about the lower sections of moulins. While some are known to extend over 200 m to the ice-rock interface, others may not penetrate more than a few tens of metres into the glacier at such a steep angle. Low angle tunnels or conduits have been observed within and beneath relatively thin glacier ice and it is assumed that these represent the lower reaches of the englacial streams which begin at the moulins. In the vadose zone above the piezometric surface (which varies quite widely because of seasonal variations in

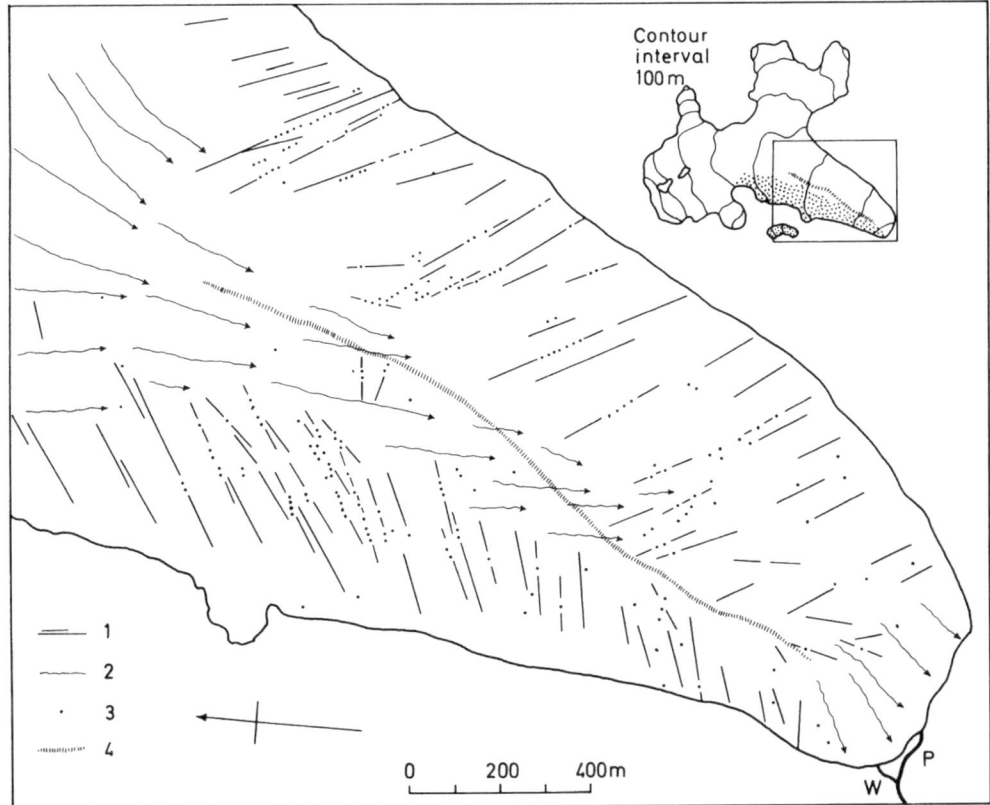

FIG. 5.57. RELATIONSHIP BETWEEN SUPERGLACIAL RUNOFF AND GLACIER STRUCTURES ON MIKKAGLACIÄREN, NORTHERN SWEDEN
1 – crevasses; 2 – surface runoff; 3 – glacier mill (Moulin); 4 – boundary between areas of crevasses striking east and west respectively. (After Stenborg, 1973).

meltwater supply), conduits may be relatively steeply inclined, water moving under gravity. Below the piezometric surface, meltwater pressures increase with depth and locally high water pressures may be expressed by water spouts or turbulent stream flow issuing from beneath glacier margins (Figure 5.58).

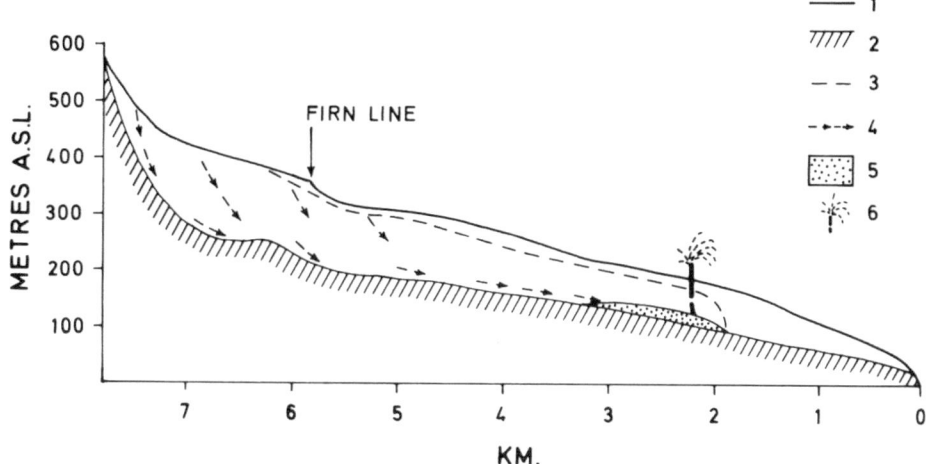

FIG. 5.58. LONGITUDINAL CROSS-SECTION OF WERENSKIOLDBREEN, VESTSPITSBRGEN, SHOWING SUPPOSED ENGLACIAL-SUBGLACIAL WATERFLOW RELATIONSHIPS
1 – glacier surface; 2 – supposed glacier bed; 3 – supposed lower level of 'cold' glacier ice; 4 – supposed water routes; 5 – supposed subglacial lake; 6 – water spout. (After Baranowski, 1973).

While the precise mode of initiation of water-carrying tubes within glaciers is still not clear, it appears likely that they are supplied by water moving down a network of intergranular veins which become progressively enlarged, preferential growth of larger ones being favoured by the exponential increase in the ratio between the heat in the meltwater and the area of the tube walls. Englacial water pressures are commonly ten times greater than the approximation of the yield stress for glacier ice (~ 1 bar) so that, once initiated, conduits may be expected to be maintained within actively moving glaciers. Enlargement may also occur by thermal erosion. The circular cross section of conduits in the zone of saturation becomes modified in the vadose zone as open channel flow within the tubes becomes dominant and a variety of entrenched and flat-floored tube forms develop. One result of this is the development and perpetuation of meandering englacial tubes and their eventual intersection with the glacial bed. In the zone of saturation, englacial water flows in conduits down a pressure gradient which broadly parallels the ice surface slope (Figure 5.31), so that the meltwater routes tend to be convergent where glacier surfaces are concave-upward (as in the accumulation zone), and divergent beneath convex ice profiles so that the ablation zone of a valley glacier may be drained by two essentially discrete systems. The great bulk of englacial water in a glacier ultimately flows out in one or a series of subglacial tubes at the ice-bed interface.

Subglacial water in thin films, sheets and streams varies in volume and pressure with variation in glacier thickness and velocity. Particularly deep depressions beneath very thick polar glaciers may give rise to thick water bodies called subglacial lakes which are contained by the thinner, cold polar ice up- and down-glacier. This situation may also occur in sub-polar glaciers where the cold ice of the upper surface and thinner snout zone prevents meltwater issuing subglacially at the snout. Basal meltwater bodies are characteristic of depressions within the undulating rock bed beneath thick temperate glaciers. At relatively low water pressures such water bodies may be isolated from water in adjacent cavities but as pressure rises or as the glacier moves, the cavity may develop an outlet to the other cavities and water will flow down the pressure gradient.

The thin film of water at the base of a temperate glacier plays a critical role in the overcoming of small-scale obstacles in the bed, water produced at high pressure (up-glacier) locations being transferred to low pressure sites in the process of regelation (see Section 5.4b). In addition, meltwater may be transferred along the base of a glacier as a thin sheet. In the regelation process, the thickness is of the order of $1 \mu m$ but for downglacier flow of basal meltwater by sheetflow, very much greater water thicknesses (of, say, 1 mm) are required, which would interfere with the regelation process. This has led to the suggestion by J. F. Nye, that the meltwater exists in a separate system contained in channels. These may be cut in the ice, in which case they will be likely to suffer local and periodic closure, or in the underlying rock head. Such rock-cut channels *(Nye channels)* provide a more permanent avenue for meltwater and may provide for the dominant mode of subglacial meltwater flow.

The discharge of glacial meltwater streams is typically variable, even over periods of only a few hours, a fact which probably represents release of meltwater from a wide variety of cavities as the glacier slides on its bed. The best documented variation is the diurnal one which is best developed in summer and is superimposed on a relatively steady seasonal curve. Tracing experiments have made it clear that the through-flow time for meltwater entering a moulin and exiting at the ice terminus varies from one day for small glaciers to over a week for large glaciers. Thus the diurnal oscillation must represent outflow peaks from an englacial reservoir system which is topped up in the manner of a cistern, rather than direct outflow of the daily input (Figure 5.59). The lag time between meltwater input and the outflow peak in the meltwater streams tends to become shorter towards the end of summer, suggesting that the conduits and tunnels reach their maximum dimensions at this time. The rate at which meltwater travels within and beneath a glacier varies from tens to some hundreds of metres per hour in open and closed conduit flow, to a few tenths of a metre per hour in intergranular vein and basal thin film flow.

Occasionally, meltwater stored in closed englacial conduits and in subglacial and proglacial lakes may be released rapidly as a result of very rapid melting (as in the case of subglacial volcanic activity), catastrophic glacier surging, or flotation of an ice tongue acting as a dam for lake water as water depth reaches a critical value. Such glacier bursts or *jökulhlaups* are common in Iceland and involve very high discharges ($\sim 50\ 000\ m^3 s^{-1}$ or more). In the case of most glaciers, however, it is evident that variation in meltwater discharge is a direct reflection of the state of the glacier's hydrological budget as controlled by weather and climate.

The periodically very high discharge and velocity of subglacial meltwater flowing within channels and as thick sheets during *jökulhlaups* are a source of energy which is expended in the transport of huge quantities of sediments and in locally severe erosion of bedrock surfaces and previously deposited proglacial sediment.

The most impressive characteristic of glacial meltwater streams, apart from their turbul-

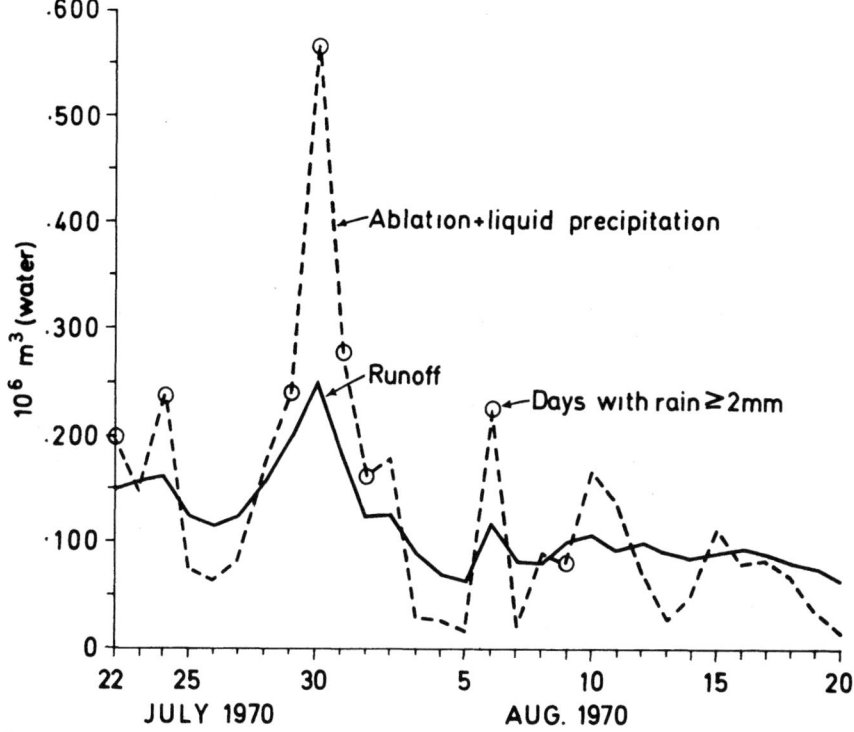

FIG. 5.59. GRAPH SHOWING DAILY AMOUNTS OF ABLATION PLUS LIQUID PRECIPITATION AND RUNOFF OVER A 35-DAY
PERIOD JULY–AUGUST 1970 FOR PARTLY GLACIER-COVERED ARCTIC WATERSHED IN ALASKA
(After Wendler, Trabant and Benson, 1973).

ence and high velocities, is their turbid nature. They transport large concentrations of
suspended sediment of silt grade, together with a saltation load of sands and fine gravels
and bed load of gravels and boulders often of impressive proportions. The abundant
saltation and bed load sediments enhance the erosive potential of glacial meltwater
streams. High velocity impact of pebbles, cobbles and boulders on hard rock surfaces
results in brittle failure of the 'percussion cone' type (Figure 5.33). The high angle of
impact produces circular cracks in the rock surface in contrast to the half-circle crescentic
fractures produced by stones held in a moving glacier sole (Section 5.4b).

The saltation and bed loads of glacial rivers play a major role in abrasion of rock surfaces
known as corrasion. Indirect evidence of this is provided by the high degree of edge-
rounding of the particles transported. Rock surfaces swept by meltwaters show a high
degree of smoothing and edge-rounding. Some such surfaces have been traced beneath
modern glaciers and it is thus not surprising that there is room for debate as to the relative
importance of meltwater flow and glacier sliding in the generation of abraded bedrock
forms.

Several kinds of observation suggest that chemical processes play a role in the erosion of
the glacier bed. Deposits of iron and manganese hydroxides on the lee faces of *roches
moutonnées* observed beneath actively flowing glaciers indicate that processes of weathering have affected the aluminium silicate minerals with associated hydration of iron,

transportation in subglacial water films and final deposition. Rather basic subglacial meltwater, especially when enriched with certain salts (e.g., common salt, magnesium sulphate and potassium chloride) may produce a type of cavernous weathering in granitic rocks beneath glacier ice. Examples are known from Europe and Antarctica. Because of the solubility of calcareous rocks such as limestones, dolomites and calcareous shales, solution is regarded as an important erosional process especially beneath temperate glaciers. High concentrations of carbon dioxide in snow may enhance the erosional potential of meltwaters derived from it, although this has not been conclusively demonstrated. Nevertheless there is some indirect evidence of the role of subglacial solution in the comparatively greater symmetry and depth-width ratios of nivation cirques and small glacial cirques cut in limestones.

Direct evidence of solution effects beneath glaciers is provided by measurements of the minerals in solution in glacial meltwaters. On calcareous rocks, calcium may account for between eighty and ninety per cent of all dissolved ions with magnesium between 10 and 20 per cent and sodium and potassium less than one per cent in a total concentration of, say, 15–35 mg/l. On volcanic rocks, calcium and quartz (SiO_2) have been recorded at concentrations greater than 8 mg/l and 7 mg/l respectively in a solution load with a concentration of almost 200 mg/l. The peak solution loads coincide with the low meltwater discharges of winter, the peak in the suspension load occurring in the summer.

5.7a Meltwater erosion

Water flow within and beneath a glacier is controlled by a pressure gradient which broadly conforms to the ice-surface gradient. While a great deal of water flow beneath temperate glaciers occurs as very thin films and sheets, small amplitude depressions in an irregular subglacial surface provide conditions in which concentrated (channel) flow may develop and persist. Incision of meltwater streams into the rock head may be essential for the evacuation of the great volume of meltwater produced by some glaciers. Evacuation of such high discharges by basal ice-walled tubes or by means of basal film and sheet flow would interfere with the regelation process and produce very high, perhaps surging velocities. It is for such reasons that J. F. Nye (1952) regards a separate system of meltwater flow by way of channels in the rock floor as probable beneath most temperate glaciers.

The presence of rock-cut channels beneath present-day glaciers is difficult to demonstrate, but many thousands of channels of Pleistocene age have been ascribed to meltwater erosional processes. Given that subglacial meltwater flow is controlled by hydrostatic rather than morphological gradients, channels eroded by subglacial waters are aligned parallel to the ice surface slope. Channels cut into ridge crests sometimes by many tens of metres show a consistent trend. Such channels are steep-sided and, when bedrock structure is conducive and subsequent slope modification slight, flat floored so that channels of Pleistocene age appear as discordant elements in the landscape. The importance of hydrostatic pressure gradients in their origin may be seen in some channels cut across spurs, cols and ridges which were disposed transversely to the former gradient of the ice sheet. Many channels cut in such conditions have a 'humped' longitudinal profile, the bed rising sometimes for some hundreds of metres before declining in the direction of meltwater flow. Such 'spur end' and 'col' channels are often the largest members of the meltwater channel population. The great size of some subglacial meltwater channels may have several explanations. In some cases, there is little doubt that the channel cross-sectional areas reflect the peak discharges.

Other factors must be considered, however. These include the probability of prolonged incision once a trunk meltwater channel is established, the periodic occupation of part of the channel by basal ice and the use of the channels during more than one phase of glaciation. Meltwater channels draped with till and having striated walls may be used in support of these inferences. Some groups of channels have consistent gradients and great continuity suggesting that their location was determined by the presence of an ice margin. Indeed, sometimes the features are replaced by sinuous rock-cut benches, the missing channel wall having been the glacier margin. Elsewhere, the channels may be more sinuous (Figure 5.60), some anastomosing may occur, and short transverse channel elements ('chutes') may be present. Variations of this kind in suites of marginal channels probably reflect variations in glacier regime, notably ice temperature as it affects the permeability of the ice margin. Complex ice-marginal channel morphology may reflect permeable, and therefore temperate, glacier conditions while simple channel morphology, and especially the presence of half-channels or benches, may be the product of impermeable polar ice. To some extent, both may be present in the same region either because of seasonal changes in meltwater-glacier relationships or because the regime of the ice sheet changed from polar in the early part of deglaciation to temperate as ice wasted toward the valley bottoms.

FIG. 5.60. GLACIAL DRAINAGE CHANNELS ALIGNED APPROXIMATELY PARALLEL TO FORMER ICE-SHEET MARGIN IN CENTRAL LABRADOR-UNGAVA
Channels cut partly in till and partly in bedrock. Meltwater flow away from camera. (Photograph: E. Derbyshire).

Meltwater channels are also cut across ridges and spurs and along hillsides as a result of the overspilling of lake waters impounded in side valleys and re-entrants by glacier ice. For many years, it was believed that all meltwater channels were the product of ice-marginal lake overflow and so were subaerial. Detailed studies in the period 1950–1970 demonstrated that a great many meltwater channels of Pleistocene age are the product of subglacial erosion. Nevertheless, ponding of lake waters by glaciers is widespread today and was probably so during the Pleistocene glaciations. While marginal and col type channels occasionally serve to drain glacial lake waters, the commonest mechanism of lake drainage is probably by subglacial tube and channel flow and by periodic flotation of the ice-edge, producing a very high discharge of subglacial waters known as a glacier flood or *jökulhlaup*. Where the slope of the ground is in the same sense as the surface slope of the glacier, meltwaters of subaerial type may incise channels into the valley floor. Some of these channels may have started as subglacial features.

The regional pattern of meltwater channels is often strikingly consistent, providing in outline, a record of dwindling ice volume, detachment of ice bodies and general ice wastage. However, the interpretation of the precise origin of channels produced by glacial meltwater is made difficult because of the many variables influencing their form and because some of these variables are indeterminable in many cases. Glacial meltwater channels remain, therefore, a challenging subject of study.

Small irregularities produced on a rock surface beneath flowing water by cavitation, impact or detachment rapidly become smoothed and rounded along their rims by the process of abrasion. Within the cavity, abrasion may be concentrated as local vortexes develop in the fluid and particles enlarge the cavity by attrition. Once entrenched, these potholes may be eroded into sound rock by the constant rotation of pebbles, cobbles or boulders trapped within them. Some potholes in glacial meltwater stream beds reach many metres in depth and diameter.

Frequently associated with potholes is a group of bedrock erosional features known as plastically-moulded surfaces or 'p-forms' (Dahl, 1965) because of their smooth, streamlined shapes. They include sichelwannen, grooves and cavettos. Sichelwannen are sickle-shaped troughs disposed transversely to glacier flow with the 'horns' pointing downglacier. Grooves and cavettos are channel-like forms disposed parallel to the ice flow lines, the former having rounded edges and the latter sharp edges. Unlike sichelwannen and grooves which are found on flat, gently sloping surfaces, cavettos are cut into steep rock faces so that they often form an overhang. Crescentic gouges and striae have been seen in them. There is no agreement as to their precise origin. However, their frequent association with bedrock potholes has led to a strong body of opinion favouring the view that they are the result of the flow of subglacial water, or a subglacial slurry of water, debris and ice crystals moving under high pressure. Sichelwannen have been likened to certain flute and ripple marks made by running water on unresistant beds and, on the basis of this analogy, it has been suggested that they may result from differential corrasion associated with areas of separated flow in subglacial stream channels (Allen, 1970).

The presence of fine striations in some of these p-forms suggest that debris-rich ice plays an essential role at least in the final stages of their formation. G. S. Boulton, for example, has suggested that they result from 'streaming' of basal ice into ice bands rich in debris and bands poor in debris as a result of plastic deformation around bedrock obstacles or large boulders. As a groove develops, normal stress will decrease as the ice tends to bridge the deepening grooves, so that there may be an optimal or equilibrium depth for given conditions of ice thickness, debris concentration, water pressure and rock resistance.

5.7b Glaci-aquatic sedimentation

Glacial debris which has been entrained, transported and deposited by water or has been laid down in bodies of standing water may be called glaci-aquatic. Within this broad category may be recognized *glacifluvial* sediments which are deposited by meltwater streams; materials brought into and deposited within ice marginal lakes, the *glacilacustrine* deposits; and *glacimarine* sediments which are derived from ice melting into the sea. The process of settling of rock particles through the water body (with the coarser, heavier particles settling more quickly than the finer material), together with variations in flow velocity and water temperature which affects the viscosity of the fluid, results in changes in particle size with depth to produce *bedding*. Thus, glaci-aquatic sediments are often referred to as *stratified drift*.

A great proportion of the water produced by glacier melting runs off in streams and sheets beneath and in front of the glacier. As flow velocities, and hence stream competence (Chapter 2) vary according to diurnal, seasonal and local factors (such as the collapse of ice tunnels and the removal of ice dams), glacifluvial deposition can be expected to occur over a wide range of particle sizes. The turbid nature of ice-front streams on the one hand, and their unstable bed and banks made up of rather coarse, well-rounded gravels, pebbles, cobbles and boulders on the other, provide a graphic reminder of the important distinction between suspension load and bedload. Much of the rock flour produced by glacial grinding is of clay or silt grade and is finally deposited only in bodies of slack water provided by ponds and lakes (see below). The material making up the bed and banks of proglacial streams is non-cohesive for it lacks a cohesive matrix and a binding mat of vegetation. Bed material may be selectively removed and packed in an upstream-dipping, overlapping fashion (upstream *imbricated fabric*) so that the bed becomes relatively stable and the critical competence level of the stream is thus raised. With non-cohesive banks, however, material is undermined on the flanks and proglacial streams have rather low hydraulic radii (high width:depth ratios). Transverse or diagonal bars result in abrupt deflection of streamflow and so play an important part in the lateral erosion of channel banks. This tendency to channel widening results in abundant lateral channel shifting. Central spool-shaped or linguoid (literally tongue-shaped) gravel bars develop, splitting the channel into two. This process is a major factor in the development of the typically braided traces of glacial meltwater streams.

An associated factor is high debris loads, especially bedloads. As the peak flood passes, channel flow is maintained by preferential development of deeper scour channels within the braided pattern. Thus braided systems may contain a wide range of meander amplitudes. Outflow of debris-charged water at the margins of ice sheets and piedmont glaciers may thus build up impressively large glacial outwash plains or *sandar* (singular *sandur*). When the zone of deposition is constrained by valley walls, as it is in the case of alpine glaciers, the linear ribbon of glacifluvial outwash is called a *valley train*.

The surfaces of sandar are dominated by pebble, cobble and boulder-sized particles. The apparent sparsity of sand is a product of post-depositional surface sorting especially by wind, for sand is the dominant grain size below the surface. Infilling with sand of the voids in the gravel framework is usual in low discharge conditions, but occasionally areas without such a matrix *(openwork fabric)* may be seen. Sand is mobilized at much lower threshold stresses than coarser material and continues in motion during late summer flows, settling differentially in depressions or 'slacks'. The diurnal and seasonal variability in meltwater discharge produces a characteristically wide range of particle size units within a sandur, silt or sand layers and beds of pebbles and cobbles with some boulders commonly being juxtaposed. The sedimentary structures, notably those arising from the bedforms in the sand layers include ripples, dunes and plane bed and these can be used in the analysis of former meltwater flow directions *(palaeocurrents)*. Outwash gravels, on the other hand, are frequently poorly sorted and show only rudimentary bedding. Where bedding occurs in sandur gravels it is usually horizontal. Factors affecting this are the dominance of upper flow regimes in meltwater systems (Froude number greater than 0.7) and the inhibition of development of cross bedding by the prevalence of shallow water flow. However, upflow imbrication is usually clearly visible in outwash gravels and this provides evidence of palaeocurrents. On the distal (downstream) sides of migrating bars, some cross bedding may develop in the gravels. In side channels and slacks, sorting is better, and cross bedding in sands together with a general tendency towards fining – upward sequences is to be expected.

Outwash gravels are characteristically well-rounded, the degree of edge-rounding increasing rapidly away from the ice front. However, the tendency for gravel size to decline and for degree of edge-rounding to increase in this direction differs according to whether the material is within channels or on the only occasionally-flooded sandur surface. For example, many boulders on sandar were deposited by the glacier and, because they are larger than the competence of the sandar sheetflood, have suffered little transport, although edge-rounding may have been increased by stream bed abrasion. Orientation of gravel clasts produces a bimodal fabric pattern, long axes lying parallel to flow in high discharge conditions but becoming transverse to flow as a result of rolling in lower flow stages. With stabilization of the sandur surface, and the eventual incision of one or two channels, dissection progresses and may result in the development of paired valley-train terraces.

When meltwaters approach and enter a body of standing water such as a lake or the sea, flow velocities are reduced and aggradation begins. The first element of deposition is the bedload in the streams at the head of the lake or arm of the sea. In the gravel-bedded streams of sandar, forward growth of the stream bed together with the tendency of gravel bed streams to shift laterally, produces a composite stream-mouth depositional form known as a *digitate delta*. The finer, suspended load is carried beyond the advancing front of the delta, the distance it is transported before it finally settles out on the lake floor depending on a variety of factors, particularly the particle size, the temperature and salinity of the water, and the presence or absence of lake-bottom currents. In particular, turbidity currents, which consist of high density flows of debris-charged water (a highly concentrated suspended load), may be of considerable importance in transporting debris long distances beyond the delta front.

As a delta advances into a standing water body, deposition by avalanching on the advancing face is discontinuous and gives rise to relatively steeply dipping *foreset beds*. The maintenance of distributory channels on the growing delta surface gives rise to an upper zone of gently dipping sediment units known as the *topset beds*. On the lake bed or sea floor in front of the foreset slope, slow settling out of the finer suspended load is essentially uninfluenced by streamflow. These fine-grained sediments make up the *bottomset beds* (Figure 5.61). In some high-energy meltwater flow situations producing rapid sedimenta-

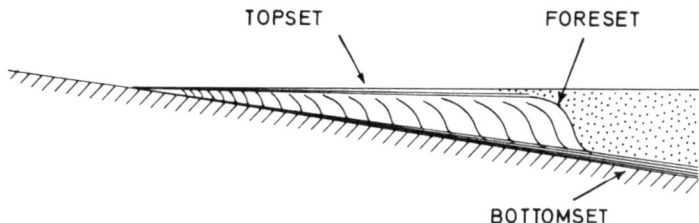

FIG. 5.61. STRUCTURE OF SIMPLE SUBAQUATIC DELTA, SHOWING RELATIONSHIP BETWEEN TOPSET, FORESET AND BOTTOM-SET BEDS. STANDING WATER INDICATED BY STIPPLE

tion, a more steeply inclined wedge of glacifluvial sediments may be laid down over the delta topsets to produce an *overwash delta*. Sometimes, when lake waters are very shallow and sedimentation rates high, a steeply inclined *supra-aquatic delta* may develop with the appearance of a rather steep, fan-shaped outwash surface.

The fine sediments found on glacial lake floors are commonly finely bedded or laminated. The regular repetition of series of laminae has resulted in the term *rhythmites* being applied to them. Relatively coarse and fine laminae make up a couplet clearly differentiated from couplets above and below. Sometimes these units appear to consist of a series of graded laminae rather than sharply defined couplets but this is illusory for few rhythmites are strictly single graded bedding units. The thickness of the couplets may be as little as 0.025 mm but, near the point of discharge, units with coarser mean grain size reach 2 or 3 cm in thickness. Since sedimentation in an ice-front lake is broadly controlled by seasonal freeze-up and melting phases, the coarser part of the rhythmite is deposited in summer, the finer part settling out slowly over the winter. Some classic studies have used this annual model to establish rates of Pleistocene ice retreat in Scandinavia and eastern Canada, counting the annual rhythmites (or *varves*) left behind in former ice-marginal lakes. Not all couplets are of annual origin, however, for pulses of sedimentation may occur as a result of temperature variations in any season, especially as they may trigger the release of turbidity suspensions. Turbidity currents may also be produced at any time by slumping of delta foresets.

When sediment-laden meltwaters enter the sea, the presence of salts in the water facilitates coagulation of the fine clays into aggregates or flocs (the process of flocculation) which settle more rapidly because of their increased size. The thickness of glacimarine rhythmites may thus more accurately reflect sediment input than is the case with freshwater rhythmites. By no means all glacimarine sediments are rhythmically laminated, however. Glacimarine sediments derived from melting of large icebergs in the Ross Sea, Antarctica, have particle size distributions very similar to terrestrial tills for example, despite deposition in seawater depths of over 600 m. Some limited sorting is evident but it can be explained as resulting from the action of low-intensity traction currents on the sea bed. Elsewhere in the Ross Sea, and in the Beaufort Sea between the north coast of Alaska and Banks Island in the Canadian arctic, glacimarine sediments remain poorly sorted but are better sorted than terrestrial tills (Figure 5.62). A more diffuse clast fabric may be anticipated in glacimarine (ice-rafted) sediments because of the tendency for 'dropstones' to embed themselves into the soft sediment surfaces at a high angle, but fabric studies of such sediments are rare.

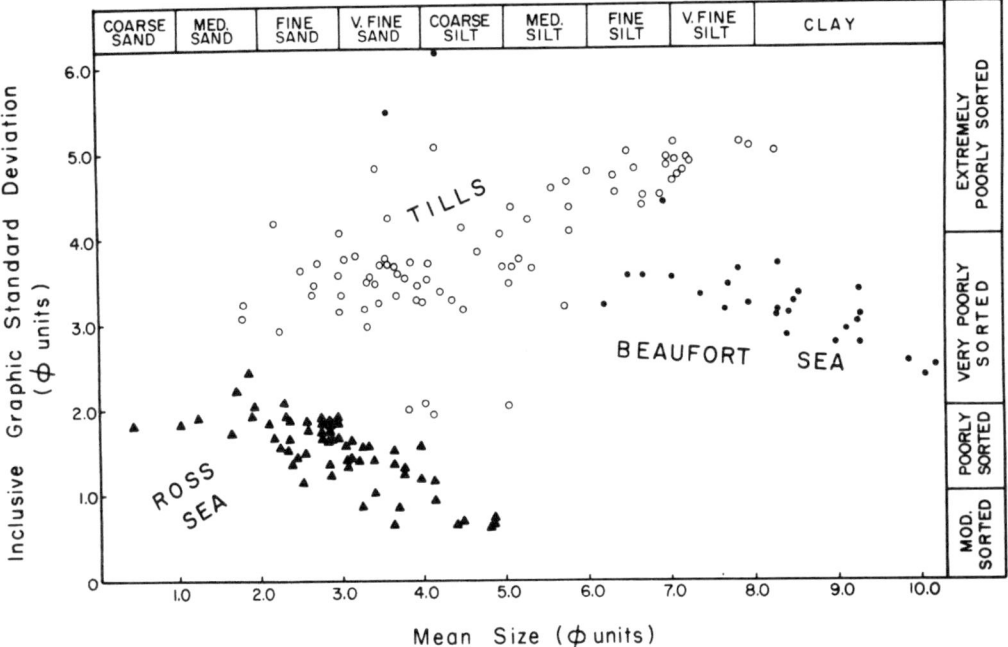

FIG. 5.62. MEAN PARTICLE SIZE PLOTTED AGAINST INCLUSIVE GRAPHIC STANDARD DEVIATION FOR TILLS AND FOR GLACIMARINE SEDIMENTS IN BEAUFORT SEA AND FOR ROSS SEA, ANTARCTICA (After Frakes and Crowell, 1975).

Glaci-aquatic deposition occurs in cavities and tubes beneath active glaciers and, while many depositional episodes and forms are strictly temporary, there is good evidence for the accumulation of a variety of sedimentary forms beneath glaciers and ice sheets. Meltwater sorting of till and rippled sand bodies has been observed under modern glaciers.

Distinctive forms of ice-contact stratified drift include *kames* and *eskers*. Kames are accumulations of glacial debris of all grain sizes but in which sands, gravels and cobbles usually dominate. Characteristic surface forms include cones, ridges and small plateaux. Bedding may be present, particularly at high angles parallel to the surface slopes, suggesting that the bedding originated by segregation during successive periods of surface sliding. Many kames have a heterogeneous structure with no apparent bedding. The morphology of kames may be constructional as for example, when they are disposed as a kame terrace along a hillside marking the former bed of an ice marginal stream or, less certainly, when they occur as discrete conical mounds suggestive of deposition in holes within thin stagnant ice.

Many kames and kame complexes are undoubtedly the result of the collapse of buried ice. The original planar bedding of the superglacial sandur is destroyed, to be replaced by a chaotic fabric and some local gravity (slide) bedding. Slow downmelting may allow the preservation of bedding in finer-grain-sized units, although these are frequently faulted. A

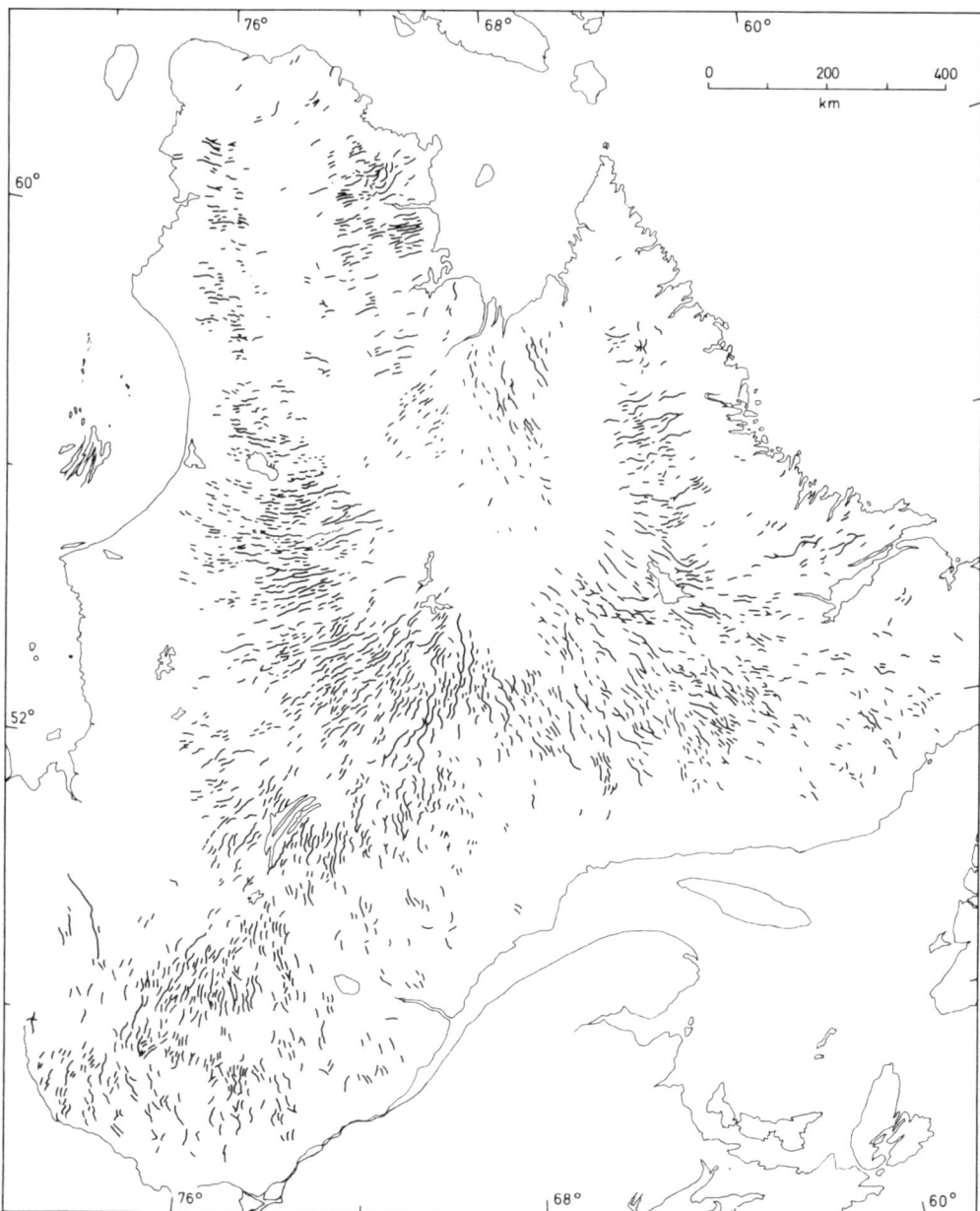

FIG. 5.63. PATTERN OF ESKERS LEFT BEHIND AS LAST ICE SHEET WASTED IN LABRADOR PENINSULA
(After Douglas and Drummond, 1953).

gradation of landforms may be produced as buried ice melts beneath sandur gravels. These range from 'pitted outwash' (a series of ice-melt depressions or 'kettle holes') to a conical kame complex, the area covered by either mounds or depressions depending on the amount and distribution of buried ice. In spatial terms, the distribution of kame hummocks may be random or it may contain linear elements related to former glacier structures.

Eskers are distinctive sinuous ridges composed mainly of sands and gravels disposed roughly parallel to the direction of glacier flow. They range in size from a few metres to many tens of metres in height and from a few tens of metres to hundreds of kilometres in length. Characteristically, they are made up of lens-like bodies of stratified sands and gravels which are often cross bedded. The great length and commonly arched bedding in eskers led to their early recognition as the deposits of englacial streams flowing in ice tunnels. Eskers are aligned down the englacial water-pressure gradient and hence down the ice surface gradient. As a result, they may be remarkably consistent in direction, trending across the landscape with little regard for morphology and gradient, except for a tendency to converge on saddles where the depositional esker may give way for short distances to glacifluvial channelling. Former ice drainage and ice thicknesses may be faithfully reflected in the remarkably consistent regional patterns of esker trend and distribution found in some areas of Pleistocene glaciation (Figure 5.63). That some eskers accumulated in a series of increments is suggested by their sedimentary structures and the occurrence of regularly spaced summits and sags along their crests in the so-called *beaded eskers*. Some eskers are draped with till, suggesting an origin in a subglacial tunnel, while some in process of formation are ice-cored and can be seen to be englacial in type.

Preservation of primary sedimentary structures is most likely in eskers of subglacial origin, although collapse of containing ice walls will induce steep lateral dips so that gable structures (upward converging faults) and arched beds are to be expected. Preservation of the primary sedimentary structures of these ice-enclosed streambed deposits is unlikely in englacial eskers or in eskers developed along entrenched superglacial stream channels. A complex series of events including terrace deposition, incision by thermal and mechanical erosion and re-entrainment and re-deposition of sediment along the line of englacial drainage tubes has been observed by the writer in some temperate glaciers. The presence of small-amplitude meanders in eskers does not preclude an englacial origin with the subsequent development of an ice-core, for englacial tunnels have been seen to meander in both the vertical and the horizontal plane within Blåisen, Southern Norway.

5.8 POST-DEPOSITIONAL PROCESSES AFFECTING GLACIAL SEDIMENTS

Modification of glacial sediments leading to the alteration of almost all primary properties may be effected by the action of wind, water, gravity, temperature oscillations and organisms. Water sorting may begin on superglacial meltout tills long before their final deposition and on subglacial lodgement and meltout tills before they emerge from beneath the glacier. Following deposition, till and glaci-aquatic sediment characteristics are frequently modified by washout, rainbeat and overland flow affecting, in particular, the particle size, fabric and density of the surface layers as well as the morphology of the deposits. Washing of fines into depressions followed by drying out and rainbeat produces a silt-rich crust within five years of deposition of some tills in Iceland and Norway. Removal of the silty crust by wind erosion produces a new surface truncating the thin, gently-undulating lacustrine beds. Wind action is locally very important in the proglacial zones of many glaciers and ice sheets and, in extreme polar conditions, it is a transportation and sorting mechanism of the first importance, removing surface fines to produce lag gravels, and altering particle shape by abrasion.

Percolation of meltwater and rainwater may result in the downward translocation of fine-grained particles and the residual concentration of coarse material at the surface. Such changes in the particle size distribution and fabric with depth below the surface are the first (physical) stages in the development of a soil profile.

Cyclic drying and wetting of glacial deposits is a cause of stress. Pore suction stresses in silty-clay tills, for example, rise from 10^2 kN/m^2 at about 18 per cent moisture content to 10^4 kN/m^2 at about 12 per cent, such moisture content variations being well within the range experienced in freshly deposited proglacial sediments. Thus, as till dries out, its void ratio falls, i.e. its degree of consolidation increases.

Organisms, especially the post-pioneer arctic-alpine vegetation species usually enhance changes due to wetting and drying and freeze-thaw particularly by penetration of roots and by trapping sediment grains. Rooting vegetation localizes weathering and redeposition and alters the local water-balance, bulk density and fabric of the sediment.

5.9 LANDFORM AND SEDIMENT ASSOCIATIONS IN SPACE AND TIME

It has been shown that certain landforms and sediments are to be found occurring together in particular environmental (climate) conditions in both glacial and periglacial regions. Such suites of landforms and sediments are known as land systems.

The efficacy of cryonival processes varies in response to climate. In the *polar desert land system*, ice-wedge polygons are the most widespread form, sometimes associated with surface boulder pavements (lag gravels) and ridges, dunes and sheets of sand or interbedded sand and snow (niveo-aeolian deposits). Local factors such as aspect may result in some development of thermokarst forms, even in some of the driest areas of Antarctica. Downslope translation of debris is dominated by gravity, frost creep and plastic deformation, gelifluction forms and sediments being rare.

In the *polar continental land system*, ice-wedge polygons are widespread but nivation, slushflow and gelifluction are more in evidence with sorted patterned ground, gelifluction lobes, nivation erosion and protalus accumulation, rock glaciers, stream erosional and depositional landforms and thermokarst forms being more active and widespread.

In *polar maritime cryonival land systems* the action of meltwater (derived from the abundant snow as well as ground ice) is concentrated in the short but often warm summers. Aeolian landforms and sediments are rare. Perennial frozen ground is often discontinuous. Ice-wedge polygons, pingos and thermokarst are widespread as are sorted and non-sorted patterned ground and gelifluction lobes and sheets. On sloping ground, altiplanation terraces, nivation hollows and blockstreams are common. The thermokarst morphology may be particularly complex on the warmer margins of this land system type as a result of several phases of permafrost degradation and aggradation.

In the *subpolar cryonival land system* the landforms produced reflect the high frequency of the frost cycle in rock and debris with a high moisture content, as well as forms associated with heavy, wet snowfalls. Small ground patterns of both sorted and unsorted type are characteristic as is a wide range of gelifluction forms. Interaction of cryonival and streamflow processes is widespread, mudflow forms occurring widely. On mountain slopes, scree and avalanche boulder tongues are well developed. Despite the often similar annual temperature range, the high solar radiation receipts and more severe radiative cooling exclude very high mountains in low latitudes from this land system category. In such areas, processes such as baring of the ground by the mechanical action of the wind (turf exfoliation), blowing of dust on to snow which then melts to form snow *pénitentes* and, especially, pipkrake-induced creep, produce a distinctive collection of small-scale landforms. Miniature soil stripes, small sorted stone polygons, turf-banked terracettes and turf mounds (*thufur* or *buttes gazonnées*) are a common association in the *low latitude cryonival land system*.

A series of glacial land systems has also been recognized (Boulton and Paul, 1976) on the basis of recurrent landform and sediment types occurring in particular conditions of glacier type (valley, piedmont, ice sheet) and glacier regime (temperate, subpolar, polar).

In the *glaciated valley land system*, cirques, horns and arêtes in the watershed region give way to steep valley side slopes. These provide a large source area for the dominant sediment type, the superglacial debris, which becomes deposited as predominantly coarse, angular medial and lateral moraines. Damming of meltwaters in re-entrants and in unglaciated tributary valleys localizes the accumulation of meltwater deposits as kame terraces and valley trains (Figure 5.64a).

The *subglacial-proglacial land system* revealed by retreat of piedmont lobes and ice sheets is characterized by streamlined forms such as *roches moutonnées* and flutes and drumlins developed in lodgement till. Moraines are predominantly of till and tend to be of the push type. A suite of glacifluvial sediments and landforms locally overlies these forms and may include eskers and kettled outwash plains (Figure 5.64b). The widespread and abundant englacially derived debris which mantles marginal zones of glaciers particularly in subpolar conditions gives rise to the distinctive landform and sediment association called the *superglacial land system*. There is an intimate association of flow and meltout tills and glacifluvial accumulations in the form of kame moraines, and sedimentary structures may be very complex (Figure 5.64c).

Such associations of landforms and sediments vary not only spatially but also through time. The climatic belts have shifted widely over the past five million years, especially in the middle latitudes of the earth. As a result, the efficacy of the cold climate geomorphic processes has varied and the distribution of the cold climate land system types has reflected this variation. Multiple glaciation has affected great areas of the northern landmasses, periglacial conditions preceding and sometimes following each glacial episode as well as extending for great distances beyond the ice sheets during glacial maxima. Conditions within the glaciations fluctuated from maritime to high polar with consequent changes in the erosional and depositional style of the ice sheets over time, as well as in the cryonival regime in unglaciated areas. The widespread subglacial-proglacial land system of the plains and low plateaux produced by advance and retreat of thick, active ice was frequently replaced or overlain by the superglacial land system produced by a shift to slow wastage in

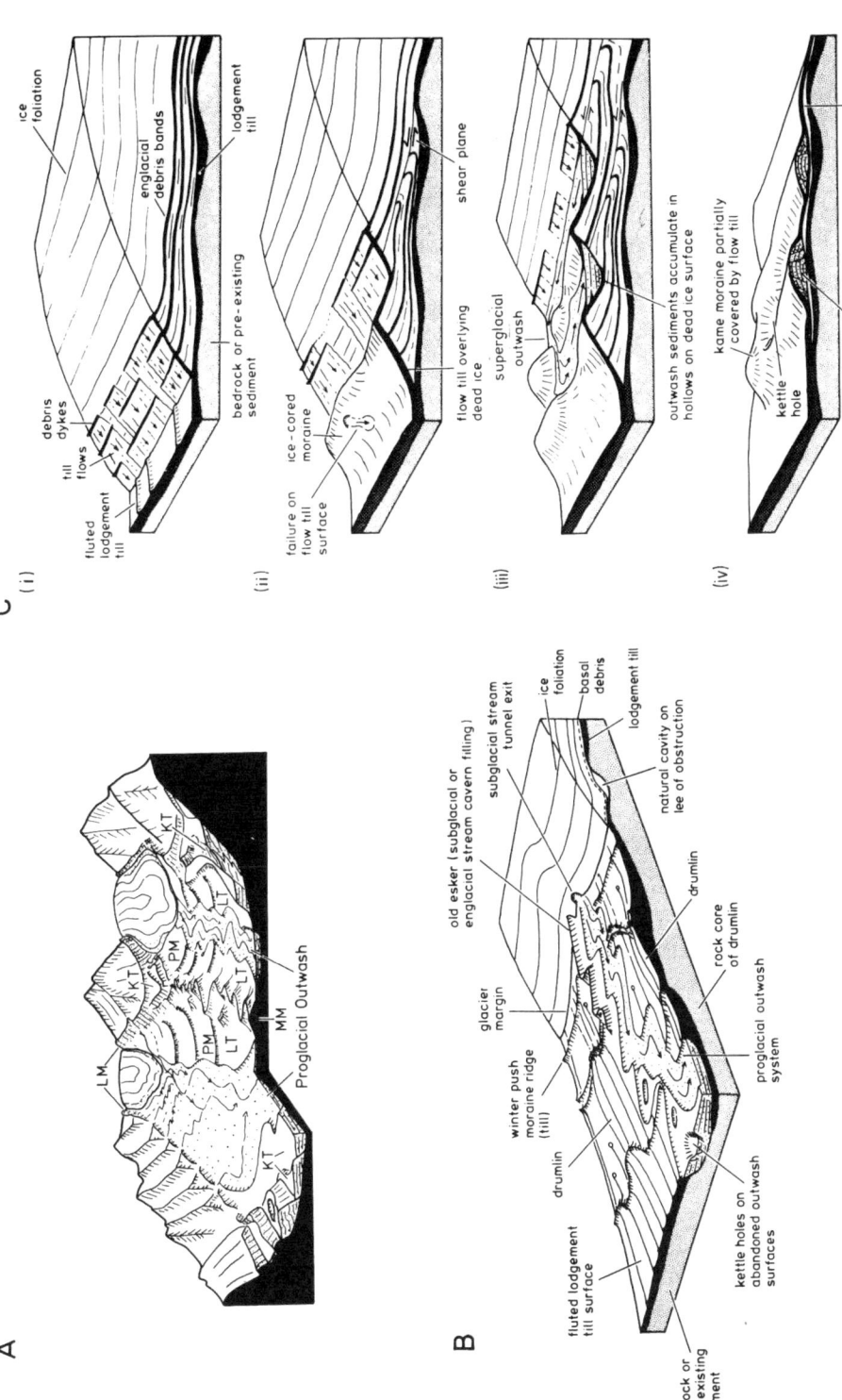

FIG. 5.64. SCHEMATIC BLOCK DIAGRAMS SHOWING THREE SEDIMENT ASSOCIATIONS AND LAND SYSTEMS

A – glaciated valley system. LM – lateral moraine; MM – medial moraine (superglacial); LT – lodgement till; PM – push moraine; KT – kame terrace.
B – subglacial/proglacial system. Simple stratigraphy illustrated consists of outwash deposits on top of till resulting from single glacial episode of advance followed by retreat.
C – superglacial system. Progressive differential downwasting of ice margin produces ice-cored moraines between which meltwater streams flow and ultimately, an inversion of relief occurs as ice cores finally melt out. Note intimate relationship between flow till and outwash sediments. Collapse structures are common in outwash materials. (After Boulton and Paul, 1976).

a subpolar environment. Adjacent to mountain areas, the effect of relief may be expressed by the dominance of the glaciated valley land system.

All glacial land systems are subject to modification by subaerial processes but prolonged periglacial episodes produce distinctive sediments and landforms which may punctuate the landform evidence of glacial type. Moreover, the cryonival land systems themselves may change their distribution over time. Slope deposits which have retained sedimentary characteristics of alternating polar desert and polar maritime conditions during the last glaciation have been described from south-facing slopes in Yakutia by Gravis (1969). In the vegetation-free polar desert conditions, accumulation of slope deposits was concentrated on south-facing slopes, preferentially-oriented ice-wedge polygons developing within them (epigenetic ice wedges). In more humid polar phases, syngenetic ice-wedge growth was stimulated, producing low-centre polygons which during mild temperature phases, became nodes for seasonal meltwater and injected ice lenses. Periods of more active gelifluction obliterated the surface expression of the ice-wedge polygons but the pattern of syngenetic ice-wedge growth is clearly related to the humid polar phases as their deposits contain a rich variety of organic remains including those of woody species (Figure 5.65).

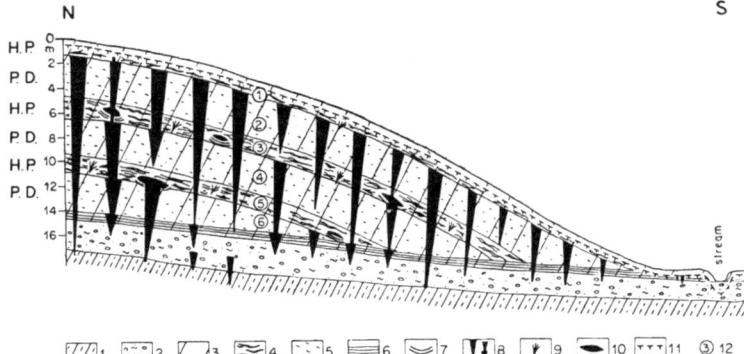

FIG. 5.65. SCHEMATIC CROSS-SECTION THROUGH SLOPE DEPOSITS IN ASYMMETRICAL VALLEY IN NORTHERN KULAR RANGE, NORTHERN YAKUTIA, USSR
1 – bedrock; 2 – river pebbles; 3 – silt; 4 – indistinct and discontinuous ice layers; 5 – thin ice bands in solifluction deposits; 6 – horizontal ice layers in alluvial sediments; 7 – arcuate ice layers in lacustrine sediments; 8 – ice veins; 9 – 'ramified' ice veins; 10 – injection-ice lenses; 11 – permafrost table; 12 – numbered slope-deposit layers. (After Gravis, 1969).

Although generally an average of only a few metres thick, gelifluctates are probably the most widely occurring cold climate sediment type in Britain, for example, especially when it is considered that the upper metre or so of many tills shows evidence of downslope transport, probably as a result of gelifluction. Indeed, the implications of the complex interstratification produced by alternating subglacial, superglacial, proglacial and cryonival depositional phases has come to be appreciated only relatively recently.

Most cold-climate landforms are quite vulnerable to modification by other subaerial processes. Erosional basins carved by ice are destined to become infilled, glaciated pavements to break down by weathering, and deposits to undergo major alterations by soil forming processes. Many drumlins and moraines of the last glaciation have already been truncated by coastal wave action.

5.10 PROSPECTS FOR COLD CLIMATE PROCESS STUDIES

It has been implied throughout this chapter that, in order to attain an acceptable level of understanding of the evolution of landforms, the relationship between environmental variables, processes and sedimentological and other properties of the materials making up the landforms must be established. This may be achieved by means of practical (laboratory) modelling, long-term field observation, mathematical modelling and the establishment of general theories. On this basis, the relationships so established may have predictive value. This becomes important in applied geomorphology (Cooke and Doornkamp, 1974; Hails, 1977) at a time when the hand of man is becoming increasingly felt in arctic and alpine regions. Active cryonival surfaces are particularly fragile and many activities, from tourism to burial of oil pipelines, are testing that fragility.

In formerly glaciated lands and shallow seas, road, reservoir, construction, pipeline and oil rig engineering requires a high level of knowledge and prediction of the properties of cold climate sediments and landforms. Even glaciers themselves have been used in the service of man. Before the days of electrical refrigeration, large tonnages of the Folgefonn ice-cap in southern Norway were shipped across the North Sea to Britain. Nowadays, tapping of subglacial meltwaters assists in the regulation of water flow for electricity generation. It has recently been suggested that large masses of ice might be used in subsurface mines in northern Sweden to prevent collapse during mining and to increase ore recovery. It is argued that, as mining proceeds, the ice would fill voids by its plastic deformation and so support walls and roof.

Man's increasing interference with natural processes must be matched by a willingness to sustain the study of natural processes and forms and to apply the principles adduced in a responsible way.

FURTHER READING

An attractive introduction to the properties and behaviour of glaciers is: SHARP, R. P., 1960, *Glaciers* (Oregon Univ. Press, Eugene) and the best introduction to glaciology is: PATERSON, W. S. B., 1969, *The Physics of Glaciers* (Pergamon, Oxford). An excellent summary of the field of periglacial landform study is provided by: EMBLETON, C. E. and KING, C. A. M., 1975, *Periglacial Geomorphology* (Arnold, London). A more detailed consideration of processes and periglacial forms may be found in: WASHBURN, A. L., 1973, *Periglacial processes and environments* (Arnold, London). Applied aspects of periglacial geomorphology are summarized in the following: COOKE, R. U., 1974, 'Freeze-thaw and periglacial environments.' Chapter 9 in Cooke, R. U. and Doornkamp, J. C., *Geomorphology in Environmental Management* (Clarendon, Oxford). DERBYSHIRE, E., 1977, 'Periglacial environments.' Chapter 7 in Hails, J. R., *Applied Geomorphology* (Elsevier, Amsterdam). A stimulating consideration of glaciers, processes and landforms is: SUGDEN, D. E. and JOHN, B. S., 1976, *Glaciers and Landscape* (Arnold, London). Fundamental contributions to the theory of glacial erosional and depositional processes are contained in the papers by G. S. Boulton (1972, 1974, 1975).

6
Conclusion: A Postscript

Emphasis throughout the foregoing chapters has been on geomorphological processes and upon how they work rather than upon their effects. The intention has been to provide an introduction to processes which is indicative rather than exhaustive so that further implications might be followed in the signposted reading at the end of each chapter. One *raison d'être* for this book is the conviction that few previous texts have been devoted concisely and exclusively to exogenetic processes and this conviction was prompted by a criticism often levelled at geomorphology in the first half of the twentieth century that it had neglected process study. By way of a postscript to the preceding chapters we may therefore ask whether greater attention to geomorphological processes is beginning to remedy the deficiencies of the past. Evidence of recent improvements in knowledge of four aspects of process study, namely rates, mechanics, models and applications, suggests that there are grounds for believing that this question may be answered in the affirmative.

One of the basic justifications for process study is to obtain a knowledge of the *rates of operation* of geomorphological processes. In the last two decades this has been achieved so that the rates of processes operating at different spatial and temporal scales and under contrasting climatic conditions is now known in much greater detail (e.g. Derbyshire, 1976). This improved knowledge necessarily refines our understanding of geomorphological systems and provides basic information which is pertinent to our understanding of change in landscape over time. Thus, in a summary of work on erosion rates achieved by diverse methods, Young (1969) concluded that known rates of land erosion are compatible with earlier estimates of rates of land surface planation over millions of years of cyclic time. One exception to this confirmation that has been elaborated is the magnitude of human activity. Work on rates of operation of geomorphological processes has necessarily highlighted those ways in which control systems have been amplified and this has provided a necessary appreciation of the way in which contemporary landscape systems operate at rates different from those of the past. Such an appreciation provides a pointer to ways in which processes may continue in the future under the direct and indirect influence of human activity (Gregory and Walling, 1979).

A further justification of the study of geomorphological processes is found in the way in which attention has been devoted to the fundamentals of these processes or to their *mechanics*. Whereas the initial attack upon the study of processes was based upon identification of the variables involved as formulated in process-response systems and upon the independent variables which accounted for variation in process rates as dependent variables, this approach was later complemented by the attention devoted to the mechanics of erosion. This is illustrated by the appearance of important books devoted to the fundamentals underlying the functioning of landscape processes (e.g. Carson, 1970; Statham, 1977). The development of such an approach is not new, of course, as is witnessed by the work of Bagnold (1941) who provided a foundation which has remained fundamental for all subsequent studies of the action of wind and of the development of sand and desert dunes. However, much more recently, geomorphologists have been seeking to formulate a framework based upon mechanics which will serve the whole range of geomorphological process study.

Processes are the means rather than the end of geomorphological enquiry. Utilization of our enhanced knowledge of process rates and mechanics requires the building of refined *models* because these provide the essential conceptual framework of a successful and reliable geomorphology. Models of many types have been employed successfully in the study of slope, drainage basin, coastal, aeolian, glacial and periglacial processes as illustrated in the preceding chapters. However, many past models have depended upon the statistical relationships established between empirically measured variables or upon the adoption of methods developed in other sciences. In this way, fundamental contributions to the understanding of glacier flow have been made by physicists, to the movement of water and sediment in channels by hydraulic engineers, and to coastal dynamics by further engineering approaches. Of course, it is imperative that geomorphological enquiry should advance in the light of such progress in allied fields. However, much of the geomorphologist's work is concerned with problems which are at the meso-scale (1 m^2–1000 km^2) in terms of spatial and temporal distribution, and it is desirable, therefore, that geomorphological processes should be viewed in the future, at least in part, in the context of models which are generated within the subject rather than in terms of those adopted and adapted from allied disciplines.

A paradigm which might become basic to the investigation of all geomorphological processes involves viewing power and landscape change as stimulus and response respectively. Power is interpreted simply as the capacity for doing work and the geomorphologist needs to know the extent to which any part of the earth's land surface possesses that capacity. As geomorphologists, therefore, we should be concerned with the potential power of an area and proceed to differentiate between areas according to their potential power. This kind of approach, based on studies of energetics, has been employed in ecology. Energy systems have been classified in terms of the magnitude of their energy flows including the natural flows and those directed by human activity (e.g. Simmons, 1978). In the context of geomorphological processes an appropriate objective might be the investigation of the power available to operate the geomorphological equation (Chapter 1) in a specific area at a particular time. Power inputs occur in all areas of the world's land surface as a result of the controls exerted by the intrinsic characteristics of the system. Such intrinsic power can be viewed as deriving from the annual pattern of climatic events or from the relief which is directly proportional to the power available. Extrinsic power can be seen as embracing those sources of energy which operate irregularly and spasmodically including endogenetic movement, human activity and catastrophic events.

Some progress in this direction has already been achieved. Thus, on a world scale, coasts have been considered in terms of their energy pattern and may be categorized into low-energy, medium-energy and high-energy types (Tanner, 1960; King, 1974). Potential energy provides a basis for the interpretation of stream channel morphology and Yang (1971) argued that nature provides potential energy to raindrops falling on the slopes of a basin according to position above a datum such as sea level. It is then possible to argue that the shape of all present stream channels is the accumulated result of the distribution and expenditure of such potential energy over time. Stream power has been utilized as a concept appropriate to the understanding of transport processes in stream channels (see 2.3b) but it is also possible to apply the concept of potential power to the drainage network (Gregory, 1979). By considering the total volume of channels comprising the drainage network in relation to relief an expression of potential network power can be obtained. Because water flow in channels is greater than it is over slopes and because water velocities are directly proportional to relief, the potential network power can be expected to be related directly to the flood potential of an area. A further example is provided by Andrews' (1972) analysis of glacier power. Total glacier power (W_T) is expressed as the product of the basal shear stress and the average velocity. Effective power (W_E) is determined by the proportion of the total average velocity resulting from basal sliding. The ratio W_T/W_E varies largely according to the proportion of basal slip to internal ice deformation, therefore. Andrews proposed that W_T/W_E is small, between zero and 0.2, for polar and sub-polar glaciers, but tends towards 1.0 and is between 0.5 and 0.8 for temperate glaciers. The implication that the glacial erosional forms produced by arctic and by temperate glaciers differ in terms of size and geometry receives some support from the literature of glacial geomorphology. Such approaches merit further exploration and testing and it is upon such foundations that ice sheet models can be established as a basis for the interpretation of patterns of glacial erosion and deposition. Thus a steady-state model of the British ice sheet of late Devensian time was constructed (Boulton, Jones, Clayton and Kenning, 1977) with a summit height of 1800 m and velocities in marginal areas ranging from 150–500 m/yr^{-1}, the basal ice being cold in the central area but temperate near to the margins. This model, used to interpret patterns of glacial erosion and deposition over Britain, provides one illustration of the way in which improved modelling may be facilitated by improved understanding of process mechanisms and this can then be employed to lead to a refined general interpretation of the glacial patterns which are already known.

Applications of knowledge of geomorphological process to practical problems provides further evidence of the value of the dynamic approach. Such applications arise for a number of reasons. Reliable information on rates of operation of geomorphic processes are an important prerequisite in decision making, especially when efforts are being made to control processes by the construction of coastal protection features, river training structures or wind breaks. Such applications are seldom straightforward because of the difficulty of determining critical thresholds of stimulus and response and because of the problem of establishing the probability of occurrence of events of a specific size. Nevertheless, applications can be extended and the future incidence of events predicted on the basis of probability analysis of past events of known magnitude and frequency. It is now appreciated that geomorphological processes present hazards to man and only by means of a well-founded knowledge of the interaction of the geomorphological process system with the human use system can the hazards be identified as a basis for management and planning. It has been suggested therefore that the aim of applied geomorphology is not to inhibit or minimize the development of resources but rather to optimize their use by reducing costs and minimizing overall landscape impacts (e.g. Clark, 1978).

Knowledge of contemporary landscape processes is essential to an understanding of landscape change, past, present and future. In further studies of processes it should be possible to develop an increasingly pertinent geomorphological approach and this remains an outstanding challenge for future researchers.

Bibliography

AARIO, R., 1977, 'Classification and terminology of morainic landforms in Finland.' *Boreas*, 6, pp 87–100.

AGASSIZ, L., 1840, *Étude Sur Les Glaciers* (Neuchatel).

AKILI, W., 1970, *Materials, Research and Standards*, MTRSA, 10, p 16.

ALLEN, J. R. L., 1968a, 'The nature and origin of bedform hierarchies.' *Sedimentology*, 10 pp, 161–82.

— 1968b, *Current Ripples: Their Relation to Patterns of Water and Sediment Motion* (North Holland, Amsterdam).

— 1969, 'Erosional current marks of weakly cohesive mud beds.' *J. Sediment. Petrol*, 39 pp, 607–23.

— 1970, *Physical Processes of Sedimentation* (Unwin).

— 1977, 'Changeable rivers: some aspects of their mechanics and sedimentation.' *River Channel Changes*, ed. K. J. Gregory (Wiley), pp 15–45.

AMSTUTZ, G. C. and CHICO, R., 1958, 'Sand size fractions of South-Peruvian barchans and a brief review of the genetic grain shape function.' *Bull. Ver. Schweiz. Petrol. Geol. Ing.*, 24(67), pp 47–52.

ANDREWS, J. T., 1972, 'Glacier power, mass balances, velocities and erosion potential.' *Zeits. für Geomorph.*, 13, pp 1–17.

— 1975, *Glacial Systems: An Approach to Glaciers and Their Environments* (Duxbury Press).

ATKINSON, T. C., 1978, 'Techniques for measuring subsurface flow on hillslopes.' *Hillslope Hydrology* ed. M. J. Kirkby (Wiley), pp 73–120.

AUFRÈRE, L., 1930, 'L'orientation des dunes continentales.' *Rept. Proc. Intern. Geogr. Congr.*, 12th, Cambridge, 1928, pp 220–31.

BAGNOLD, R. A., 1940, 'Beach formation by waves; some model experiments in a wave tank.' *J. Inst. Civil Eng.*, 15, pp 27–52.

— 1941, *The Physics of Blown Sand and Desert Dunes* (Methuen). [Second edn., 1954].

— 1953, 'The surface movement of blown sand in relation to meteorology.' *Desert Research, Proc. Int. Symp.*, Jerusalem, 1952, sponsored by Research Council of Israel and UNESCO, *Res. Counc. of Israel Spec. Pub.*, 2, pp 89–93.

— 1954, 'Experiments on a gravity dispersion of large solid spheres in a Newtonian fluid under stress.' *Proc. Roy. Soc. London*, 225, pp 49–63.

— 1956, 'The flow of cohesionless grains in fluids.' *Phil. Trans. Roy. Soc. London*, Ser. A, 249, pp 235–97.

BARANOWSKI, S., 1973, *Intl. Assoc. Sci. Hydrol.*, Pub. 95, p 133.

BARNES, P., TABOR, D. and WALKER, J. C. F., 1957, *Proc. R. Soc. Lond.*, Ser. A, 324, pp 127–55.

BASCOM, W. N., 1951, 'The relationship between sand-size and beach-face slope.' *Trans. Am. Geophys. Union*, 32, pp 866–74.

— 1964, *Waves and Beaches* (Doubleday, Garden City, New York).

BEATY, C. B., 1956, 'Landslides and slope exposure.' *Journal of Geology*, 64, pp 70–74.

— 1974, 'Needle ice and wind in the White Mountains of California.' *Geology*, 2(11), pp 565–7.

BENEDICT, J. B., 1970, 'Downslope soil movement in a Colorado alpine region: rates, processes, and climatic significance.' *Arctic and Alpine Research*, 2, pp 165–226.

BESKOW, G., 1935 (1947), *Soil Freezing and Frost Heaving with Special Reference to Roads and Railways* (Northwestern Univ., Technol. Inst., Evanston) (trans. J. Osterberg).

BILLAM, J., 1972, 'Some aspects of the behaviour of granular materials at high pressure.' *Stress-strain Behaviour of Soils* ed. R. H. G. Parry (Foulis).

BLACK, R. F. and BERG, T. E., 1963, 'Glacier fluctuations recorded by patterned ground, Victoria Land.' *Antarct. Geol.*, (S.C.A.R. Proc., 1963), 3(1), pp 107–22.

BLANFORD, W. T., 1876, 'On the physical geography of the Great Indian Desert with especial reference to the former existence of the sea in the Indus Valley, and on the origin and mode of formation of the sandhills.' *J. Asiatic Soc. Bengal (Calcutta)*, 45(2), pp 86–103.

BLENCH, T., 1972, 'Morphometric changes.' *River ecology and man* ed. R. T. Oglesby, C. A. Carlson and J. A. McCann (Academic Press), pp 287–308.

BLUCK, B. J., 1967, 'Sedimentation of beach gravels: examples from south Wales.' *J. Sediment. Petrol.*, 37, pp 128–56.

BOTCH, S. G., 1946, *Bull. Soc. Geogr. USSR*, 78, pp 201–34.

BOULTON, G. S., 1968, 'Flow tills and related deposits on some Vestspitzbergen glaciers.' *J. Glaciol.*, 7(51), pp 391–412.

— 1972, 'The role of thermal regime in glacial sedimentation.' *Polar Geomorphology, Inst. Brit. Geogr. Spec. Pub.*, ed. R. J. Price and D. E. Sugden 4, pp 1–19.

— 1974, 'Processes and patterns of glacial erosion.' *Glacial Geomorphology*, ed. D. R. Coates (State University of New York, Binghampton), pp 41–87.

— DENT, B. L. and MORRIS, E. M., 1974, *Geografiska Annaler.*, 56a, p 141.

— 1975, 'Processes and patterns of subglacial sedimentation: a theoretical approach.' *Ice Ages Ancient and Modern,* ed. H. E. Wright and F. Moseley, (Seal House Press, Liverpool) pp 7–42.

— JONES, A. S., CLAYTON, K. M. and KENNING, M. J., 1977, 'A British ice sheet model and patterns of glacial erosion and deposition in Britain.' *Brit. Quart. Studies: recent advances*, ed. F. W. Shotton, (Clarendon, Oxford), pp 231–46.

BOULTON, G. S. and PAUL, M. A., 1976, 'The influence of genetic processes on some geotechnical properties of glacial tills.' *Q. J. Geol. Soc.*, 9(3), pp 159–94.

BOULTON, G. S. and VIVIAN, R., 1973, 'Underneath the glaciers,' *Geogr. Mag.*, 45(4), pp 311–19.

BOWEN, A. J., 1969a, 'Rip currents, 1: Theoretical investigations.' *J. Geophys. Res.*, 74, pp 5467–78.

— 1969b, 'The generation of longshore currents on a plane beach.' *J. Mar. Res.*, 37, pp 206–15.

BOWEN, A. J. and INMAN, D. L., 1969, 'Rip currents, 2: Laboratory and field observations,' *J. Geophys. Res.*, 74, pp 5479–90.

BROOKFIELD, M., 1970, 'Dune trends and wind regime in central Australia.' *Zeit. für Geom. Suppl.*, 10, pp 121–53.

BROWN, E. H., 1975, 'The content and relationships of physical geography.' *Geographical Journal*, 141, pp 35–48.

BRUUN, P., 1962, 'Sea level as a cause of shore erosion, *J. Waterways and Harbours Div.*, *Am. Soc. Civ. Eng.*, 88, pp 117–30.

BÜDEL, J., 1969, *Erkunde*, 23, pp 165–82.

BUTZER, K. W., 1973, 'Pluralism in geomorphology.' *Proc. Assoc. Am. Geogr.*, 5, pp 39–43.

CAINE, N., 1968, *The Block Fields of Northeastern Tasmania* (Aust. Nat. Univ. Canberra, Dept. Geogr.), Publ. G/6.

CARSON, M. A., 1970, *The Mechanics of Erosion* (Pion).

— 1971, 'Application of the concept of threshold slopes to the Laramie mountains, Wyoming.' *Trans. Inst. Brit. Geog.*, Spec. Pub. 3, pp 31–48.

CARSON, M. A. and PETLEY, D., 1970, 'The existence of threshold slopes in the denudation of the landscape.' *Trans. Inst. Brit. Geog.*, 49, pp 71–96.

CARSON, M. A. and KIRKBY, M. J., 1971, *Hillslope Form and Process* (Cambridge University Press).

CHAMBERLIN, T. C. and CHAMBERLIN, R. T., 1911, 'Certain phases of glacial erosion.' *J. Geol.* 19, pp 193–216.

CHAMBERS, M. J. G., 1967, 'Investigations of patterned ground at Signy Island, South Orkney Islands, III: Miniature patterns, frost heaving and general conclusions.' *Brit. Antarctic Survey Bull.*, 12, pp 1–22.

CHANDLER, R. J., 1977, 'The application of soil mechanics methods to the study of slopes.' *Applied Geomorphology* ed. J. R. Hails (Elsevier), pp 157–82.

CHEPIL, W. S. and WOODRUFF, N. P., 1963, 'The physics of wind erosion and its control.' *Advances in Agronomy*, 15, pp 211–302.

CHIEN-NING, 1961, 'The braided stream of the lower Yellow River.' *Scientia Sinica*, 10, pp 734–54.

CHILINGARIAN, G. V. and WOLF, K. H., 1975, *Compaction of Coarse-grained Sediments–1* (Elsevier, Amsterdam).

CHORLEY, R. J., 1962, 'Geomorphology and general systems theory.' *U.S. Geol. Survey Prof. Paper 500-B*.

— 1971, 'The role and relations of physical geography.' *Progress in Geography*, 2, pp 87–109.

— 1978, 'Bases for theory in geomorphology.' *Geomorphology: Present Problems and Future Prospects* ed. C. Embleton, D. Brunsden and D. K. C. Jones (OUP), pp 1–13.

CHORLEY, R. J. and KENNEDY, B. A., 1971, *Physical Geography: A Systems Approach* (Prentice Hall).

CHORLEY, R. J., DUNN, A. J. and BECKINSALE, R. P., 1964, *The History of the Study of Landforms: Vol. 1. Geomorphology before Davis* (Methuen).

— 1973, *The History of the Study of Landforms: Vol. 2. The Life and Work of William Morris Davis* (Methuen).

CLANCY, E. P., 1969, *The Tides* (Anchor, Garden City, New York).

CLARK, M.d., 1978, 'Geomorphology in coastal zone environment management', *Geography*, 63, pp 273–82.

CLAYTON, K. M., 1974, 'Progress in geomorphology'. *Inst. Brit. Geogr.*, Spec. Pub. 7.

CLOS-ARCEDUC, A., 1966, 'L'Application des méthodes d'interpretation des images à des problèmes géographiques, exemples et résultats, méthodologie.' *Rev. de l'Inst. français du Pétrole*, 21, pp 1783–1800.

COATES, D. R., 1973 (ed.), *Coastal Geomorphology* (Publications in Geomorphology, State University of New York, Binghamton).

— 1976, *Geomorphology and Engineering* (State University of New York, Binghamton).

COOKE, R. U. and WARREN, A., 1973, *Geomorphology in Deserts* (Batsford).

COOPER, W. S., 1958, 'Coastal sand dunes of Oregon and Washington.' *Geol. Soc. Am. Mem.*, 72.

CORNISH, V., 1897, 'On the formation of sand dunes.' *Geog. J.*, 9, pp 278–302.

— 1900, 'On desert sand dunes bordering the Nile Delta.' *Geog. J.*, 15, pp 1–30.

CORTE, A. E., 1962, 'Vertical migration of particles in front of a moving freezing plane.' *J. Geophys. Res.*, 67(3), pp 1085–90.

— 1966a, 'Experiments on sorting processes and the origin of patterned ground.' Permafrost International Conference, Lafayette, Indiana, Nov. 1963, *Proc. Natl. Acad. Sci.*, Natl. Res. Council Publ. 1287.

— 1966b, 'Particle sorting by repeated freezing and thawing.' *Biuletyn Peryglacjalny*, 15, pp 175–240.

— 1972, 'Laboratory formation of extrusion features by multicyclic freeze-thaw in soils.' *Caen, Centre de Géomorphologie, Bull.* 13–15, pp 157–180.

COSTIN, A. B., JENNINGS, J. N., BAUTOVICH, B. C. and WIMBUSH, D. J., 1973, 'Forces developed by snowpatch action, Mt Twynam, Snowy Mountains, Australia.' *Arctic and Alpine Research*, 5, pp 121–6.

CRICKMORE, M. J., WATERS, C. B. and PRICE, W. A., 1972, 'The measurement of offshore shingle movement.' *Proc. Conf. Coastal Eng.*, 13, *Am. Soc. Civ. Eng.*, 2, pp 1005–25.

CURTIS, W. F., CULBERTSON, J. K. and CHASE, E. B., 1973, 'Fluvial-sediment discharge to the oceans from the conterminous United States.' *U.S. Geol. Surv. Circ.*, 670.

CZUDEK, T. and DEMEK, J., 1970, 'Thermokarst in Siberia and its influence on the development of lowland relief.' *Quaternary Research*, 1, pp 103–20.

DAHL, R., 1965, 'Plastically sculptured detail forms on rock surfaces in northern Nordland, Norway.' *Geografiska Annaler*, 47, pp 83–140.

DALRYMPLE, B., CONACHER, A. J. and BLONG, R. J., 1968, 'A hypothetical nine-unit landsurface model.' *Zeitschrift für Geomorphologie*, 12, pp 60–76.

DAVIES, J. L., 1964, 'A morphogenic approach to world shorelines.' *Zeits. für Geomorph.*, 8, pp 127–42.

DAVIS, R. A. and ETHINGTON, R. L. 1976 (eds.), *Beach and Nearshore Sedimentation* (*Soc. Econ. Palaeont. and Minerol.*), Spec. Publ. 24.

DEAN, R. G., 1973, 'Heuristic models of sand transport in the surf zone.' *Conf. on Eng. Dyn. in the Surf Zone*, Sydney, Australia.

DE LA NOË, G. and MARGERIE, E. DE, 1888, *Les Formes du Terrain* (Imprimerie Nationale).

DERBYSHIRE, E., 1972, 'Tors, rock weathering and climate in southern Victoria Land, Antarctica.' *Polar Geomorphology* ed. D. E. Sugden and R. J. Price, *Inst. Brit. Geog., Spec. Pub.*, 4, pp 93–105.

— 1973 (ed.), *Climatic Geomorphology* (Macmillan).

— 1975, 'The engineering behaviour of glacial materials.' *Midland Soil Mechanics and Foundation Engng. Soc.*, pp 6–17.

— 1976 (ed.), *Geomorphology and Climate* (Wiley).

DERBYSHIRE, E., McGOWN, A. and RADWAN, A., 1976, '"Total" fabric of some till landforms.' *Earth Surface Processes*, 1, pp 17–26.

DERBYSHIRE, E. and PETERSON, J. A., 1978, 'A photo-geomorphic map of the Mt Menzies nunatak, Prince Charles Mountains, Australian Antarctic Territory.' *Zeits. für Gletscherk. und Glazialgeol.*, 14(1), pp 17–26.

DOBKINS, J. E. and FOLK, R. L., 1970, 'Shape development on Tahiti-Nui.' *J. Sediment. Petrol.*, 40, pp 1167–1203.

DOLAN, R. and FERM, J. C., 1966, 'Swash processes and beach characteristics.' *Prof. Geogr.*, 18, pp 210–213.

DOUGLAS, I., 1977, *Humid Landforms* (Australian National University Press).

DREIMANIS, A. and VAGNERS, U. J., 1971, *Till–A Symposium*, ed. R. P. Goldthwait (Ohio State University Press), p 243.

— 1971, *ibid.*, p 24.

DOUGLAS, M. V. and DRUMMOND, R. N., 1953, *Trans. R. Soc. Canada*, 47, pp 11–16.

DURY, G. H., 1966, 'The Concept of Grade.' *Essays in Geomorpholgy*, ed. G. H. Drury, (Heinemann), pp 211–33.

DUNCAN JR., J. R., 1964, 'The effects of water table and tide cycle on swash-backwash sediment distribution and beach profile development.' *Mar. Geol.*, 2, pp 186–97.

DUNNE, T., 1978, 'Field studies of hillslope flow processes.' *Hillslope Hydrology*, ed. M. J. Kirkby (Wiley).

DUNNE, T. and LEOPOLD, L. B., 1978, *Water in Environmental Planning* (W. H. Freeman).

DURY, G. H., 1972, 'Some recent views on the nature, location, needs and potential of geomorphology.' *Professional Geographer* 24, pp 199–202.

DYLIK, J., 1967, 'Solifluxion, congelifluxion and related slop processes.' *Geografiska Annaler*, 49A, pp 167–77.

ECKART, C., 1951, *Surface Waves in Water of Variable Depth* (Univ. of California, Scripps Inst. of Oceanography, Wave Report No.100, S10, Ref. 51–12, La Jolla).

EMBLETON, C. E. and KING, C. A. M., 1975, *Periglacial geomorphology* (Arnold).

EMMETT, W. W., 1970, 'The hydraulics of overland flow on hillslopes,' *U.S. Geol. Survey Prof. Paper* 662a.

ENQUIST, F., 1932, 'The relation between dune form and wind direction.' *Geol. Fören Stockholm Forh.*, 54, pp 19–59.

ERNST, W. G., 1969, *Earth Materials* (Prentice Hall, Englewood Cliffs).

ESCHER, B. G., 1937, 'Experiments on the formation of beach cusps.' *Leid. Geol. Meded.*, 9, pp 79–104.

EVANS, I. S., 1970, 'Salt crystallization and rock weathering: a review.' *Rev. Géomorphol. Dyn.*, 19, pp 153–77.

FAIRBRIDGE, R. W., 1964, 'African ice-age aridity.' *Problems in Palaeoclimatology*, ed. A. E. M. Nairn (Interscience, New York, N.Y.), pp 356–9.

FEDOROVICH, B. A., 1972, 'Recent and ancient, cold and warm loesses and their relationship with glaciations and deserts.' *Acta Geol. Acad. Sci. Hung.*, 16, pp 371–81.

FINKEL, H. J., 1959, 'The barchans of southern Peru.' *J. Geol.*, 67, pp 614–47.

FLINT, R. F., 1959, 'Pleistocene climates in eastern and southern Africa.' *Geol. Soc. Am. Bull.*, 70, pp 343–74.

FLINT, R. F. and BOND, G., 1968, 'Pleistocene sand ridges and pans in western Rhodesia.' *Geol. Soc. Am. Bull.*, 79, pp 299–314.

FOLK, R. L., 1966, 'A review of grain-size parameters.' *Sedimentology*, 6, pp 73–93.

— 1971a, 'Longitudinal dunes of the northwestern edge of the Simpson Desert, Northern Territory, Australia, 1–Geomorphology and grain size relationships.' *Sedimentology*, 16, pp 5–54.

— 1971b, 'Genesis of longitudinal and oghurd dunes elucidated by rolling upon grease.' *Bull. Geol. Soc. Am.*, 82, pp 3461–8.

FOX, H. L., 1976, 'The urbanizing river: A case study in the Maryland Piedmont.' *Geomorphology and Engineering*, ed. D. R. Coates (State University of New York, Binghamton), pp 245–71.

FRAKES, L. A. and CROWELL, J. C., 1975, *Gondwana Geology*, ed. K. S. W. Campbell (Aust. Natl. Univ. Press), p 374.

FRANCIS, P., 1976, *Volcanoes* (Penguin Books, Harmondsworth).

FRENCH, H. M., 1971, 'Slope asymmetry of the Beaufort plain, Northwest Banks Island, N.W.T. Canada.' *Can. J. Earth Sci.* 8, pp 717–31.

FRIEDMAN, G. M., 1961, 'Distinction between dune, beach, and river sands from their textural characteristics.' *J. Sediment. Petrol.*, 31, pp 514–29.

— 1967, 'Dynamic processes and statistical parameters compared for size, frequency and distribution of beach and river sands.' *J. Sediment. Petrol.*, 37, pp 327–54.

GALVIN, J. C., 1967, 'Longshore current velocity: a review of theory and data.' *Revs. in Geophys.*, 5, No. 3, pp 287–304.

GILBERT, G. K., 1905, *Bull. Geol. Soc. Am.*, 17, p 306.

GLASS, H. D., FRYE, J. C. and WILMAN, H. B., 1968, 'Clay mineral composition, a source indicator of Midwest loess.' *The Quaternary of Illinois*, (Univ. Illinois College of Agric.), Spec. Pub. 14, pp 35–40.

GLENNIE, K. W., 1970, *Desert Sedimentary Environments (Developments in Sedimentology, 14* (Elsevier, Amsterdam).

GOLDSMITH, V., 1976, 'Continental shelf wave climate models: Critical links between shelf hydraulics and shoreline processes.' *Beach and Nearshore Sedimentation* ed. R. A. Davies, Jr. and R. L. Ethington.

GOLDTHWAIT, R. P., 1976, 'Frost sorted patterned ground: a review.' *Quaternary Research*, 6(1), pp 27–35.

GRAF, W. F., 1977, 'The impact of suburbanization on fluvial geomorphology.' *Water Resources Research*, II, pp 690–2.

GRAVIS, R. F., 1969, 'Fossil slope deposits in the northern Arctic asymmetrical valleys.' *Biuletyn Peryglacjalny*, 20, pp 239–57.

GREENWOOD, B., 1969, 'Sediment parameters and environmental discrimination: an application of multivariate statistics.' *Con. J. Earth Sci.*, 6, pp 1347–58.

GREGORY, K. J., 1976, 'Changing drainage basins.' *Geog. J.*, 142, pp 237–47.

— 1976, 'Drainage basin adjustments and man.' *Geographica Polonica*, 34, pp 155–73.

— 1978a, 'A physical geography equation.' *National Geographer*, 12, pp 13–141.

— 1978b, 'Valley carved in the Yorkshire moors.' *Geographical Magazine*, 50, pp 276–79.

— 1979, 'Drainage network power,' *Water Resources Research*, in press.

GREGORY, K. J. and PARK, C. C., 1976, 'The development of a Devon gully and man.' *Geography*, 61, pp 77–82.

GREGORY, K. J. and Walling, D. E., 1973, *Drainage Basin Form and Process* (Edward Arnold).

GREGORY, K. J. and WALLING, D. E., 1979 (eds.), *Man and Physical Landscape Processes* (Wm. Dawson, Folkestone).

GROVE, A. T., 1958, 'The ancient erg of Hausaland and similar formations on the south side of the Sahara.' *Geog. J.*, 124, pp 528–33.

GUILLIEN, Y. and LAUTRIDOU, J.-P., 1970, 'Recherches de gélifraction expérimentale du Centre de Géomorphologie–1. Calcaire de Charentes.' *Centre de Géomorphologie Bull.*, Caen, 5.

GUZA, R. T. and INMAN, D. L., 1975, 'Edge waves and beach cusps.' *J. Geophys. Res.*, 80, No. 21, pp 2977–3012.

HACK, J. T. and GOODLETT, J. C., 1960, 'Geomorphology and forest ecology of a mountain region in the central Appalachians.' *U.S. Geol. Survey Prof. Paper* 347.

HAILS, J. R., 1964, 'A reappraisal of the nature and occurrence of heavy mineral deposits along parts of the east Australian coast.' *Aust. J. Sci.*, 27(1), pp 22–23.

— 1967, 'Significance of statistical parameters for distinguishing sedimentary environments in New South Wales, Australia.' *J. Sediment. Petrol.*, 37, pp 1059–69.

— 1969, 'The nature and occurrence of heavy minerals in three coastal areas of New South Wales.' *Proc. R. Soc., N.S.W.*, 102, pp 21–39.

— 1972, 'The significance and limitations of statistical parameters for recognizing sedimentary environments.' *Proc. Soc. Analyt. Chem.*, 9, pp 115–18.

— 1974, 'A review of some current trends in nearshore research.' *Earth Sci. Rev.*, 10, pp 171–202.

— 1975, 'Submarine geology, sediment distribution and Quaternary history of Start Bay, Devon.' *J. Geol. Soc. Lond.*, 131, pp 19–35.

HAILS, J. R., 1977 (ed.), *Applied Geomorphology* (Elsevier, Amsterdam).

HAILS, J. R. and CARR, A. P. (eds.), 1975, *Nearshore Sediment Dynamics and Sedimentation* (John Wiley and Sons, New York).

HAILS, J. R. and HOYT, J. H., 1969, 'The significance and limitations of statistical parameters for distinguishing ancient and modern sedimentary environments of the Lower Georgia Coastal Plain.' *J. Sediment. Petrol.*, 39, pp 559–80.

HAILS, J. R., SEWARD-THOMPSON, B. and CUMMINGS, L., 1973, 'An appraisal of the significance of sieve intervals in grain size analysis for environmental interpretation.' *J. Sediment. Petrol.*, 43, pp 889–93.

HARRIS, T. F. W., 1967, *Field and Model Studies of the Nearshore Circulation*, Ph.D. dissertation, Department of Physics, Univ. of Natal, South Africa.

HARRISON, W., 1969, 'Empirical equations for foreshore changes over a tidal cycle.' *Mar. Geol.*, 7, pp 529–51.

— 1972, 'Changes in foreshore sand volume on a tidal beach: role of fluctuations in water table and ocean still-water level.' *Proc. Int. Geol. Congr. 24th, Montreal*, Sect. 12, pp 159–66.

HASTENRATH, S. L., 1967, 'The barchans of the Arequipa region, southern Peru.' *Zeit. für Geom.*, 11, pp 300–331.

HAYES, M. O., 1967, 'Hurricanes as geologic agents: case studies of hurricanes Carla, 1961, and Cindy, 1963.' *Bur. Econ. Geol. Rep. Inv.*, (Univ. Texas, Austin, No.61).

HERRON, W. J. and HARRIS, R. L., 1966, 'Littoral bypassing and beach restoration in the vicinity of Port Hueneme, California.' *Proc. Conf. Coastal Eng.*, 10th, *Am. Soc. Civ. Eng.*, 1, pp 651–75.

HEWITT, K. and BURTON, I., 1976, *The Hazardousness of a Place: A Regional Ecology of Damaging Events* (University of Toronto Press).

HINO, M., 1972, 'Theory on formation of shore current system and systematic deformation of coastal topography,' *Tech. Rep.*, 13, pp 99–113. (Department of Civ. Eng., Tokyo Inst. of Technol., Tokyo).

— 1973, 'Hydrodynamic instability theory on the formation of system of shore current and coastal topography, 2.' *Tech. Rep.*, 14, pp 27–41 (Department of Civ. Eng., Tokyo Inst. of Technol., Tokyo).

HOEKSTRA, P., 1969, 'Water movement and freezing pressures.' *Soil Sci. Am. Proc.*, 33, pp 512–18.

HOLM, D. A., 1960, 'Desert Geomorphology in the Arabian Peninsula.' *Science*, 132, pp 1369–70.

HOLMES, P., 1975, 'Wave conditions in Start Bay.' *J. Geol. Soc. Lond.*, 131, pp 57–62.

HOM-MA, M. and SONU, C. J., 1963, Rhythmic pattern of longshore bars related to sediment characteristics, *Proc. Conf.* Coastal Engng. 8th, Am. Soc. Civ. Engng., 248–278.

HORIKAWA, K. and SHEN, H. W., 1960, 'Sand movement by wind.' *U.S. Army Corps of Engineers, B.E.B., Tech. Memo.*, 119.

HORTON, R. E., 1933, 'The role of infiltration in the hydrological cycle.' *Transactions American Geophysical Union*, 14, pp 446–60.

— 1945, 'Erosional development of streams and their drainage basins: hydrophysical approach to quantitative morphology.' *Bull. Geol. Soc. Am.* 56, pp 275–370.

HOUBOLT, J. J. H. C., 1968, 'Recent sediments in the Southern Bight of the North Sea.' *Geol. en Mijnbouw*, 47, pp 245–73.

HOYT, J. H., 1966, 'Air and sand movements to the lee of dunes.' *Sedimentology*, 7, pp 137–44.

HUNTLEY, D. A. and BOWEN, A. J., 1975a, 'Field observations of edge waves and their effect on beach material.' *J. Geol. Soc. Lond.*, 131, pp 69–81.

— 1975b, 'Comparison of the hydrodynamics of steep and shallow beaches.' *Nearshore Sediment Dynamics and Sedimentation*, ed. J. R. Hails and A. P. Carr (John Wiley and Sons), pp 69–109.

298 *Bibliography*

HUTCHINSON, J. N., 1968, 'Mass movement.' *Encyclopedia of Geomorphology* ed. R. W. Fairbridge (Reinhold), pp 688–95.

INGLE, J. C., 1966, *The Movement of Beach Sand* (American Elsevier, New York).

INMAN, D. L. and BAGNOLD, R. A., 1963, 'Beach and nearshore processes, Part II: Littoral processes.' *The Sea, Vol. 3, The Earth Beneath the Sea History*, ed. M. N. Hill (Intersci-ence, New York), pp 529–53.

IWAGAKI, Y. and NODA, H., 1962, 'Laboratory study of scale effects in two-dimensional beach processes.' *Proc. Conf. Coastal Eng.*, 8th, *Am. Soc. Civ. Eng.*, pp 194–210.

JENNINGS, J. N., 1973, ' "Any milleniums today, Lady?" The geomorphic bandwaggon parade.' *Australian Geographical Studies*, 11, pp 115–33.

JOHNSON, D. W., 1919, *Shore Processes and Shoreline Development* (Wiley, New York).

JOHNSON, J. P., 1973, *Research in Polar and Alpine Geomorphology*, ed. B. D. Fahey and R. D. Thompson (Geo Abstracts, Norwich), p 88.

JOHNSON, J. W., 1949, 'Scale effects in hydraulic models involving wave motion.' *Trans. Am. Geophys. Union*, 30, pp 517–25.

JONES, T. A., 1970, 'Comparison of the descriptors of sediment grain-size distributions.' *J. Sediment. Petrol.*, 40, pp 1204–15.

KAWAMURA, R., 1951, 'Study on sand movement by wind.' *Inst. Sci. and Tech., Tokyo, Rep.*, 5(3–4), pp 95–112.

— 1953, 'Mouvement du sable sous l'effet du vent.' *Actions Éoliennes, Cent. Nat. de Rech. Sci., Paris, Coll. Int.*, 35, pp 117–51.

KELLER, E. A., 1971, 'Areal sorting of bed-load material: the hypothesis of velocity reversal.' *Bull. Geol. Soc. Amer.*, 82, pp 753–6.

X KENNEDY, B. A., 1976, 'Valley-side slopes and climate.' *Geomorphology and Climate*, ed. E. Derbyshire (Wiley), pp 171–202.

KIDSON, C., 1961, 'Movement of beach materials on the east coast of England.' *East. Mid. Geogr.*, 16, pp 3–16.

KING, C. A. M., 1972, *Beaches and Coasts*, Second edn (Edward Arnold).

— 1974, 'Coasts,' *Geomorphology in Environmental Management*, ed. R. U. Cooke and J. C. Doornkamp (Clarendon Press, Oxford), pp 188–222.

KING, C. A. M. and McCULLAGH, M. J., 1971, 'A simulation model of a complex recurved spit.' *J. Geol.*, 79, pp 22–37.

KING, L. C., 1959, 'Denudational and tectonic relief in southeastern Australia.' *Trans. Geol. Soc. South Africa*, 62, pp 113–38.

KIRKBY, M. J., 1969, 'Infiltration, throughflow and overland flow.' *Water, Earth and Man*, ed. R. J. Chorley (Methuen), pp 215–27.

— 1978 (ed.), *Hillslope Hydrology* (Wiley).

KOMAR, P. D., 1971b, 'Nearshore cell circulation and the formation of giant cusps.' *Geol. Soc. Am. Bull.*, 82, pp 2643–50.

— 1972, 'Nearshore currents and the equilibrium cuspate shoreline.' Oregon State Uni-versity Department of Oceanography, *Tech. Rep.* 239.

— 1976, *Beach Processes and Sedimentation* (Prentice-Hall, Englewood Cliffs, New Jersey).

KOMAR, P. D. and INMAN, D. L., 1970, 'Longshore sand transport on beaches.' *J. Geophys. Res.*, 75, pp 5194–5927.

KRUMBEIN, W. C., 1936, 'Application of logarithmic moments to size frequency distribution of sediments.' *J. Sediment. Petrol.*, 6, pp 35–47.

KRUMBEIN, W. C. and GRAYBILL, F. A., 1965, *An Introduction to Statistical Models in Geology* (McGraw Hill, New York).

KUENEN, Ph. H., 1948, 'The formation of beach cusps.' *J. Geol.*, 56, pp 34–40.

LACHENBRUCH, A. H., 1962, 'Mechanics of thermal contraction cracks and ice-wedge polygons in permafrost.' *Geol. Soc. Amer., Spec. Paper*, 70.

LAM-KIN-CHE, 1977, 'Patterns and rates of slopewash on the badlands of Hong Kong.' *Earth Surface Processes*, 2, pp 319–32.

LANDON, R. E., 1930, 'An analysis of beach pebble abrasion and transportation.' *J. Geol.*, 38, pp 437–46.

LANE, F. W., 1966, *The Elements Rage* (David and Charles).

LANGBEIN, W. B. and LEOPOLD, L. B., 1964, 'Quasi-equilibrium states in channel morphology.' *Am. J. Sci.*, 262, pp 782–94.

LANGBEIN, W. B. and SCHUMM, S. A., 1958, 'Yield of sediment in relation to mean annual precipitation.' *Trans. Am. geophys. Un.* 39, pp 1076–84.

LAUTRIDOU, J. P., 1972, 'Bilan des recherches de gélivation expérimentale effectuées au centre de Géomorphologie.' *Centre de Géomorphologie, Caen, Bull.*, 13, 14, 15, pp 63–73.

LE BLOND, P. H. and TANG, C. L., 1974, 'On energy coupling between waves and rip currents.' *J. Geophys. Res.*, 79, pp 811–16.

LEFFINGWELL, E. DE K., 1915, 'Ground ice wedges.' *J. Geol.*, 23, pp 635–54.

LEOPOLD, L. B. and MADDOCK, T., 1953, 'The hydraulic geometry of stream channels and some physiographic implications.' *U.S. Geol. Survey Prof. Paper*, 252.

LEOPOLD, L. B., EMMETT, W. W. and MYRICK, R. W., 1966, 'Channel and hillslope processes in a semi-arid area, New Mexico.' *U.S. Geol. Survey Prof. Paper*, 352G, pp 193–253.

LETTAU, K. and LETTAU, H., 1969, 'Bulk transport of sand by the barchans of the Pampa La Joya in southern Peru.' *Zeit. Geom.*, N.F., 13, pp 182–95.

LEWIS, A. N., 1925, 'Notes on a geological reconnaissance of the Mt La Perouse Range.' *Papers and Proc. Royal Soc. Tasmania, 1924*, pp 9–44.

LEWIS, W. V., 1960 (ed.), 'Norwegian cirque glaciers.' *R. Geogr. Soc. Res. Ser.*, 4.

LINDSAY, J. F., 1973a, 'Ventifact evolution in Wright Valley, Antarctica.' *Geol. Soc. Am. Bull.*, 84(5), pp 1791–8.

— 1973b, 'Reversing barchan dunes in lower Victoria Valley, Antarctica.' *Geol. Soc. Am. Bull.*, 84(5), pp 1799–1806.

LISITZIN, E., 1974, *Sea Level Changes* (Elsevier, Amsterdam).

LONG, J. T. and SHARP, R. P., 1964, 'Barchan-dune movement in the Imperial Valley, California.' *Geol. Soc. Am. Bull.*, 75, pp 149–56.

LONGWELL, C. K., FLINT, R. F. and SANDERS, J. E., 1969, *Physical Geology* (John Wiley).

LONGUET-HIGGINS, M. S. and STEWART, R. W., 1962, 'Radiation stress and mass transport in gravity waves.' *J. Fluid Mech.*, 13, pp 481–504.

— 1963, 'A note on wave setup.' *J. Mar. Res.*, 21, pp 4–10.

— 1964, 'Radiation stresses in water waves, a physical discussion, with applications.' *Deep-Sea Res.*, 11, pp 529–62.

MACKAY, J. R. and MATHEWS, W. H., 1967, 'Observations on pressure exerted by creeping snow, Mount Seymour, British Columbia, Canada.' *Physics of Snow and Ice, Proc. Intl. Conf. Low Temp Sci.*, ed. H. Oura (Sapporo, Japan, 1966), pp 1185–97.

MACKIN, J. H., 1948, 'Concept of the graded river.' *Bull. Geol. Soc. Am.* 59, pp 463–512.

MADIGAN, C. T., 1930, 'An aerial reconnaissance into the southeastern portion of central Australia.' *Proc. R. Geogr. Soc. Aust., Session 1928–1929*, 30, pp 83–108.

— 1936, 'The Australian sand-ridge deserts.' *Geogr. Rev.*, 26, pp 205–27.

MAKEEV, O. W. and KERZHENTSEV, A. S., 1974, 'Cryogenic processes in the soils of northern Asia.' *Geoderma*, 12, pp 101–9.

MARK, D. M., 1974, *Geology*, 2, p 102.

MASON, C. C. and FOLK, R. L., 1958, 'Differentiation of beach, dune, and aeolian flat environments by size analysis, Mustang Island, Texas.' *J. Sediment. Petrol.*, 28, pp 211–16.

MATTHES, F. E., 1900, 'Glacial sculpture of the Bighorn Mountains, Wyoming.' *U.S. geol. Surv. 21st Ann. Rep. (1899–1900)*, pp 167–90.

McGOWN, A. and DERBYSHIRE, E., 1977, 'Genetic influences on the properties of tills.' *Q. Jour. Engng. Geol.*, 10, p 391.

McKEE, E. D., 1966, 'Structures of dunes at White Sands National Monument, New Mexico (and a comparison with structures of dunes from other selected areas).' *Sedimentology*, 7(1), pp 1–69.

McKEE, E. D. and TIBBITTS, G. C., JR., 1964, 'Primary structures of a seif dune and associated deposits in Libya.' *J. Sediment. Petrol.*, 34(1), pp 5–17.

McLEISH, W., 1968, 'On the mechanism of wind-slick generation.' *Deep Sea Res.*, 15, pp 461–89.

McROBERTS, E. C., 1975, 'Some aspects of a simple secondary creep model for deformation of permafrost slopes.' *Can. Geotech. J.*, 12, pp 98–105.

MEADE, R. H., 1969, 'Landward transport of bottom sediments in estuaries of the Atlantic Coastal Plain.' *J. Sediment. Petrol.*, 39, pp 222–34.

MEYER, R. E., 1972, *Waves on Beaches and Resulting Sediment Transport* (Academic Press, New York).

MILLER, M. M., 1973, *Research in Polar and Alpine Geomorphology*, ed. B. D. Fahey and R. D. Thompson (Geo Abstracts, Norwich), p 137.

MORISAWA, M. E., 1968, *Streams: Their Dynamics and Morphology. (McGraw-Hill, New York).*

MÜLLER, F., 1959, 'Beobachtungen über Pingos. Detailuntersuchungen in Ostgrönland und in der Kanadischen Arktis.' *Meddr. Grönland*, 153 (3).

MUNK, W. H., 1949, 'The solitary wave theory and its applications to surf problems.' *Acad. Sci. Am.*, New York, 51, pp 376–424.

NERSCOVA, Z. A. and TSYTOVICH, N. A., 1966, Permafrost Int. Conf., Lafayette, Indiana, 1963, *Proc. Natl. A. Sci.*, Natl. Res. Council Pub. 1287, p 230.

NORRIS, R. M., 1966, 'Barchan dunes of Imperial Valley, California.' *J. Geol.*, 74, pp 292–306.

NYE, J. F., 1952, 'The mechanics of glacier flow.' *J. Glaciol*, 2(12), pp 82–93.

NYE, J. F. and MARTIN, P. C. S., 1968, 'Glacial erosion.' *Intnl. Assoc. Scient. Hydrol., Publ.*, 79, pp 78–86.

OLLIER, C. D., 1969, *Volcanoes* (M.I.T. Press).

— 1969, *Weathering* (Oliver and Boyd, Edinburgh)5

OTVOS, E. G., 1964, 'Observations of beach cusps and beach ridge formation on the Long Island Sound.' *J. Sediment. Petrol.*, 34, pp 554–60.

OUTCALT, S. L., 1971, 'An algorithm for needle ice growth.' *Water Resources Research*, 7, pp 394–400.

PARKER, W. R., 1975, 'Sediment mobility and erosion on a multibarred foreshore (southwest Lancashire, UK).' *Nearshore Sediment Dynamics and Sedimentation*, ed. J. R. Hails and A. P. Carr (John Wiley & Sons), pp 151–79.

PATERSON, W. S. B., 1969, *The Physics of Glaciers* (Pergamon, Oxford).

PELTIER, L. C., 1950, 'The geographic cycle in periglacial regions as it is related to climatic geomorphology.' *Ann. Assoc. Am. Geogr.*, 40, pp 214–36.

PENCK, W., 1924, *Morphological Analysis of Land Forms*. Trans. H. Czech and K. C. Boswell (Macmillan, 1953).

PENMAN, H., 1963, *Vegetation and Hydrology*. Technical Communication No. 53 (Commonwealth Bureau of Soils, Farnham Royal).

PENNER, E., 1966, 'Pressures developed during the unidirectional freezing of water-saturated porous materials: experiment and theory.' *Physics of snow and ice, Intnl. Conf. Low Temp. Sci.*, ed. H. Oura (Sapporo, Japan), 1(2), pp 1401–12.

PICKUP, G. and WARNER, R. F., 1976, 'Effects of hydrologic regime on magnitude and frequency of dominant discharge.' *Journal of Hydrology* 29, pp 51–75.

PILGRIM, D. H., 1966, 'Radioactive tracing of storm runoff on a small catchment.' *Journal of Hydrology* 4, pp 289–326.

PRICE, W. A., 1968, 'Variable dispersion and its effects on the movements of tracers on beaches.' *Proc. Conf. Coastal Eng., 11th, Am. Soc. Civ. Eng.*, 1, pp 329–34.

PRIOR, D. B. and HO, C., 1971, 'Coastal and mountain slope instability on the islands of St Lucia and Barbados.' *Engng. Geol.*, 6, pp 1–18.

PRIOR, D. B. and STEPHENS, N., 1972, 'Some movement patterns of temperate mudflows: examples from north eastern Ireland.' *Bull. Geol. Soc. of Am.* 33, pp 2533–44.

RADOK, R. and RAUPACH, M., 1977, 'Sea level and transport phenomena in St Vincent Gulf.' *Inst. Engrs. 3rd Aust. Conf. Coastal and Ocean Eng.*, pp 103–9.

RAPP, A., 1960, 'Recent development of mountain slopes in Karkevagge and surroundings, northern Scandinavia.' *Geografiska Annaler*, 42, pp 65–200.

RECTOR, R. L., 1954, *Laboratory Study of the Equilibrium Profiles of Beaches* (U.S. Army Corps of Engrs., Beach Erosion Board Tech. Memo. No. 41).

REES, A. I., 1968, 'The production of preferred orientation in a concentrated dispersion of elongated and flattened grains.' *J. Geol.*, 76, pp 457–65.

RILEY, S. J., 1976, 'Aspects of Bankfull geometry in a distributary system of eastern Australia.' *Hydrol. Sci. Bull.*, 21, 545–560.

RIM, M., 1958, 'Simulation by dynamical model of sand track morphologies occurring in Israel.' *Bull. Res. Council of Israel*, 7-G(2/3), pp 123–133.

RITTMAN, A., 1962, *Volcanoes and Their Activity* (John Wiley Interscience, New York).

ROBIN, G. DE Q., 1974, *J. Glaciol.*, 16, p 185.

ROSE, J. and LETZER, J. M., 1977, *J. Glaciol.*, 18, p 478.

RUSSELL, R. J., 1958, 'Long straight beaches.' *Ecol. Geol. Helv.*, 51, pp 591–8.

RYCKBORST, H., 1975, 'On the origin of pingos.' *J. Hydrol.*, 26(3/4), pp 303–14.

SAVIGEAR, R. A. G., 1965, 'A technique of morphological mapping.' *Annls Assoc. Am. Geog.* 55, pp 514–38.

SCHEIDEGGER, A. E., 1970, *Theoretical Geomorphology* (Springer-Verlag, Berlin), 2nd edn.

SCHICK, A. P., 1977, 'A tentative sediment budget for an extremely arid watershed in the southern Neger.' *Geomorphology in Arid Regions*, ed. D. O. Doehting (State University of New York), pp 139–63.

SCHMERTMANN, J. H. and TAYLOR, R. S., 1965, 'Quantitative data from a patterned ground site over permafrost.' *U.S. Army Cold Reg. Res. and Eng. Lab. Res. Rept.*, 96.

SCHUMM, S. A., 1963, 'The disparity between present rates of denudation and orogeny.' *U.S. Geol. Surv. Prof. Paper*, 454H.

— 1973, 'Geomorphic thresholds and complex response of drainage systems.' *Fluvial Geomorphology*, ed. M. E. Morisawa (State University of New York Binghamton), pp 299–310.

— 1977, *The Fluvial System* (Wiley, New York).

SCHUMM, S. A. and LICHTY, R. W., 1965, 'Time, space and causality in geomorphology.' *Am. J. Sci.*, 263, pp 110–19.

SCHUMSKII, P. A., 1964, *Principles of structural glaciology* (Dover, New York).

SCOTT, T., 1954, *Sand Movement By Waves* (U.S. Army Corps of Engrs., Beach Erosion Board Tech. Memo., No. 48).

SELBY, M. J., 1977, 'Transverse erosional marks on ventifacts from Antarctica.' *N.Z. Journal of Geology and Geophysics*, 20(5), pp 949–69.

SELBY, M. J., RAINS, R. B. and PALMER, R. W. P., 1974, 'Eolian deposits of the ice-free Victoria Valley, southern Victoria Land, Antarctica.' *N.Z. Journal of Geology and Geophysics*, 17(3), pp 543–62.

SHARP, R. P., 1948, *J. Glaciol.*, 1, pp 182–9.

— 1963, 'Wind ripples.' *J. Geol.*, 71, pp 617–36.

— 1966, 'Kelso dunes, Mojave desert, California.' *Geol. Soc. Am. Bull.*, 77, pp 1045–74.

SHARPE, C. F. S., 1938, *Landslides and Related Phenomena* (Columbia University Press).

SHAW, J., 1977, 'Tills deposited in arid polar environments.' *Can. J. Earth Sci.*, 14(6), pp 1239–45.

SHEPARD, F. P., 1952, 'Revised nomenclature for depositional coastal features.' *Bull. Am. Assoc. Petrol. Geologists*, 36, pp 1902–12.

— 1973, *Submarine Geology*, Third edn (Harper and Row).

SHEPARD, F. P., EMERY, K. O. and LaFOND, E. C., 1941, 'Rip currents: a process of geological importance.' *J. Geol.*, 49, pp 337–69.

SHEPARD, F. P. and INMAN, D. L., 1950, 'Nearshore circulation related to bottom topography and wave refraction.' *Trans. Am. Geophys. Union*, 31, 555–65.

SHEPARD, F. P. and YOUNG, R., 1961, 'Distinguishing between beach and dune sands.' *J. Sediment. Petrol.*, 31, pp 196–214.

SILVESTER, R., 1974, *Coastal Engineering*, vols 1 and 2 (Elsevier Scientific Publishing Company).

SIMMONS, I. G., 1978, 'Physical geography in environment science.' *Geography*, 63, pp 314–23.

SIMONETT, D. S., 1967, 'Landslide distribution and earthquakes in the Bewain and Tarricelli, Mountains, New Guinea: A statistical analysis.' *Landform Studies from Australia and New Guinea*, ed. J. N. Jennings and J. A. Mabbutt (Australian National University Press) pp 64–84.

SIMONS, F. S., 1956, 'A note on Pur-Pur Dune, Viru Valley, Peru.' *J. Geol.*, 64, pp 517–21.

SMALLEY, I. J., 1964, 'Flow-stick transitions in powders.' *Nature*, 201, pp 173–234.

SMALLEY, I. J. and UNWIN, D. J., 1968, 'The formation and shape of drumlins and their distribution and orientation in drumlin fields.' *J. Glaciol.*, 7(51), pp 377–90.

SMITH, D. I. and ATKINSON, T. C., 1976, 'Process, landforms and climate in limestone regions.' *Geomorphology and Climate*, ed. E. Derbyshire (Wiley), pp 367–409.

SMITH, H. T. U., 1965, 'Dune morphology and chronology in central western Nebraska.' *J. Geol.*, 73(4), pp 557–8.

— 1973, 'Photogeologic study of periglacial talus glaciers in northwestern Canada.' *Geografiska Annaler*, 55A(2), pp 69–84.

SO, C. L., 1971, 'Mass movements associated with the rainstorms of June 1966 in Hong Kong.' *Trans. Inst. Brit. Geographers*, 53, pp 55–65.

SONU, C. J., 1973, 'Three dimensional beach changes.' *J. Geol.*, 81, pp 42–64.

STATHAM, I., 1977, *Earth Surface Sediment Transport* (Clarendon Press).

STEARNS, S. R., 1966, 'Permafrost.' *U.S. Army Cold Regions Res. Eng. Lab. Tech. Rep.*, 1, (C.R.R.E.L.).

STENBORG, T., 1973, *Intnl. Assoc. Sci. Hydrol.*, Pub. 95, p 121.

STEWART, J. H. and LAMARCHE, V. C., 1967, 'Erosion and deposition produced by the floods of December 1964 on Coffee Creek, Trinity County California.' *U.S. Geol. Survey Prof. Paper*, 422-K.

STRAHLER, A. N., 1952, 'Dynamic basis of geomorphology.' *Bull. Geol. Soc. Am.*, 63, pp 923–38.

— 1966, 'Tidal cycle of changes on an equilibrium beach.' *J. Geol.*, 74, pp 247–68.

SUGDEN, W., 1964, 'Origin of facetted pebbles in some recent desert sediments of Southern Iraq.' *Sedimentology*, 3, pp 65–74.

SUGDEN, D. E. and JOHN, B. S., 1976, *Glaciers and Landscape* (Edward Arnold).

SUNDBORG, A., 1956, 'The river Klaralven, a study of fluvial processes.' *Geografiska Annaler*, 38, pp 127–316.

TABER, S., 1929, 'Frost heaving.' *J. Geol.*, 37, pp 428–61.

— 1930, 'The mechanics of frost heaving.' *J. Geol.*, 38, pp 303–17.

TANNER, W. F., 1958, 'The equilibrium beach.' *Trans. Am. Geophys. Union*, 39, pp 889–91.

—, 1960, 'Bases of coastal classification.' *South Eastern Geology*, 2, pp 13–22.

TARLING, D. H. and TARLING, M. P., 1971, *Continental Drift* (Bell).

THORNES, J. B. and BRUNSDEN, D., 1977, *Geomorphology and Time* (Methuen).

TISCHENDORF, W. G., 1969, Quoted in Chorley, R. J., 1978.

TRASK, P. D., 1952, *Source of Beach Sand at Santa Barbara, California, as Indicated by Mineral Grain Studies* (U.S. Army Corps of Engrs., Beach Erosion Board Tech. Memo., No. 28).

TRICART, J., 1974, *Structural Geomorphology*, trans. S. H. Beaver and E. Derbyshire (Longmans).

TRICART, J. and CAILLEUX, A., 1962, *Le modelé glaciaire et nival* (Sedes, Paris).

TUCKER, M. J., 1950, 'Surf beats: sea waves of 1 to 5 min period.' *Proc. R. Soc. London*, Series A, 202, pp 565–73.

UGOLINI, F. C., 1975, 'Ice-rafted sediments as a cause of some thermokarst lakes in the Noatak River delta, Alaska.' *Science*, 188 (4183), pp 51–3.

UNITED STATES ARMY, 1973, *Shore Protection Manual* (U.S. Army Corps of Eng. Coastal Eng. Res. Cent.), Vols. 1–111.

VEEVERS, J. J. and WELLS, A. T., 1961, 'The geology of the Canning Basin, western Australia.' *Commonwealth Aust. Bur. Min. Res. Geol. Geophys. Bull.*, 60, p 323.

VERLAQUE, C., 1958, 'Les dunes d' in Salah.' *Trav. Inst. Rech. Sahariennes, Travaux*, 17, pp 12–58.

VERSTAPPEN, H. TH., 1977, *Remote sensing in Geomorphology* (Elsevier, Amsterdam).

WALLING, D. E., 1971, 'Sediment dynamics of small instrumented catchments in south east Devon.' *Trans. Devonshire Assoc.*, 103, pp 147–65.

— 1975, 'Solute variations in small catchment streams: some comments.' *Trans. Inst. Brit. Geog.*, 64, pp 141–7.

WALLING, D. E. and FOSTER, I. D. L., 1975, 'Variations in the natural chemical concentration of river water during flood flows and the lag affect: some further comments.' *J. Hydrol.*, 13, pp 325–37.

WARREN, A., 1971, 'The dunes of the Ténéré Desert.' *Geogr. J.*, 137, pp 458–61.

— 1972, 'Observations on dunes and bi-model sands in the Ténéré Desert.' *Sedimentology*, 19, pp 37–44.

WASHBURN, A. L., 1973, *Periglacial processes and environments* (Edward Arnold).

WATTS, G. M., 1954, *Laboratory Study on the Effect of Varying Wave Periods on the Beach Profiles* (U.S. Army Corps of Engrs., Beach Erosion Board Tech. Memo., 53).

WEERTMAN, J., 1961, 'Mechanism for the formation of inner moraines found near the edge of cold ice caps and ice sheets.' *J. Glaciol.*, 3, pp 965–78.

WENDLER, G., TRABANT, D. and BENSON, C., 1973, *Intnl. Assoc. Sci. Hydrol.*, Pub. 107, p 433.

WHALLEY, W. B., 1974, 'The mechanics of high-magnitude, low-frequency rock failure.' *Department of Geography, Univ. Reading, Geog. Paper*, 27.

— 1976, *Properties of Materials and Geomorphological Explanation* (OUP).

WHIPKEY, R. Z., 1965, 'Subsurface stormflow on forested slopes.' *Bull. Intnl. Assoc. Sci. Hydrol.*, 10, pp 74–85.

WHITNEY, M. I. and DIETRICH, R. V., 1973, 'Ventifact sculpture by windblown dust.' *Geol. Soc. Am. Bull.*, 84, 2561–81.

WIEGEL, R. L., 1964, *Oceanographical Engineering* (Prentice-Hall, Englewood Cliffs, New Jersey).

WINGATE, O., 1934, 'In search of Zerzura.' *Geogr. Rev.*, pp 281–308.

WILLIAMS, G., 1964, 'Some aspects of the aeolian saltation load.' *Sedimentology*, 3, pp 257–87.

WILLIAMS, P. J., 1957, 'Some investigations into solifluction features in Norway.' *Geogr. J.*, 123, pp 42–58.

WILSON, I. G., 1971, 'Desert sandflow basins and a model for the development of ergs.' *Geogr. J.*, 137, pp 180–99.

— 1972, 'Aeolian bedforms – their development and origins.' *Sedimentology*, 19, pp 173–210.

— 1973, 'Ergs.' *Sediment. Geol.*, 10, pp. 77–106.

WILSON, L., 1973, 'Relationships between geomorphic processes and modern climates as a method in paleoclimatology.' *Climatic Geomorphology*, ed. E. Derbyshire (Macmillan), pp 269–84.

WOLMAN, M. G. and MILLER, J. P., 1960, 'Magnitude and frequency of forces in geomorphic processes.' *J. Geol.*, 68, pp 54–74.

YATSU, E., 1971, 'Landform materials science – rock control in geomorphology.' *Proc. First Guelph Symp. Geomorph.* (University of Guelph), pp 49–56.

YOUNG, A., 1969, 'Present rate of land erosion.' *Nature*, 224, pp 851–2.

— 1972, *Slopes* (Oliver and Boyd, Edinburgh).

— 1974, 'The rate of slope retreat.' *Progress in Geomorphology*, ed. E. H. Brown and R. S. Waters, Spec. Pub. No. 7 (Institute of British Geographers), pp 65–78.

YANG, CHIH TED, 1976, 'Minimum unit stream power and fluvial hydraulics.' *J. Hydraul. Div. Proc. Am. Assoc. Civ. Eng.*, 102, pp 919–34.

ZAKRZEWSKA, B., 1967, 'Trends and methods in landform geography.' *Annls Assoc. Am. Geog.*, 57, pp 128–65.

ZENKOVICH, V. P., 1967, *Processes of Coastal Development* (Oliver and Boyd, Edinburgh).

ZOLTAI, S. C. and TARNOCAI, C., 1971, 'Properties of a wooded palsa in northern Manitoba.' *Arctic and Alpine Res.*, 3, pp 115–29.

— 1975, *Can. J. Earth Sciences*, 12, pp 28–43.

Subject Index

Geographical Index